Smart Ways of Biomaterial Designing Synthesis and Characterization

Prospects of Enhanced Application From Labs to Clinics

Editors

Arvind K. Singh Chandel
Assistant Professor
Department of Chemical System Engineering
The University of Tokyo
Tokyo, Japan

Arpana Parihar
Department of Translational Medicine
All India Institute of Medical Sciences
Bhopal, Madhya Pradesh, India

Raju Khan
Industrial Waste Utilization, Nano & Biomaterials
CSIR-Advanced Materials & Processes Research Institute (AMPRI)
Bhopal, Madhya Pradesh, India

CRC Press
Taylor & Francis Group
Boca Raton London New York

CRC Press is an imprint of the
Taylor & Francis Group, an **informa** business
A SCIENCE PUBLISHERS BOOK

Cover credit: Cover image provided by the editor, Dr. Arvind K. Singh Chandel. This is a copyright free image downloaded from Pixabay.

First edition published 2025
by CRC Press
2385 NW Executive Center Drive, Suite 320, Boca Raton FL 33431

and by CRC Press
4 Park Square, Milton Park, Abingdon, Oxon, OX14 4RN

CRC Press is an imprint of Taylor & Francis Group, LLC

Library of Congress Cataloging-in-Publication Data (applied for)

ISBN: 978-1-032-30678-0 (hbk)
ISBN: 978-1-032-30683-4 (pbk)
ISBN: 978-1-003-30624-5 (ebk)

DOI: 10.1201/9781003306245

Typeset in Times New Roman
by Prime Publishing Services

Preface

Various natural and synthetic polymeric materials and composites are extensively used as biomaterials for diverse biomedical applications, such as drug delivery, adhesion barriers, implantable devices, drug delivery, gene delivery, tissue engineering, bioimaging, bioadhesive, antimicrobial coating, sealant, and so on. Advances in polymer chemistry have led to the generation of a comprehensive range of biomaterials based on polymers and composites for different biomedical applications, reflecting the unique nature of polymers in their usage. The versatility and simplicity of modification of the chemical, physical, surface, and biomimetic properties of polymers have made them much more convenient to researchers working in the biomedical field. Applications of biomaterials have driven the development of polymers that are biocompatible, biodegradable, and/or resorbable. A variety of research results on polymers and composites are published almost daily, aimed at enhancing their properties as biomaterials to meet the ongoing and evolving challenges in the biomedical field. The designing of the materials is important to achieve the desired goals. This book deals with the material designing strategies for specific biomedical applications and characterization techniques of biomedical materials. It intends to provide detailed updates and information to students, technicians, researchers, scientists, and teachers working in the biomedical field. By drawing exclusively from the latest research-based studies, it presents an extensive summary of the various polymeric materials used in biomedical applications.

This book focuses on biomaterial design aspects, synthesis, and characterization from the point of view of its extended applications, along with technical details that will make it easy to achieve the accurate application and desired results. Slight changes in the composition, such as molecular weight, degree of substitution, or impurity in the processed biomaterials, make huge changes in the results of any experimental application. The details of research-based techniques and their flaws and tips to overcome the limitations will be included in this book, which will be really beneficial to research students and professionals to overcome the problems in their research. This book is going to cover all important real-world practical experiences of material chemists, biomedical experts, surgeons, and physicians, focusing on how biomaterials behave in actual applications. In order to achieve the desired results from any biomaterials in real *in vivo* application, it is crucial to characterize it thoroughly and create a close memetic system before application. Also, the characterization of biomaterials is an important and obligatory step in the development of any biomedical device, implant, drug delivery system and imaging probes.

This book would help researchers, polymer chemists, biomedical engineers, pharmaceutical dosage forms designers, and clinicians for the development of biomaterials-based drug delivery, tissue engineering, antiadhesion sealants, implants, stents, hemostatic, antimicrobial sutures, designing and characterization-based knowledge. This book also covers several troubleshooting steps-based tips and in-depth knowledge of experienced researcher's real experiences, which they generally face in their biomaterials-based research practices. Therefore, it is expected that this book will play a conclusive role and open new pathways for biomaterials-based biomedical materials research.

Contents

1

Biomaterials

An Overview for Future Utility in the Biomedical Domain

Vedika Khare,[1] *Arpana Parihar,*[2,*] *Raju Khan*[3] and *Arvind Singh Chandel*[4]

1. Introduction

The domain of biomaterials holds great significance in the biomedical field because of its potential to revolutionise healthcare and enhance the quality of life for individuals all over the world. Biomaterials are artificial or natural compounds that interact with biological systems ranging from cells and tissues to whole organs (Basu 2020). They are employed in a wide range of medical applications, such as implantable devices, tissue engineering scaffolds, and drug delivery systems. Biomaterials have a long history, dating back many decades when researchers saw the need for materials that could merge smoothly with the human body, aiding healing and restoring functionality. The discovery of biocompatible and bioactive materials has opened the path for medical breakthroughs, such as joint replacements, artificial organs, and regenerative treatments (Mubarak et al. 2023). The value of biomaterials is found in their capacity to replicate the characteristics and operations of natural biological tissues, thereby providing compatibility, toughness, and biointegration. They are essential in dealing

[1] Michigan Technological University, Houghton, MI, USA.

[2] Department of Translational Medicine, All India Institute of Medical Sciences, Bhopal, Madhya Pradesh, India.

[3] Industrial Waste Utilization, Nano and Biomaterials, CSIR-Advanced Materials and Processes Research Institute (AMPRI), Hoshangabad Road, Bhopal - 462026, MP, India; Academy of Scientific and Innovative Research (AcSIR), Ghaziabad - 201002, India.

[4] Center for Disease Biology and Integrative Medicine, Faculty of Medicine, The University of Tokyo, Tokyo 113-8655, Japan.

* Corresponding author: arpana_parihar@yahoo.com

with medical issues, such as chronic illnesses, tissue damage, and organ failure (Freedman and Mooney 2019). With the use of biomaterials, personalised medicine is made possible, allowing for customised approaches to interventions and therapies that are distinct for each patient. Additionally, biomaterials promote the development of targeted medicines, drug discovery, and diagnostic methods. They offer platforms for controlled medication release, enabling the administration of exact dosages and localised therapy. Biomaterials encourage innovation and improve healthcare technology by promoting the development of innovative medical devices (Mohd Isa et al. 2022).

Additionally, materials science, engineering, biology, and medicine are all involved in the research and development of biomaterials. In order to advance this profession, it is crucial for scientists, engineers, and physicians to work together. Emerging health issues, including tissue regeneration, personalised medicine, and the incorporation of technology into medical devices, may be addressed with the research of biomaterials (Zhang et al. 2022a). The objective of this chapter is to give a general overview of biomaterials used in biomedical applications, including their history, characteristics, existing and new biomaterials, and their uses in diagnostic and therapeutic procedures. The purpose of this chapter is to emphasise the importance of biomaterials in transforming healthcare and enhancing patient outcomes. It paves the way for more developments and discoveries in the field of biomaterials research by discussing the difficulties and potential directions for the future.

2. Biomaterials: History

The history of biomaterials dates back centuries, with early examples of their use in medical practises. Ancient civilisations, such as the Egyptians and Romans, utilised various natural materials, including wood, ivory, and metals, for prosthetic limbs and dental restorations. These early applications laid the foundation for developing biomaterials as a distinct field of research and innovation. In the mid-20th century, significant advancements in materials science and engineering led to the emergence of biomaterials as a specialised discipline (Hildebrand 2013). The discovery and understanding of biocompatible materials, such as stainless steel and titanium alloys, opened new possibilities for medical implants and devices. These materials exhibited improved mechanical properties, corrosion resistance, and biocompatibility, setting the stage for transformative developments in healthcare (Sidambe 2014).

In the 1950s, the field of polymer science began to contribute significantly to biomaterials research. Polymethyl methacrylate (PMMA), for instance, gained prominence as a bone cement for joint replacements. This marked a shift toward using synthetic polymers in biomedical applications, driven by their versatility, tuneable properties, and processability (Bauer et al. 2013). The 1970s witnessed a surge in biomaterials research with the advent of tissue engineering and regenerative medicine. The pioneering work of researchers such as Robert Langer and Joseph Vacanti paved the way for the concept of using biomaterial scaffolds to support tissue regeneration (Sohn et al. 2020). Synthetic and natural polymers, such as poly (lactic-co-glycolic acid) (PLGA) and collagen, became vital components in tissue engineering strategies, enabling the development of three-dimensional constructs to promote cellular growth and tissue formation (Rahmati et al. 2021).

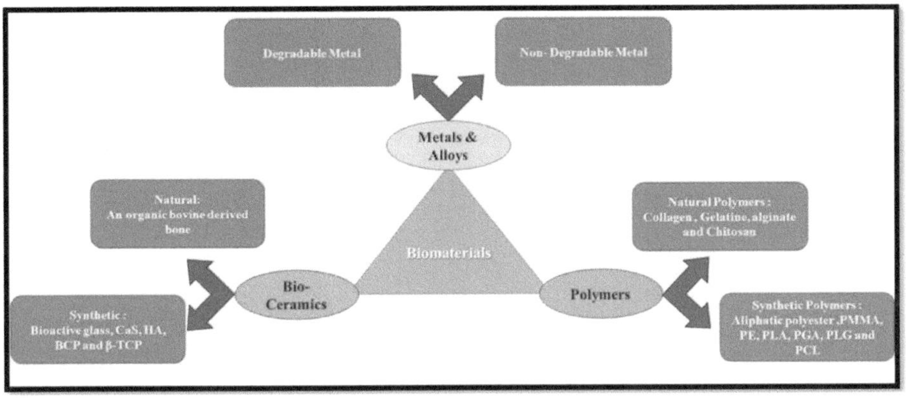

Figure 1: Classification of biomaterials.

Over the years, biomaterials research has continued to evolve and diversify. Advances in nanotechnology have enabled the fabrication of nanostructured materials with unique properties and functionalities, leading to innovations in drug delivery systems and diagnostic devices (Suwardi et al. 2022). Integrating biomaterials with biotechnology and bioengineering has further expanded the possibilities for creating biologically inspired materials and devices (Huang et al. 2018). Figure 2 depicts the evolution of biomaterials in the biomedical domain.

Today, biomaterials play an indispensable role in various fields of medicine, including orthopaedics, cardiovascular interventions, dentistry, and ophthalmology, among others. They are instrumental in developing artificial organs, biosensors, and advanced wound healing technologies (Pazhouhnia et al. 2022). Ongoing research focuses on enhancing the biocompatibility, bioactivity, and functionalisation of biomaterials, as well as exploring emerging technologies like 3D printing and bio-fabrication (Sonatkar et al. 2022).

3. Properties of Biomaterials

Biomaterials for biological applications have various desired characteristics that are now the subject of intensive study and development. These characteristics are intended to improve the efficiency of biomaterials in various biological fields and solve certain problems.

One key property of biomaterials is biocompatibility, ensuring that they interact harmoniously with biological systems without causing adverse reactions or toxicity. To increase the biocompatibility of materials and their integration with surrounding tissues, researchers are investigating innovative surface modifications, such as biomimetic coatings or bioactive compounds (Ostroff et al. 2022). Mechanical characteristics continue to be an important factor in the study of biomaterials. Researchers work to develop materials that are strong, stiff, and elastic enough to meet the biomechanical needs of particular tissues or organs. For instance,

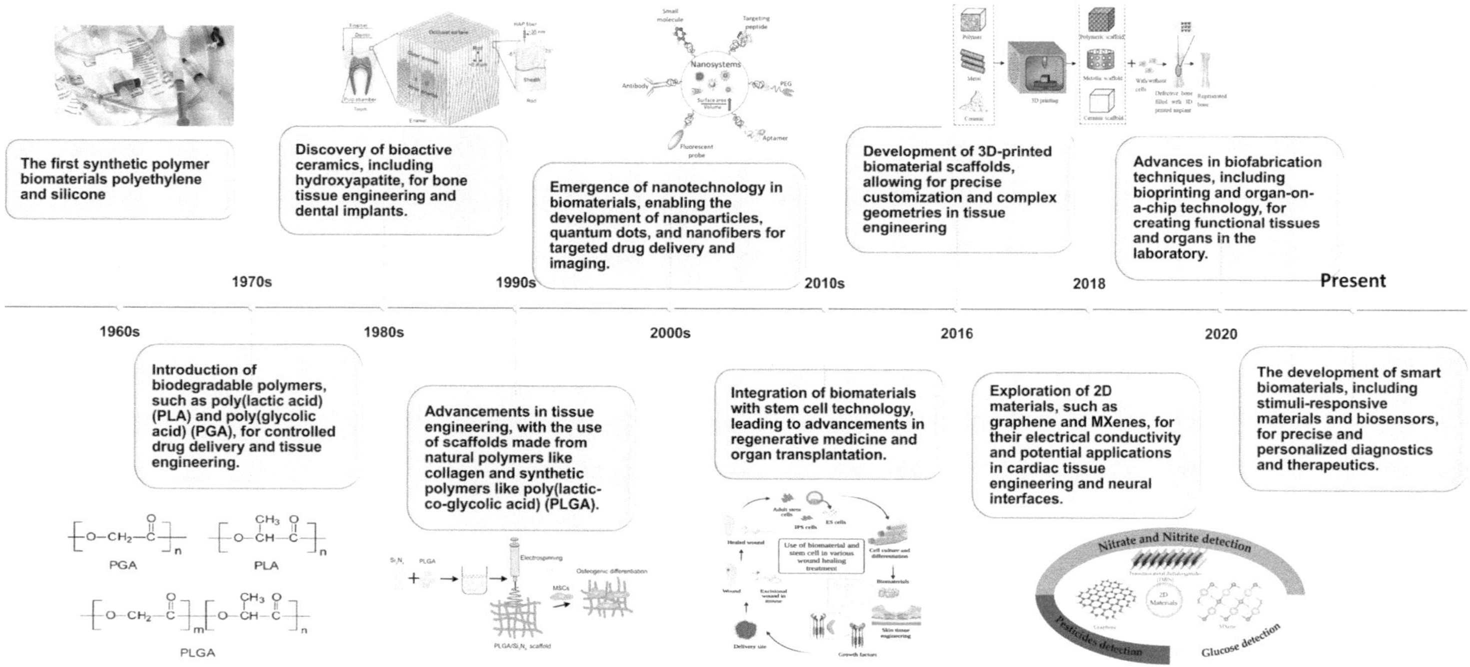

Figure 2: Timeline for biomaterials in the biomedical domain.

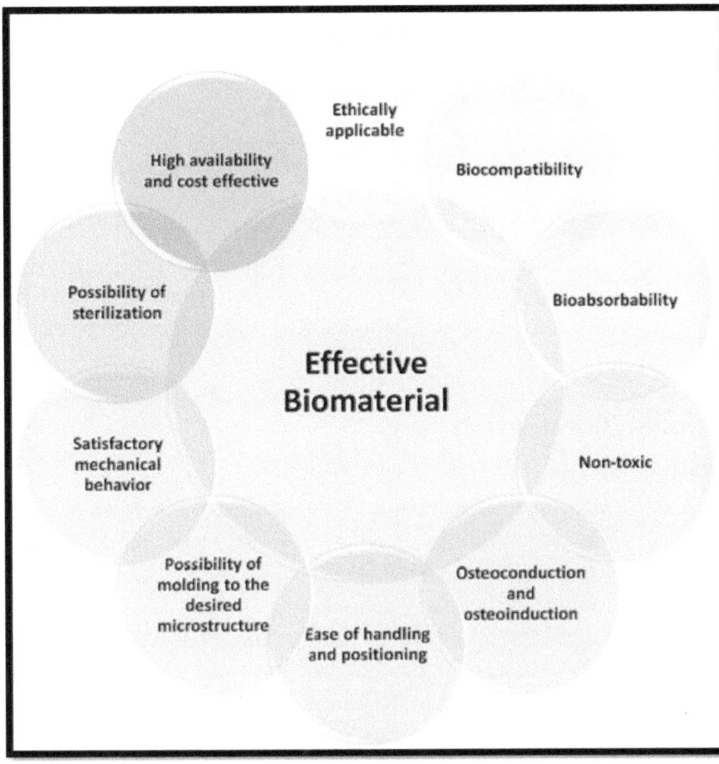

Figure 3: Characteristics of an effective biomaterial to be utilised in the biomedical field (Vaiani et al. 2023).

scaffolds used in tissue engineering must have mechanical characteristics that are similar to those of the original tissue in order to encourage correct cell migration, proliferation, and differentiation (Wang et al. 2020). Biodegradable or bioresorbable materials have drawn a lot of interest recently. These biomaterials do not require surgical removal since they gradually break down and resorb within the body over time. They are thoroughly investigated for uses including temporary implants, medication delivery mechanisms, and scaffolds for tissue engineering (Mukherjee et al. 2023). In order to maximise the pace of resorption and guarantee proper tissue repair, researchers are looking into innovative biomaterial compositions, processing methods, and degradation kinetics. The current research also focuses on the surface characteristics of biomaterials (Zhang and King 2020). To improve cell adhesion, manage protein interactions, stop bacterial adherence, or enable targeted medication delivery, researchers are looking at surface changes. The performance of biomaterials in biomedical applications will be enhanced by exploring surface patterning, functionalisation, and coating approaches to customise the interactions between biological entities and biomaterials (Khan and Alamry 2021; Yang et al. 2021). Furthermore, a rising field of research is the integration of biomaterials with cutting-edge technology. To generate biomaterials with exact structures, controlled release qualities, and adaptable designs, researchers are experimenting with the inclusion of nanomaterials, 3D printing technologies, and bio-fabrication approaches (Khalilov 2023). This connection creates opportunities for patient-specific tissue engineering techniques, on-demand medication administration, and personalised treatment. In a nutshell, the ongoing search for biocompatible, bioactive, mechanically appropriate, resorbable, and surface-engineered materials is the driving force behind contemporary research in biomaterials for biomedical applications (Vasanthan et al. 2022). In order to revolutionise healthcare in areas like tissue engineering, regenerative medicine, drug delivery, and medical device technologies, researchers are constantly improving these qualities. Their goal is to create biomaterials that can smoothly integrate with the human body and promote healing.

4. Traditional Biomaterials

Conventional biomaterials have been employed extensively in the biomedical field for many years. These materials have a demonstrated track record of reliability, effectiveness, and body compatibility. Traditional biomaterials most frequently come in the form of metals and alloys, ceramics, and polymers (Egbo 2021), as shown in Figure 4. Orthopaedic implants, cardiovascular devices, and dental applications have all made substantial use of metals and alloys like titanium, stainless steel, and cobalt-chromium alloys. Excellent mechanical strength, longevity, and corrosion resistance are all features of these materials (Thanigaivel et al. 2022). Metals are appropriate for load-bearing applications because they can endure physiological stresses and offer structural support. It is common practise to use surface coatings and alterations to increase biocompatibility and tissue integration (Bandyopadhyay et al. 2022). Dental implants, bone graft alternatives, and coatings for metal implants all use ceramic materials, such as alumina, zirconia, and hydroxyapatite. High biocompatibility, exceptional wear resistance, and chemical stability are all characteristics of ceramics (Choi et al. 2023). In order to encourage osseointegration and establish solid contact between the implant and the surrounding tissue, they can closely match the characteristics of natural bone. Biomedical uses for polymers, both natural and artificial, are many. Natural polymers with strong biocompatibility and biodegradability include collagen, chitosan, and alginate, which are produced from biological sources (Arif et al. 2023). They are often utilised in medication delivery systems, wound dressings, and tissue engineering scaffolds. Synthetic polymers with a variety of mechanical and degrading qualities include polyethylene, polyurethane, and poly (lactic-co-glycolic acid) (PLGA) (Mukherjee et al. 2023). The development, testing, and optimisation of conventional biomaterials have been substantial. They are easily accessible for clinical usage because of the regulatory framework that has existed for a long time (Basu et al. 2022). They have a lengthy history of usage, yielding useful clinical information on their effectiveness, safety, and long-term consequences. Traditional biomaterials, however, also have drawbacks. Despite being robust and long-lasting, metals and ceramics are relatively stiff and can result in implant-related issues such as stress shielding (Losic 2021). Polymers may have constraints regarding mechanical strength and degradation control while being more

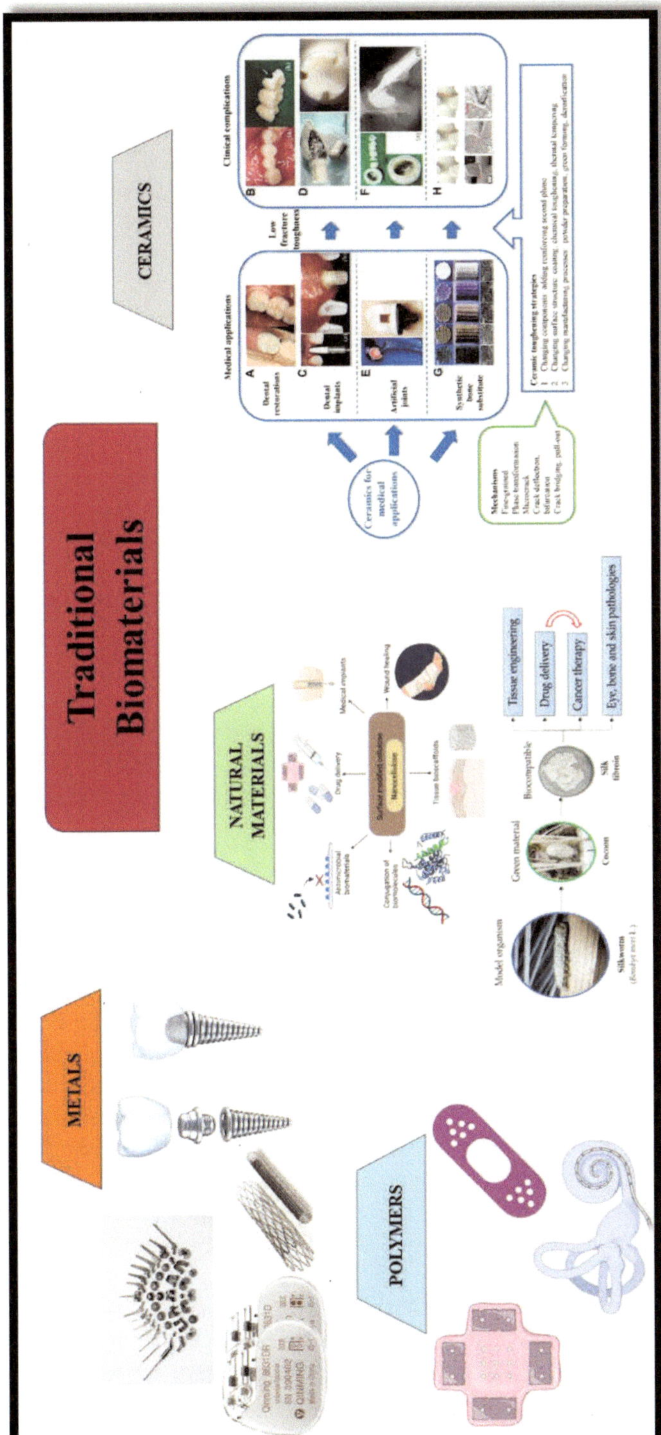

Figure 4: Traditional biomaterials along with biomedical application (Nguyen et al. 2019; Troy et al. 2021; Vaiani et al. 2023).

flexible and versatile. Additionally, conventional biomaterials might not be bioactive enough to stimulate a particular biological response, necessitating additional alterations or coatings (Mondschein et al. 2017).

5. Emerging Biomaterials

Due to their distinct qualities and potential to completely change how healthcare is provided, emerging biomaterials have attracted much attention in the biomedical community. These materials, which are frequently at the cutting edge of research and development, provide cutting-edge capabilities and functions that can solve present problems and open up new applications (Wang et al. 2021). Natural polymers, synthetic polymers, biomimetic materials, nanomaterials, and smart materials are some important subcategories of developing biomaterials. In applications for tissue engineering, wound healing, and drug administration, natural polymers generated from biological sources, such as collagen, silk, and hyaluronic acid, are becoming increasingly popular (Liu et al. 2023). These biomaterials display excellent biocompatibility, biodegradability, and bioactivity. Natural polymers can enhance cell proliferation, provide a bioactive environment, and encourage tissue regeneration (Zhang et al. 2022).

When it comes to adjusting their characteristics for particular biomedical applications, synthetic polymers provide great adaptability. Examples include poly(caprolactone), poly(ethylene glycol), and poly(ethyleneimine) (Radu et al. 2023). Mechanical characteristics, degradation rates, and surface functions of synthetic polymers may all be precisely controlled through engineering. They are used in medical equipment, tissue engineering scaffolds, and drug delivery systems (Zielińska et al. 2023). The goal of biomimetic materials is to replicate the composition and operation of biological tissues or organs. Decellularized extracellular matrices (ECMs) made of tissues, for instance, offer a scaffold that closely replicates the milieu of the genuine tissue (Brown et al. 2022). These materials can all aid the adhesion, migration, and tissue regeneration of cells. Using biomimetic materials in tissue engineering and regenerative medicine is extremely promising (Anjum et al. 2022).

At the nanoscale level, nanomaterials such as nanoparticles, nanofibers, and nanocomposites provide distinctive features. They are useful in biological applications because of their high surface-to-volume ratio, adjustable characteristics, and adaptable surface changes (Hassan et al. 2021). Targeted medication delivery, biosensing, bioimaging, and tissue engineering may all be accomplished with nanomaterials. Gold nanoparticles, quantum dots, and carbon nanotubes are a few examples (Harish et al. 2022). Smart materials, often referred to as stimulus-responsive materials, have dynamic characteristics that may react to outside stimuli like temperature, pH, light, or magnetic fields. Hydrogels, shape memory polymers, and drug-releasing systems are a few examples of intelligent materials (Malekmohammadi et al. 2021). When certain triggers operate on these materials, reversible alterations can occur, allowing regulated medication release, tissue actuation, or on-demand therapeutic treatments.

Innovative drug delivery methods, personalised treatment, tissue regeneration, and diagnostic technologies can all benefit from emerging biomaterials, as shown

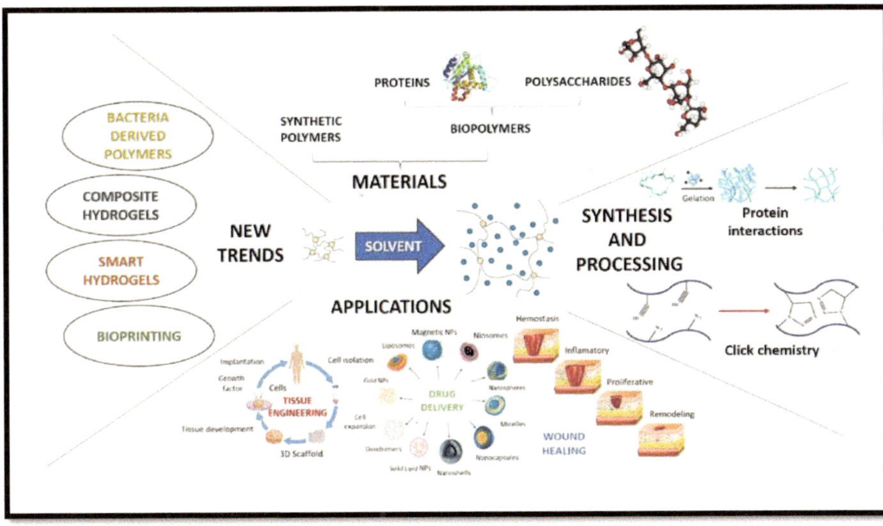

Figure 5: An overview of recent and current biomaterials research trends in the biomedical area (Sánchez-Cid et al. 2022).

in Figure 5. They make it possible to manipulate the characteristics of materials precisely, how they interact with biological systems, and how treatments work (Han et al. 2022). However, to completely comprehend their long-term biocompatibility, regulatory issues, and scalability for widespread clinical usage, more study is required. Researchers want to overcome current constraints, create novel solutions, and influence the future of biomedical applications by continually investigating and optimising developing biomaterials (Øvrebø et al. 2022). These resources have a huge potential to boost treatment plans, enhance patient outcomes, and progress healthcare.

6. Biomaterials in Biomedical Diagnostic Applications

Biomaterials are essential for breakthroughs in disease diagnosis, monitoring, and personalised treatment in diagnostic biomedical applications. In order to improve sensitivity, specificity, and accuracy, these materials are included in a variety of diagnostic instruments and processes. Applications for biomaterial-based diagnostics are numerous and include microfluidic devices, imaging agents, and biosensors.

In order to identify and measure certain biomolecules or analytes, analytical tools called biosensors combine a biomaterial with a transducer. Antibodies, enzymes, nucleic acids, and synthetic receptors that specifically bind to target molecules can all be utilised as biomaterials in biosensors (Khan et al. 2021, 2022; Parihar et al. 2022a, 2022c, 2023b, 2023a). To identify and measure certain antigens or antibodies in biological samples, immunosensors, for instance, use antibodies immobilised on a biomaterial surface (Huang et al. 2021). This makes detecting illnesses, including autoimmune disorders, quickly and accurately, cancer biomarkers, and infectious microorganisms possible. A "turn-on" magnetic fluorescence biosensor based on graphene quantum dots (GQDs), iron oxide (Fe_3O_4), and molybdenum disulfide

(MoS_2) nanosheets is an illustration of a natural biomaterial utilised in a biosensing application. This biosensor is made to separate and identify circulating tumour cells (CTCs) quickly, effectively, and sensitively. The biosensor uses photoluminescent GQDs functionalised with an aptamer that specifically targets receptors for the adhesion molecule for epithelial cells (EpCAM). A potential biosensing nanocomposite for fluorescence-guided tumour cell enrichment and bioimaging, the biosensing system exhibits minimal cytotoxicity, high capture efficiency, quick detection, and the capacity to identify as little as ten tumour cells in whole blood (Cui et al. 2019). Likewise, an example of a biopolymer used in an immunosensor is a label-free electrochemical strategy employing a glassy carbon electrode (GCE) modified with a biocompatible and conducting biopolymer functionalised molybdenum disulfide-reduced graphene oxide (CS–MoS_2/rGO) nanohybrid. This immunosensor detects specific antibodies against the SARS-CoV-2 virus through the interaction of recombinant SARS-CoV-2 Spike RBD protein (rSP) with the antibodies, resulting in decreased current responses measured by differential pulse voltammetry (DPV) (Yadav et al. 2023a).

Medical imaging techniques include magnetic resonance imaging (MRI), computed tomography (CT), positron emission tomography (PET), and ultrasound, all employ imaging agents based on biomaterials.

These substances, which are frequently nanoparticles or contrast agents, are made to make certain tissues, organs, or disease signs more visible. Lignocellulosic-based nanoparticles made from rose biomass waste illustrate a biomaterial utilised as an imaging agent. When stimulated, these nanoparticles' inherent luminescence causes them to produce light in the near-infrared (NIR) spectrum. They provide a strong photoluminescence signal without the necessity for encapsulation or imaging agent functionalisation. With an average size of 156 nm, they exhibit little *in vitro* toxicity (IC50 approximately 3 mg/mL), and at dosages as high as 57 mg/kg, they exhibit no *in vivo* toxicity (Arcentales et al. 2023). Due to their favourable properties, including simple manufacturing, biodegradability, and biocompatibility, biopolymers have drawn much interest. In addition, early transition metal nanoclusters known as inorganic polyoxometalates (POMs), a type of metal-oxygen anionic compounds, have a variety of useful applications in both the biomedical and industrial sectors (Mousavi et al. 2022). Similarly, Shaheen et al. (2023) produced composite hydrogel beads using biopolymers and ammonium metavanadate (AMV) to develop a biomaterial-based imaging agent. Through electrostatic interactions, cationic chitosan, chitosan/gelatine, or chitosan/methylcellulose were crosslinked with AMV to create the hydrogel beads. These hydrogel beads were yellow in colour and showed strong mechanical properties. The biomaterial-based hydrogel mixture demonstrated outstanding luminescence and adsorption abilities, good swelling behaviour (up to 432% maximum swelling), and adjustable physicochemical characteristics. The hydrogel beads demonstrated specific ascorbic acid sensing behaviour, making them appropriate for imaging applications. Furthermore, the hydrogels' antibacterial efficacy and cytotoxicity against embryonic kidney cells were evaluated, which revealed the dose-dependent cytotoxicity of the embedded imaging agents (Shaheen et al. 2023).

Biomaterials are used in microfluidic devices as microchannels, surfaces, or coatings to manipulate, analyse, and categorise biological samples precisely. These gadgets provide portable, high-throughput, and miniature diagnostic platforms (Parihar et al. 2020, 2021b, 2021a, 2023e; Yadav et al. 2023; Parihar and Dube 2023; Parihar et al. 2023d; Parihar and Khan 2023). Wearable contact lenses for non-invasive diagnostics are being developed, thanks to biomaterials revolutionising microfluidic devices. Yang et al. (2020) have developed flexible contact lenses with biocompatibility, hydrophilicity, and elasticity using a UV-curable biomaterial. Tears may now travel via capillary networks and reservoirs thanks to tear inlets built into the lenses. Biomarkers, such as glucose, chloride, and urea, elicited colorimetric changes by interacting with chemical substrates embedded into the lenses. In order to calculate the concentrations of biomarkers, external equipment was used to take pictures and analyse RGB data. *In vitro* investigations employing a synthetic microfluidic hydrogel eyeball showed reliable lens functioning. The tear storage and diagnostic capabilities of these biomaterial-based microfluidic contact lenses represent an interesting development for point-of-care diagnostics (Yang et al. 2020). In addition to this, Wang et al. (2023) created a point-of-care (POC) detection tool for salmon rotting utilising biomaterials. They combined bromothymol blue (BTB) and silk fibroin methacryloyl (SilMA) to generate a pH-responsive colorimetric microneedle (MN) patch (BTB/SilMA@MN patch). The MN patch included needle points that could cut through cling film on food and draw fluids from the fish tissue. The MN patch's colour changed as a result of the extracted fluids being transferred to the backing layer. The salmon's freshness could be seen in the colour transition from yellow to yellowish green to green. With the use of this POC equipment, salmon spoilage may be quickly and non-destructively determined, giving customers a simple and reliable way to gauge the quality of their food (Wang et al. 2023). Biomaterials, including hydrogels, quantum dots, gold nanoparticles, DNA/RNA probes, and fluorescent dyes, are a few examples that are employed in diagnostic applications. With the help of certain biological ligands, hydrogels may be functionalised to collect and concentrate desired biomarkers from difficult samples (Liu et al. 2022). Improved sensitivity, quick analysis, smaller sample quantities, and the possibility for point-of-care testing are some benefits of biomaterial-based diagnostics. They make it possible to diagnose diseases early, track treatment outcomes, and use personalised medical techniques (Gačanin et al. 2020). Current research focuses on creating new biomaterials, enhanced sensing methods, and digital technology integration for more accurate, effective, and easily accessible diagnostic instruments. By enabling early detection, precise diagnosis, and focused treatments for different diseases, these developments have the potential to revolutionise healthcare, as briefly described in Table 1.

7. Biomaterials in Biomedical Therapeutic Applications

By providing a wide range of therapy and recovery options, biomaterials have transformed biomedical therapeutic applications. These materials are widely employed in many therapeutic methods, such as implanted medical devices, drug delivery systems, tissue engineering, and regenerative medicine. Biomaterials

Table 1: Brief description of emerging biomaterials in diagnostic application.

Sr No.	Application	Biomaterials	Target Analyte/ Function	Key Features	References
1.	Biosensing (CTC detection)	Graphene quantum dots	Circulating tumour cells (CTCs)	Photoluminescent, high capture efficiency, low cytotoxicity	(Cui et al. 2019)
2.	Immunosensing (COVID-19 detection)	Biopolymer-based nanohybrid	Antibodies against SARS-CoV-2 virus	Label-free, electrochemical detection, decreased current response	(Yadav et al. 2023b)
3.	Medical Imaging	Lignocellulosic nanoparticles	Tissues, organs, disease signs	Near-infrared (NIR) luminescence, low toxicity	(Arcentales et al. 2023)
4.	Biomaterial-based Imaging Agent	Composite hydrogel beads	Ascorbic acid sensing, imaging agent	Strong mechanical properties, luminescence, adjustable characteristics	(Shaheen et al. 2023)
5.	Microfluidic Devices	UV-curable biomaterial	Non-invasive tear-based diagnostics	Biocompatible, hydrophilic, elasticity	(Yang et al. 2020)
6.	Point-of-Care (POC) Detection	pH-responsive microneedle patch	Salmon freshness	Colorimetric change, non-destructive, food quality assessment	(Wang et al. 2023b)
7.	Diagnostic Applications	Hydrogels, quantum dots, gold nanoparticles, DNA/RNA probes, fluorescent dyes	Biomarker collection and concentration	Improved sensitivity, quick analysis, point-of-care testing	(Liu et al. 2022)

serve as a scaffold or matrix in tissue engineering to facilitate the development and regeneration of sick or damaged tissues. They support the development of functional tissues by fostering an environment that is favourable for cellular adhesion, migration, and differentiation (Abdollahiyan et al. 2021). Natural polymers, synthetic polymers, or hybrid materials can be used to create biomaterial scaffolds that replicate the characteristics of the target tissue. They act as a support structure for cells to proliferate and regenerate, allowing for the replacement or repair of damaged organs or tissues (Elhadad et al. 2022). Biomaterials with electroconductive qualities are gaining popularity in tissue engineering applications for regenerating tissues, including myocardial, neurological, musculoskeletal, and bone tissue—tissues that require electrical activity. 2D MXene ($Ti_3C_2T_x$) was covalently crosslinked with collagen type I to develop a biohybrid platform by Asaro et al. (2023). Together with the increased rigidity and improved electrical conductivity offered by the two-dimensional titanium carbide material, the resultant platform demonstrated the bio functionality of collagen. High biocompatibility was shown by these

MXene-based systems, which encouraged cell growth and spread while inhibiting bacterial adhesion. Additionally, the platforms improved cell spreading, survivability, and expression of cx43 when grown with cardiomyocytes, indicating their potential for *in vitro* cardiac tissue regeneration (Asaro et al. 2023). Similarly, using a combination of piezoelectric and bioactive ceramic-crystallised glass, Polley et al. (2023) have developed macroporous biomaterial scaffolds using additive manufacturing techniques. Binder jetting is used to create and thoroughly characterise scaffolds made of 45S5 bioactive glass and barium titanate. These 3D-printed composite scaffolds have good mechanical strength and bioactive behaviour, which is shown by the buildup of calcium phosphate that resembles bone on the surface (Parihar et al. 2022b, 2023c; Kumar et al. (2022)). The scaffolds' piezoelectric constants, which range from 1 to 21 pC/N, mimic or even exceed those of bone. MC3T3-E1 osteoblast precursor cells have good cytocompatibility, cell attachment, and proliferation during *in vitro* investigations. These glass composites made of 45S5 ceramic-crystallised barium titanate show promise for use in bone tissue engineering (Polley et al. 2023).

Additionally, biomaterials are essential in drug delivery systems because they improve the effectiveness, security, and targeted administration of medicinal substances. Controlled release profiles can be obtained by encapsulating medications inside biocompatible and biodegradable carriers, such as nanoparticles, liposomes, or hydrogels. Biomaterial-based drug delivery methods enable localised or targeted distribution to certain tissues or cells, reducing adverse effects, and improving therapeutic results (Fenton et al. 2018). They can also increase medication stability, solubility, and bioavailability. For illustration, Kamaci and Kaya (2023) produced biodegradable hybrid hydrogels for drug delivery applications employing polyurethane with azomethine as a synthetic polymer and chitosan as a natural polymer. The hydrogels' thermal, mechanical, wettability, water uptake, biodegradation, protein absorption, drug loading, and release characteristics were examined, and FT-IR and SEM analyses were used to characterise them. The outcomes showed that the hybrid hydrogels had good mechanical strength, hydrophilicity, ability to absorb water, and biodegradation behaviour. The hybrid hydrogels also had a 50% drug release, which along with their favourable characteristics for drug delivery applications, made them a good choice for the controlled administration of the medication 5-fluorouracil (5-FU) (Kamaci and Kaya 2023).

Biomaterials serve as the basis for the creation and operation of implantable medical devices. Implants, prosthetics, orthopaedic devices, and cardiovascular devices are all made using biomaterials with the right mechanical characteristics, biocompatibility, and corrosion resistance. For instance, bioceramics or titanium alloys are frequently utilised in orthopaedic implants, whereas biocompatible polymers are used in tissue scaffolds or cardiovascular stents. The quality of life for patients is enhanced by the support, durability, and integration with the surrounding tissues that biomaterials in implanted devices offer (Nasr Azadani et al. 2022). Biomaterial advancements have resulted in the creation of mechanically compliant craniofacial implants that prioritise the preservation of the cerebrum after trauma or neurosurgery treatments. Rahmani et al. (2022) used powder bed fusion technology and powder metallurgy to embed three different types of bioactive ceramic powders into metallic scaffolds. The study also looked at using TiNi and $Ti_{22}Al_{25}Nb$ alloys for their dynamic impact resistance (Rahmani et al. 2022). The results demonstrated reduced porosity and limited cracking in the impact region, indicating the promise of these advancements in creating more effective and durable craniofacial implants. Biomaterial-based implant technological advancements have addressed the critical concern of peri-implant infection in dental implants. However, a recent development by Wu et al. (2021) involves an N-halamine polymeric coating on the titanium surface that exhibits long-lasting, renewable antibacterial properties while maintaining stability and biocompatibility. Importantly, the coating's antibacterial activity lasts long *in vitro*, in animal models, and even in the human oral cavity during the osseointegration process. Furthermore, the coating is easily reusable using chlorine, allowing for a renewable antibacterial coating idea for dental implants (Wu et al. 2021).

The science of biomaterials is evolving and progressing, providing new options for biomedical therapeutic applications, as briefly described in Table 2. Researchers continually investigate new materials, creating novel production procedures and improving their qualities to satisfy specific therapeutic requirements. Biomaterials have enormous promise for improving patient outcomes, enabling personalised treatment, and tackling difficult medical problems, all of which contribute to improving healthcare and human well-being.

Table 2: Brief description of emerging biomaterials in therapeutic application.

Sr No.	Therapeutic Application	Biomaterials	Function	Key Features	References
1.	Tissue Engineering	2D MXene	Cardiac tissue regeneration	Biohybrid platform, electrical conductivity, biocompatibility	(Asaro et al. 2023)
2.	Tissue Engineering	Piezoelectric ceramic-glass composite	Bone tissue engineering	Good mechanical strength, bioactive behaviour	(Polley et al. 2023)
3.	Drug Delivery Systems	Biodegradable hybrid hydrogels	Controlled drug delivery	Mechanical strength, biodegradation, drug release	(Kamaci and Kaya 2023)
4.	Implantable Medical Devices	Bioceramics and titanium alloys	Orthopaedic implants	Mechanical characteristics, biocompatibility	(Nasr Azadani et al. 2022)
5.	Implantable Medical Devices	N-halamine polymeric coating	Dental implants	Long-lasting antibacterial properties, biocompatibility	(Wu et al. 2021)

8. Challenges and Future Directions

Although the field of biomaterials has made significant strides, there are still a number of obstacles to overcome and possibilities to seize for future development and use. The demand for increased biocompatibility and long-term stability of biomaterials within the physiological milieu is a significant problem. While many biomaterials initially appear compatible, extended interactions with host tissues may result in unfavourable reactions or degradation, limiting their usefulness (Winter et al. 2007; Raza et al. 2022). It is still important to create biomaterials that can operate properly and maintain their structural integrity throughout time. The ability to precisely regulate how cells and tissues interact with biomaterials is another difficulty. In order to encourage tissue regeneration and integration, biomaterials should be made to direct cellular behaviour, such as adhesion, proliferation, and differentiation (Yu et al. 2015). A greater comprehension of cell-material interactions, including surface changes, bioactive chemicals, and tissue-specific signals, is necessary to accomplish this (Joyce et al. 2021). Additionally, the creation of biomaterials with many functions is a developing field. Highly desired biomaterials can administer therapeutic agents, offer prolonged drug release, react to outside stimuli, or actively engage with cells and tissues (Fenton et al. 2018). In order to create such multifunctional biomaterials, the integration of cutting-edge technologies such as nanotechnology, 3D printing, and bio-fabrication offers promise. Another crucial factor is dealing with the biomaterials' scalability and cost-effectiveness. The production of many biomaterials in large numbers or at reasonable prices is frequently difficult, especially for those with complicated structures or advanced characteristics (Ventola 2014). The transition of biomaterials from the lab to clinical applications can be facilitated by streamlining production procedures and investigating novel fabrication techniques.

Future potential for biomaterials includes personalised and regenerative medicine, which are developing fields. Intense research is being done on producing tissue-engineered constructions for organ replacement and regeneration, as well as customising biomaterials to the demands of individual patients (Bomkamp et al. 2022). In a nutshell, even though the field of biomaterials has made great progress, issues including biocompatibility, cellular interactions, multifunctionality, scalability, and personalised medicine still need to be resolved (Wang et al. 2023). The future of biomaterials will be enabled by creative solutions for a variety of biomedical applications, eventually improving patient outcomes, as shown in Figure 6, by addressing these issues and using developing technology.

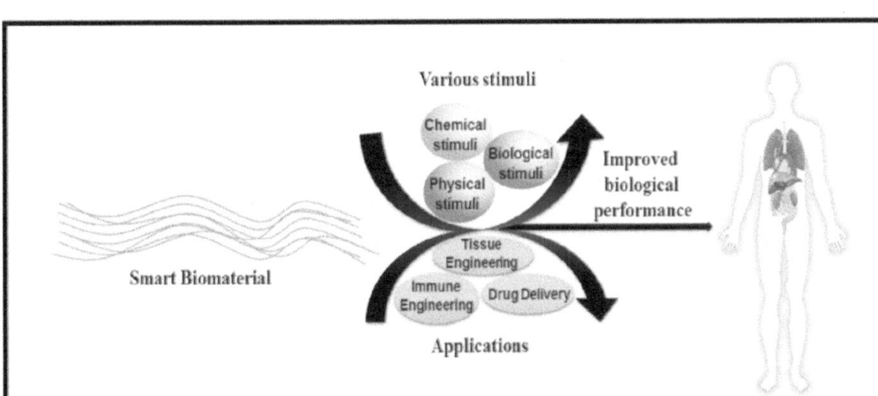

Figure 6: Smart biomaterial utilise in the biomedical domain to improve biological performance (Kowalski et al. 2018).

9. Conclusion

With their enormous potential to revolutionise healthcare and raise people's quality of life worldwide, biomaterials have become an essential part of the biomedical sector. They have a long, illustrious history, and conventional and novel biomaterials have developed significantly in recent years. The effectiveness of the use of biomaterials in diverse biological areas is greatly influenced by their features, including biocompatibility, mechanical capabilities, biodegradability, and surface changes. Both diagnostic and therapeutic applications benefit greatly from the use of biomaterials. Biomaterials are used in diagnostics to create microfluidic devices, medical imaging agents, and biosensors that enable early illness detection, accurate diagnosis, and individualised care. Biomaterials are used as scaffolds in therapeutics for tissue engineering, drug delivery systems, and implanted medical devices, promoting controlled drug release, controlled tissue regeneration, and better patient outcomes. However, biomaterials face a number of difficulties and chances in the future. The development of multifunctional biomaterials, scalability, cost-effectiveness, and precise control of cell-material interactions are continuing problems. Personalised medicine, regenerative medicine, and the incorporation of cutting-edge technologies like nanotechnology and 3D printing are the future directions. The future of biomaterials will be paved with solutions to these problems and the effective use of cutting-edge technology. It will result in novel treatments for a variety of biological issues, revolutionising medical care and eventually enhancing patient outcomes. Biomaterials will continue to influence medicine's future and help progress healthcare technology as long as scientists, engineers, and doctors collaborate on their research.

References

Abdollahiyan, P., F. Oroojalian and A. Mokhtarzadeh. 2021. The triad of nanotechnology, cell signalling, and scaffold implantation for the successful repair of damaged organs: An overview on soft-tissue engineering. Journal of Controlled Release 332: 460–492.

Anjum, S., F. Rahman, P. Pandey et al. 2022. Electrospun biomimetic nanofibrous scaffolds: a promising prospect for bone tissue engineering and regenerative medicine. International Journal of Molecular Sciences 23(16): 9206.

Arcentales, M., R. Martín-Sampedro, R. Santos-Oliveira et al. 2023. Lignocellulosic-based nanoparticles with photoluminescent properties for bioimaging. ACS Applied Materials & Interfaces.

Arif, Z. U., M. Y. Khalid, R. Noroozi et al. 2023. Additive manufacturing of sustainable biomaterials for biomedical applications. Asian Journal of Pharmaceutical Sciences: 100812.

Asaro, G. A., M. Solazzo, M. Suku et al. 2023. MXene functionalized collagen biomaterials for cardiac tissue engineering driving iPSC-derived cardiomyocyte maturation. npj 2D Materials and Applications 7(1): 44.

Bandyopadhyay, A., I. Mitra, S. B. Goodman et al. 2022. Improving biocompatibility for next generation of metallic implants. Progress in Materials Science: 101053.

Basu, B., N. Gowtham, Y. Xiao et al. 2022. Biomaterialomics: Data science-driven pathways to develop fourth-generation biomaterials. Acta Biomaterialia 143: 1–25.

Bauer, S., P. Schmuki, K. Von Der Mark et al. 2013. Engineering biocompatible implant surfaces: Part I: Materials and surfaces. Progress in Materials Science 58(3): 261–326.

Bomkamp, C., S. C. Skaalure, G. F. Fernando et al. 2022. Scaffolding biomaterials for 3D cultivated meat: prospects and challenges. Advanced Science 9(3): 2102908.

Brown, M., J. Li, C. Moraes et al. 2022. Decellularized extracellular matrix: New promising and challenging biomaterials for regenerative medicine. Biomaterials: 121786.

Choi, S.-r., J.-w. Kwon, K.-s. Suk et al. 2023. The clinical use of osteobiologic and metallic biomaterials in orthopedic surgery: the present and the future. Materials 16(10): 3633.

Cui, F., J. Ji, J. Sun et al. 2019. A novel magnetic fluorescent biosensor based on graphene quantum dots for rapid, efficient, and sensitive separation and detection of circulating tumor cells. Analytical and Bioanalytical Chemistry 411: 985–995.

Egbo, M. K. 2021. A fundamental review on composite materials and some of their applications in biomedical engineering. Journal of King Saud University-Engineering Sciences 33(8): 557–568.

Elhadad, A. A., A. Alcudia, B. Begines et al. 2022. A multidisciplinary perspective on the latest trends in artificial cartilage fabrication to mimic real tissue. Applied Materials Today 29: 101603.

Fenton, O. S., K. N. Olafson, P. S. Pillai et al. 2018. Advances in biomaterials for drug delivery. Advanced Materials 30(29): 1705328.

Freedman, B. R. and D. J. Mooney. 2019. Biomaterials to mimic and heal connective tissues. Advanced Materials 31(19): 1806695.

Gačanin, J., C. V. Synatschke and T. Weil. 2020. Biomedical applications of DNA-based hydrogels. Advanced Functional Materials 30(4): 1906253.

Han, X., A. Alu, H. Liu et al. 2022. Biomaterial-assisted biotherapy: A brief review of biomaterials used in drug delivery, vaccine development, gene therapy, and stem cell therapy. Bioactive Materials 17: 29–48.

Harish, V., D. Tewari, M. Gaur et al. 2022. Review on nanoparticles and nanostructured materials: Bioimaging, biosensing, drug delivery, tissue engineering, antimicrobial, and agro-food applications. Nanomaterials 12(3): 457.

Hassan, T., A. Salam, A. Khan et al. 2021. Functional nanocomposites and their potential applications: A review. Journal of Polymer Research 28: 1–22.

Hildebrand, H. F. 2013. Biomaterials–a history of 7000 years. BioNanoMaterials 14(3-4): 119–133.

Huang, W., S. Ling, C. Li et al. 2018. Silkworm silk-based materials and devices generated using bio-nanotechnology. Chemical Society Reviews 47(17): 6486–6504.

Huang, X., Y. Zhu and E. Kianfar. 2021. Nano biosensors: properties, applications and electrochemical techniques. Journal of Materials Research and Technology 12: 1649–1672.

Joyce, K., G. T. Fabra, Y. Bozkurt et al. 2021. Bioactive potential of natural biomaterials: Identification, retention and assessment of biological properties. Signal Transduction and Targeted Therapy 6(1): 122.

Kamaci, M. and I. Kaya. 2023. Chitosan based hybrid hydrogels for drug delivery: Preparation, biodegradation, thermal, and mechanical properties. Polymers for Advanced Technologies 34(2): 779–788.

Khalilov, R. 2023. A comprehensive review of advanced nano-biomaterials in regenerative medicine and drug delivery. Advances in Biology & Earth Sciences 8(1).

Khan, A. and K. A. Alamry. 2021. Recent advances of emerging green chitosan-based biomaterials with potential biomedical applications: A review. Carbohydrate Research 506: 108368.

Khan, R., A. Parihar, A. Kaushik et al. 2022. Advanced Biosensors for Virus Detection: Smart Diagnostics to Combat SARS-CoV-2.

Khan, R., A. Parihar and S. Sanghi. 2021. Biosensor Based Advanced Cancer Diagnostics: From Lab to Clinics

Kumar, A., Dr. A. Parihar, U. Panda et al. 2022. Microfluidics-based point-of-care testing (POCT) devices in dealing with waves of COVID-19 Pandemic: The Emerging Solution. ACS Appl. Bio. Mater 5: 2046–2068. https://doi.org/10.1021/acsabm.1c01320.

Kowalski, P. S., C. Bhattacharya, S. Afewerki et al. 2018. Smart biomaterials: recent advances and future directions. ACS Biomaterials Science & Engineering 4(11): 3809–3817.

Liu, M., L. Wang, Y. Lo et al. 2022. Aptamer-enabled nanomaterials for therapeutics, drug targeting and imaging. Cells 11(1): 159.

Liu, S., J.-M. Yu, Y.-C. Gan et al. 2023. Biomimetic natural biomaterials for tissue engineering and regenerative medicine: new biosynthesis methods, recent advances, and emerging applications. Military Medical Research 10(1): 16.

Losic, D. 2021. Advancing of titanium medical implants by surface engineering: Recent progress and challenges. Expert Opinion on Drug Delivery 18(10): 1355–1378.

Malekmohammadi, S., N. Sedghi Aminabad, A. Sabzi et al. 2021. Smart and biomimetic 3D and 4D printed composite hydrogels: opportunities for different biomedical applications. Biomedicines 9(11): 1537.

Mohd Isa, I. L., S. A. Mokhtar, S. A. Abbah et al. 2022. Intervertebral disc degeneration: biomaterials and tissue engineering strategies toward precision medicine. Advanced Healthcare Materials 11(13): 2102530.

Mondschein, R. J., A. Kanitkar, C. B. Williams et al. 2017. Polymer structure-property requirements for stereolithographic 3D printing of soft tissue engineering scaffolds. Biomaterials 140: 170–188.

Mousavi, S. M., S. A. Hashemi, S. Mazraedoost et al. 2022. Anticancer, antimicrobial and biomedical features of polyoxometalate as advanced materials: a review study. Inorganic Chemistry Communications: 110074.

Mubarak, N. M., M. Anwar, S. Debnath et al. 2023. Application to Biomaterials. Fundamentals of Biomaterials: A Supplementary Textbook, Springer: 129–139.

Mukherjee, C., D. Varghese, J. Krishna et al. 2023. Recent advances in biodegradable polymers–properties, applications and future prospects. European Polymer Journal: 112068.

Nasr Azadani, M., A. Zahedi, O. K. Bowoto et al. 2022. A review of current challenges and prospects of magnesium and its alloy for bone implant applications. Progress in Biomaterials 11(1): 1–26.

Nguyen, T. P., Q. V. Nguyen, V.-H. Nguyen et al. 2019. Silk fibroin-based biomaterials for biomedical applications: A review. Polymers 11(12): 1933.

Ostroff, E., K. Parekh, A. Prominski et al. 2022. Biocompatible and nanoenabled technologies for biological modulation. Advanced Materials Technologies 7(2): 2100216.

Øvrebø, Ø., G. Perale, J. P. Wojciechowski et al. 2022. Design and clinical application of injectable hydrogels for musculoskeletal therapy. Bioengineering & Translational Medicine 7(2): e10295.

Parihar, A., N. K. Choudhary, P. Sharma et al. 2023a. Carbon nanomaterials-based electrochemical aptasensor for point-of-care diagnostics of cancer biomarkers. Mater Today Chem. 30: 101499. https://doi.org/10.1016/J.MTCHEM.2023.101499.

Parihar, A., N. K. Choudhary, P. Sharma et al. 2023b. MXene-based aptasensor for the detection of aflatoxin in food and agricultural products. Environmental Pollution 316: 120695. https://doi.org/10.1016/J.ENVPOL.2022.120695.

Parihar, A. and A. Dube. 2023. Structural alterations in cell organelles induced by photodynamic treatment with chlorin p6-histamine conjugate in human oral carcinoma cells probed by 3D fluorescence microscopy. Luminescence 38: 1175–1184. https://doi.org/10.1002/BIO.4307.

Parihar, A. and R. Khan. 2023. Yttrium functionalized reduced graphene oxide nanocomposite-based aptasensor for ultrasensitive detection of a breast cancer biomarker. ACS Appl. Nano. Mater. 0: https://doi.org/10.1021/acsanm.3c03234.

Parihar, A., R. Khan, A. Kumar et al. 2022a. Computational Approaches for Novel Therapeutic and Diagnostic Designing to Mitigate SARS-CoV2 Infection: Revolutionary Strategies to Combat Pandemics. Computational Approaches for Novel Therapeutic and Diagnostic Designing to Mitigate SARS-CoV2 Infection: Revolutionary Strategies to Combat Pandemics 1–594. https://doi.org/10.1016/C2020-0-04145-9

Parihar, A., A. Kumar, U. Panda et al. 2023c. Cryopreservation: a comprehensive overview, challenges, and future perspectives. Adv. Biol. 7: 2200285. https://doi.org/10.1002/ADBI.202200285.

Parihar, A., V. Pandita and R. Khan. 2022b. 3D printed human organoids: High throughput system for drug screening and testing in current COVID-19 pandemic. Biotechnol. Bioeng. 119: 2669–2688. https://doi.org/10.1002/BIT.28166.

Parihar, A., V. Pandita, A. Kumar et al. 2021a. 3D Printing: advancement in biogenerative engineering to combat shortage of organs and bioapplicable materials. Regenerative Engineering and Translational Medicine 2021 1–27. https://doi.org/10.1007/S40883-021-00219-W.

Parihar, A., P. Ranjan, S. K. Sanghi et al. 2020. Point-of-care biosensor-based diagnosis of COVID-19 holds promise to combat current and future pandemics. ACS Appl. Bio. Mater 3: 7326–7343. https://doi.org/10.1021/ACSABM.0C01083.

Parihar, A., P. Sharma, N. Kumar Choudhary et al. 2023d. Green Synthesis of CdS and CdS/rGO nanocomposites: evaluation of electrochemical, antimicrobial, and photocatalytic properties. ACS Appl. Bio. Mater. 6: 3706–3716. https://doi.org/10.1021/acsabm.3c00390.

Parihar, A., R. Shrivastava and A. Dube. 2021b. Interaction of Cp6-his and Cp6 with bovine serum albumin and liver microsomes: Spectroscopic and molecular docking studies. J. Photochem. Photobiol. 5: 100013. https://doi.org/10.1016/j.jpap.2020.100013.

Parihar, A., A. Singhal, N. Kumar et al. 2022c. Next-generation intelligent MXene-Based electrochemical aptasensors for point-of-care cancer diagnostics. Nano-Micro. Letters 14: 1–34. https://doi.org/10.1007/S40820-022-00845-1.

Parihar, A., S. Yadav, M. A. Sadique et al. 2023e. Internet-of-medical-things integrated point-of-care biosensing devices for infectious diseases: Toward better preparedness for futuristic pandemics. Bioeng. Transl. Med. e10481. https://doi.org/10.1002/BTM2.10481.

Pazhouhnia, Z., N. Beheshtizadeh, M. S. Namini et al. 2022. Portable hand-held bioprinters promote in situ tissue regeneration. Bioengineering & Translational Medicine 7(3): e10307.

Polley, C., T. Distler, C. Scheufler et al. 2023. 3D printing of piezoelectric and bioactive barium titanate-bioactive glass scaffolds for bone tissue engineering. Materials Today Bio.: 100719.

Radu, E. R., S. I. Voicu and V. K. Thakur. 2023. Polymeric membranes for biomedical applications. Polymers 15(3): 619.

Rahmani, R., N. Kamboj, M. Brojan et al. 2022. Hybrid metal-ceramic biomaterials fabricated through powder bed fusion and powder metallurgy for improved impact resistance of craniofacial implants. Materialia 24: 101465.

Rahmati, M., D. K. Mills, A. M. Urbanska et al. 2021. Electrospinning for tissue engineering applications. Progress in Materials Science 117: 100721.

Raza, S., E. Ghasali, Y. Orooji et al. 2022. Two dimensional (2D) materials and biomaterials for water desalination; structure, properties, and recent advances. Environmental Research: 114998.

Sánchez-Cid, P., M. Jiménez-Rosado, A. Romero et al. 2022. Novel trends in hydrogel development for biomedical applications: A review. Polymers 14(15): 3023.

Shaheen, A., M. I. Zargar and A. A. Dar. 2023. Luminescent and self-healing hybrid ionotropic hydrogel beads of ammonium metavanadate and chitosan: promising biomaterial as an antimicrobial agent, efficient dye adsorbent and ascorbic acid sensor. Soft Matter.

Sidambe, A. T. 2014. Biocompatibility of advanced manufactured titanium implants—A review. Materials 7(12): 8168s8188.

Sohn, S., M. V. Buskirk, M. J. Buckenmeyer et al. 2020. Whole organ engineering: approaches, challenges, and future directions. Applied Sciences 10(12): 4277.

Sonatkar, J., B. Kandasubramanian and S. O. Ismail. 2022. 4D printing: Pragmatic progression in biofabrication. European Polymer Journal 169: 111128.

Suwardi, A., F. Wang, K. Xue et al. 2022. Machine learning-driven biomaterials evolution. Advanced Materials 34(1): 2102703.

Thanigaivel, S., A. Priya, D. Balakrishnan et al. 2022. Insight on recent development in metallic biomaterials: Strategies involving synthesis, types and surface modification for advanced therapeutic and biomedical applications. Biochemical Engineering Journal 187: 108522.

Troy, E., M. A. Tilbury, A. M. Power et al. 2021. Nature-based biomaterials and their application in biomedicine. Polymers 13(19): 3321.

Vaiani, L., A. Boccaccio, A. E. Uva et al. 2023. Ceramic materials for biomedical applications: an overview on properties and fabrication processes. Journal of Functional Biomaterials 14(3): 146.

Vasanthan, K. S., V. Srinivasan, V. Mathur et al. 2022. 3D Bioprinting for esophageal tissue regeneration: A review. Journal of Materials Research 37(1): 88–113.

Ventola, C. L. 2014. Medical applications for 3D printing: current and projected uses. Pharmacy and Therapeutics 39(10): 704.

Wang, J.-Y., L.-J. Chen, X. Zhao et al. 2023. Silk fibroin-based colorimetric microneedle patch for rapid detection of spoilage in packaged salmon samples. Food Chemistry 406: 135039.

Wang, L., C. Wang, S. Wu et al. 2020. Influence of the mechanical properties of biomaterials on degradability, cell behaviors and signaling pathways: current progress and challenges. Biomaterials Science 8(10): 2714–2733.

Wang, N., Y. Dai, J. Y. H. Fuh et al. 2023. Applications of triboelectric nanogenerators in bone tissue engineering. Advanced Materials Technologies: 2201310.

Wang, X., D. Wu, W. Li et al. 2021. Emerging biomaterials for reproductive medicine. Engineered Regeneration 2: 230–245.

Winter, J. O., S. F. Cogan and J. F. Rizzo. 2007. Retinal prostheses: current challenges and future outlook. Journal of Biomaterials Science, Polymer Edition 18(8): 1031–1055.

Wu, S., J. Xu, L. Zou et al. 2021. Long-lasting renewable antibacterial porous polymeric coatings enable titanium biomaterials to prevent and treat peri-implant infection. Nature Communications 12(1): 3303.

Yadav, S., M. A. Sadique, P. Ranjan et al. 2023. Synergistically functionalized molybdenum disulfide-reduced graphene oxide nanohybrid based ultrasensitive electrochemical immunosensor for real sample analysis of COVID-19. Analytica Chimica Acta 1265: 341326.

Yadav, S., A. Parihar, M. Abubakar Sadique et al. 2023. Emerging point-of-care optical biosensing technologies for diagnostics of microbial infections. ACS Applied Optical Materials 1: 1245–1262. https://doi.org/10.1021/acsaom.3c00129.

Yang, L., S. Pijuan-Galito, H. S. Rho et al. 2021. High-throughput methods in the discovery and study of biomaterials and materiobiology. Chemical Reviews 121(8): 4561–4677.

Yang, X., H. Yao, G. Zhao et al. 2020. Flexible, wearable microfluidic contact lens with capillary networks for tear diagnostics. Journal of Materials Science 55: 9551–9561.

Yu, X., X. Tang, S. V. Gohil et al. 2015. Biomaterials for bone regenerative engineering. Advanced Healthcare Materials 4(9): 1268–1285.

Zhang, F. and M. W. King. 2020. Biodegradable polymers as the pivotal player in the design of tissue engineering scaffolds. Advanced Healthcare Materials 9(13): 1901358.

Zhang, H., J. Guo, Y. Wang et al. 2022. Natural polymer-derived bioscaffolds for peripheral nerve regeneration. Advanced Functional Materials 32(41): 2203829.

Zhang, K., B. Ma, K. Hu et al. 2022. Evidence-based biomaterials research. Bioactive Materials 15: 495–503.

Zielińska, A., J. Karczewski, P. Eder et al. 2023. Scaffolds for drug delivery and tissue engineering: The role of genetics. Journal of Controlled Release 359: 207–223.

2

Properties of Biomaterial

Surender Singh Gulia,[1] *Reena Antil,*[1,*] *Navneet Kumari*[2] and *Jyoti Lathwal*[3]

1. Introduction

The term "biomaterial" refers to any material that can be employed for biomedical and therapeutic applications. Biomaterials are not a new field of study; they have been exercised in medical science over the last many decades. Non-biological elements were first introduced into the human body many ages ago, in pre-history. They are non-viable materials that are utilized in medical devices to interface with the biological system. They are employed in cardiovascular materials, neural prostheses, bone substitutes, dental implants, orthopedics prostheses, plasma and cytapheresis, and dialysis (Hanker and Giammara 1988). Different branches of sciences, viz. medicine, biology, tissue engineering, chemistry, and material science, are all incorporated under biomaterial science. In some cases, surgical repairs to the body are frequently required because of trauma, degeneration, and illness. Some organs, viz. knees, elbows, hips, finger joints, teeth, vertebrae, and other bodily vital organs such as kidney, heart, and skin are generally replaced in these procedures. All of these materials, each performing a certain function when live materials are substituted, are referred to as biomaterials. There are many definitions given to biomaterials; according to Helmus et al. (2008), biomaterials are materials employed in medical applications in medical devices, especially in situations where the device contacts are temporarily inserted and have specific design requirements. A biomaterial is a material used in the form of specialized structures to construct prostheses and biomedical devices that are designed to replace or restore impaired body functions in order to save or make life better (Gobbi et al. 2020; Miljana et al. 2020). Biomaterials

[1] Pt. Nekiram Sharma Government College Rohtak, Haryana.
[2] Department of Botany, Baba Mast Nath University Rohtak – 124001 (India).
[3] Department of Botany, Maharshi Dayanand University, Rohtak, Haryana (India).
* Corresponding author: rinantil@gmail.com

are materials that interact with biological systems to evaluate, treat, augment, or replace any organ tissues or function in the body (Williams et al. 1992). A biomaterial is a substance that has been designed to take a form that can be utilized to direct the course of any therapeutic or diagnostic operation in human beings and veterinary medicine by controlling interactions with living system components (Heimann 2020). Some commonly used biomaterials for medical implants are described as bioactive materials, synthetic and biodegradable polymers, natural macromolecules, metal, carbon, and ceramics that may include cellulose, silicon rubber, dacron, poly (methyl methacrylate), polyurithanes, hydrogels, stainless steel, alumina, titanium, hydroxyapatite and collagen (Boretos and Eden 1984; Helmus 2003; Markhoff et al. 2017). Different biomaterials are used to design different types of medical devices depending on their properties (Table 1). The major goal is to extend the lifespan of the biomaterial implanted or to ensure that it lasts to the end of life without defects or the need for reoperation, thereby improving the patient's quality of life. In nature, a variety of materials and alloys are accessible, but only a few of them are biocompatible (Jenkoa et al. 2018). As a result, the biomaterial must meet a number of basic properties particular to the application, including acceptable mechanical strength, corrosion resistance, biocompatibility, high wear resistance, low friction, and mechanical compatibility (Gobbi et al. 2020). The objective of this chapter is to introduce the reader to some basic properties of biomaterials.

2. Properties of Biomaterial

Three fundamental properties must be met for any material to be used for biomedical applications. The first requirement is that the material be compatible, meaning the organism should not perceive it as a foreign thing. Second, the material must be biodegradable; it must disintegrate harmlessly in the organism's body, allowing it to resume normal function. Third, the material must be mechanically sound; for load-bearing structural replacements, the material must have similar or superior mechanical stability to assure the graft's high reliability.

The factors that should be taken into account while selecting a biological material include a clear specification of the material's intended function, a precise description of the environment in which it must operate, as well as the influence that environment will have on the material's qualities; a clear knowledge of what it means to be safe for human usage.

Some of the important properties (Figure 1) that a biomaterial should possess in order to perform well as an implant in a biological system are given below:

2.1 Biocompatible

One of the most important properties that differentiate a biomaterial from other materials is its biocompatibility. The proper host response to biomaterials is a concept used to describe this phenomenon. Biomaterials success is inextricably linked to their acceptance by the human body. A biomaterial must not cause any adverse effect to the beneficiary, viz. any injury, cytotoxicity, mutagenicity, or carcinogenicity (Freire et al. 2015). It should perform the desired function in relation to medical therapy and should generate the most appropriate beneficial cellular and tissue responses

Table 1: Biomaterial classification and their associated properties (Wagner 2020; Todros et al. 2021; Bharadwaj 2021; Suwardi et al. 2022).

Attributes	Polymers	Metals and Alloys	Ceramics
Different types of bonds present	Covalent and wander wall forces	Metallic	Covalent and Ionic
Melting point	Low	Intermediate	High
Chemical stability	Poor	Good	Very High
Thermal conductivity	Low to intermediate	High	Low
Electrical conductivity	Very high	High	Very low but varies
Properties and advantages	Degradable, inert, akin to soft tissue and processing ability, eco-friendly to body tissues, Resilience, resistance to rust, less load, simple to construct	Hardness, high strength, enough vigor, outstanding tiredness resistance	Non-conductivity and inertness closely resemble biological bone qualities, eco-friendly to living tissues, resistant to rust and wear, less density
Mechanical deformation	Very high; plastic	High (Ductile)	Low (brittle)
Major issues	Thermally unstable; low strength	Wear and corrosion	High-density and brittle
Examples	Polymethylmethacrilate (PMMA), Polylactic acid (PLA), Poly tetrafluroethylene (PTFE), Nylone, Polyurethane, Polyethylene, Celluloid, Cellophane, Polycaprolactone (PCL), Polylactic acid (PLA), Polyglycolic (PGA)acid and Polyethylene glycol (PEG)	Stainless steel, titanium, Co CrMo, nitinol, platinum, nickel, tantalum	Alumina, hydroxyapatite zirconia, pyrolytic carbon, β tri-calcium phosphate
Biomedical applications	Drug delivery systems, orthopedic, orthodontic, cardiovascular, breast implants, scaffold for soft tissue	Especially hard tissue applications (orthopedic and dental implants)	Orthopedic, orthodontic, cardiovascular

(Naahidi et al. 2013). A biomaterial implant should be compatible in order to work properly in the body of the patient (Todros et al. 2021). Biocompatibility generally refers to biomaterial's ability to perform in a given application while exhibiting a sufficient host reaction without causing an inflammatory or allergic reaction (Davis et al. 2022). Biomaterials are required to undergo a variety of tests to determine their biocompatibility, which are carried out by following the recommendations of several regulatory organizations (Williams et al. 2008). Biocompatibility testing determines a

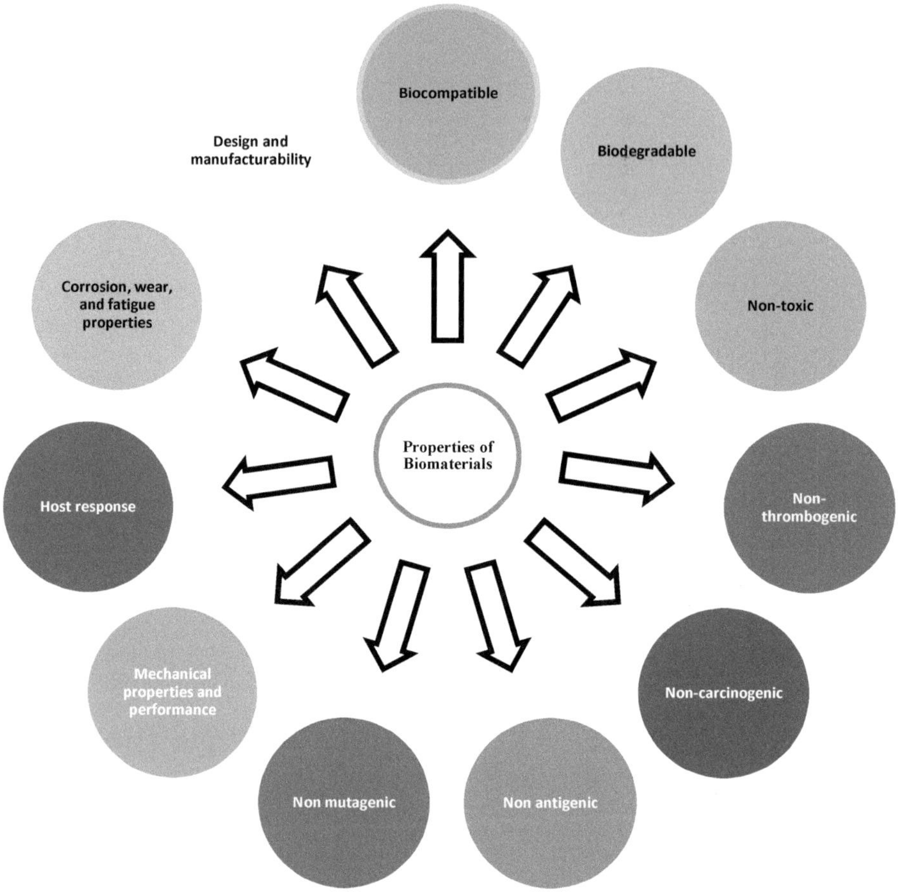

Figure 1: Different properties associated with biomaterials.

biomaterial's technical suitability for human use by scrutinizing possibly detrimental physiological effects (Davis et al. 2022). Some of the common problems associated with biomaterials mainly include premature wear and dislocation of hip and knee prostheses, infection and rejection of dental implants, restenosis and obstruction of coronary stents, etc. Furthermore, it is estimated that bacterial biofilms formed on surfaces of biomaterials, implants, or surgical devices cause 80% of infections in hospitals. These biofilms are up to thousands of times more resistant to antimicrobial agents than planktonic bacteria (Costerton et al. 1999). Henceforth, the implant's own physical, mechanical, degradation properties, sterilization (implant's surface free from all types of microorganisms), toxicological aspect (intoxicating substances), implantation site, surgical procedure, material-cell interfacial interactions, and load-bearing capacity all play a significant role in the biocompatibility of a biomedical implant's clinical performance (Figure 2) (Davis et al. 2022).

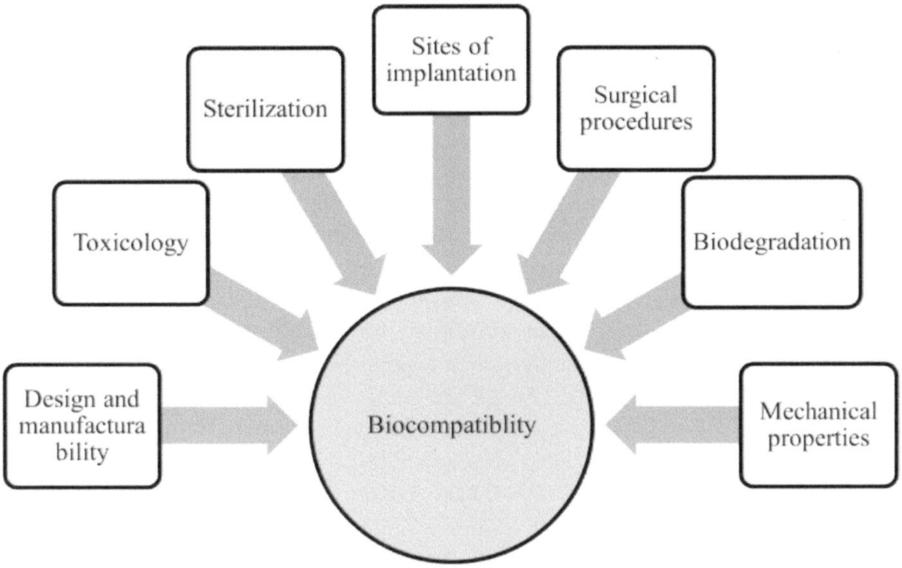

Figure 2: Important factors affecting biocompatibility of biomedical implant.

2.2 Non-Thrombogenic

In the field of biomaterials, non-thrombogenic materials and surface treatments are still needed. Biomaterial must be non-thrombogenic in blood-contacting applications to avoid thrombi and emboli-related problems. Biomaterials and tissue-engineered constructions that imitate the vasculature are being established in response to this challenge. Various biomaterials-based blood-contacting medical devices are frequently utilized to treat a variety of cardiovascular disorders. Cardiovascular device failure is frequently caused by thrombus development (Braune et al. 2019). There are presently no clinically viable biomaterials that can completely prevent thrombosis in more difficult situations (i.e., low flow in the venous system). Although some biomaterials diminish protein adsorption or cell adhesion, thrombosis and inflammation are still a concern. Elastin is an extracellular matrix macromolecule that contributes to arterial elasticity, accounting for up to 50% of the vessel's non-hydrated mass. Elastin is a widely used biomaterial because of its non-thrombogenic properties (Waterhouse et al. 2021).

2.3 Biodegradable

Biodegradable biomaterials are of considerable interest because of their potential to be absorbed gradually by the body. A biodegradable product must have a harmless breakdown product that the body's metabolic pathways can eliminate. Biopolymers are now widely recognized and employed in biomedical applications for a variety of functions. Tissue engineering, drug delivery systems, and therapeutics have all benefited from these biodegradable polymers. (1) Large implants including bone plates, bone screws, and contraceptive reservoir; (2) small implants viz. sutures, staples, and micro-sized drug delivery vehicles; (3) multifilament meshes or porous structures for tissue engineering; and (4) plain membranes for guided tissue regeneration are some of the present biomedical applications of biodegradable polymeric materials (Vert 2005). The most widely utilized biodegradable compounds are homopolymers or copolymers of alpha hydroxyl acids, such as lactic and glycolic acid. The most important significance of biodegradable implants over metal implants is that they do not need a second surgical procedure to remove the devices. Drug delivery systems or other implants made of natural polymers reduce the risk of persistent inflammation, immunological responses, and toxicity, which are common with synthetic polymers (Bostman and Pihlajamaki 2000). Polymeric materials are quickly displacing other material types in biomaterials, such as metals, alloys, and ceramics, due to their versatility. The sales of biomaterials are continuously increasing every year, and it was predicted that there will be a huge market for biodegradable biomaterials in the coming decades (Naira et al. 2007).

Biodegradable biomaterials should have some important properties, as shown below (Lloyd 2002).

- The substance must have a long shelf life.
- The breakdown product must not be hazardous, digestible and excreted from the body.
- The material should have permeability and processibility suitable for the intended use.
- The deterioration time of the material should correlate to the healing or regeneration process.
- Mechanical properties of the material must really be appropriate for the application, and changes in mechanical properties with deterioration must be compatible with the healing and regeneration process.
- The material's mechanical characteristics should be appropriate for the application, and the change in mechanical properties with deterioration should be compatible with the healing or regeneration process.

Polymeric biomaterials can be divided into two types based on how they degrade: (1) hydrolyticälly degrädäble polymers and (2) enzymäticälly degradable polymers.

1. Hydrolyticälly Degrädäble Polymers

Polymers with hydrolytically labile chemical linkages in their backbone are hydrolytically degradable. Esters, orthoesters, anhydrides, carbonates, amides, urethanes, ureas, and other functional groups are sensitive to hydrolysis (Li et al. 1999). Natural polymers were the first biodegradable biomaterials to be employed in clinical trials. When compared to enzymatically degradable polymers, hydrolytically degradable polymers are chosen as implants because there is less variation from location to location and patient to patient (Katti et al. 2002).

2. Enzymaticälly Degrädäble Polymers

The majority of polymers found in nature are degraded by enzymes. However, depending on the availability and quantity of enzymes, the rate of enzymatically degradable polymer degradation *in vivo* varies substantially depending on the implantation site. The rate of degradation of these polymers can also be influenced by chemical modifications.

2.4 Non-Toxic

Non-toxicity applies to biomaterials that are non-carcinogenic, non-pyrogenic, non-allergic, non-inflammatory, and blood-compatible. The term "toxicology of biomaterials" refers to the components that emerge from biomaterials. In the live body environment, a well-designed biomaterial should do its purpose while causing no harm to other cells. A well-designed biomaterial should fulfill its function in the living body environment while causing no harm to other cells, organs, and entire organisms. It is plausible to believe that unless it is designed to do so, a biomaterial should not release or create anything from its mass (Ratner et al. 2003; Raghavendra et al. 2015).

2.5 Mechanical Properties and Performance

The physical parameters of a biomaterial must be coordinated with the target tissue or organ in the live system where it will be transplanted. Biomaterial utilized for medical devices must meet mechanical and performance parameters comparable to those of the organ or tissue being replaced (Hans et al. 2021). Metal biomaterials are said to make up 70–80% of implant biomaterials, and they are significant in the remodeling of tissues that have been damaged, particularly hard tissues (Niinomi 2012).

The mechanical and physical characteristics of the designed biomaterials can be placed into three categories.

2.5.1 Mechanical Performance

The mechanical performance of biomaterials, as medical devices, reveals their mechanical qualities in order to function properly in the living bodily environment. The biomaterial's mechanical properties differ depending on the application spot. Some of the most important properties of biomaterial to evaluate prior to implantation include elastic modulus, tensile strength, yield strength, corrosion, creep, and hardness (Prasad et al. 2017). Mechanical characteristics are extremely important in hard tissue applications. The production of acceptable biomaterials that can assist human tissue repair and regeneration is ongoing, and biomaterial mechanical qualities are important to their effectiveness in the therapeutic context.

2.5.2 Mechanical Durability

The amount of time it takes a biomaterial to perform its principal function efficiently is referred to as durability. The mechanical durability of some of the biomaterials used in tissue implants is given in Table 2.

Table 2: Some of the primary mechanical and physical properties of biomaterial design.

Mechanical Performance	
Biomaterials	**Mechanical Characteristics**
Dialysis membrane	Strong, flexible, and non-elastomeric
Hip prosthesis	Strong, rigid
Heart valve leaflet	Flexible, tough
Articular cartilage	Soft, elastomeric
Tendon material	Strong, flexible
Endosteal root	Strong, rigid
Intramedullary rods	Strong, rigid
Dialysis membrane	Strong, flexible
Mechanical Durability	
Biomaterials	**Mechanical Durability**
Catheter	3 days
Bone plate	Greater than 6 months
Artificial heart	1–3 Years
Hip joints	10–15 years
Cochlear implant	Lifetime
Leaflets in heart valves	Lifetime
Physical Properties	
Biomaterials	**Physical Properties**
Dialysis membrane	Permeability
Articular cup of the hip joint	High lubricity
Intraocular lens	Clarity and refraction
Sutures	Absorbable
Artificial blood vessels	Ability to heal, remodel, and contract
Contact lenses	Oxygen permeability, wettability

2.5.3 Physical Properties

The primary requirement of the biomaterial is that it has physical properties that are similar to or extremely similar to those of the organism. As a result, these materials are created based on the characteristics of the tissue in which they are used. Moreover, it has become clear that both physical and chemical features of materials can influence biological responses (Mitragotri and Lahann 2009). Materials are modified for optimal adaptation to the biological environment based on physical strength, chemical features, photoreactivity, calcification potency, and surface structure.

2.6 Non-Carcinogenic

A designed biomaterial must fulfill its purpose in the living body tissues without affecting the other parts of the body. For that, biomaterial should be non-carcinogenic. Any carcinogenic or mutagenic action, physical irritation, inflammation, and toxicity should be avoided by patient body parts or tissues that come into contact with implants (Fathi et al. 2003; Freire et al. 2015). Under certain circumstances, the implantation of biomaterial-related products into body tissues is linked with the growth of tumors. There are many factors that can be responsible for carcinogenicity, viz., the nature of the material themselves, the nature of the host, and the relationship between the host and biomaterial.

Some of the properties of biomaterial that are known to have an influence on the carcinogenicity process mainly include surface chemistry, intrinsic chemistry, chemical nature of released components, if any, physical and crystallographic nature of any released particulate components, size, shape, hardness of the surface, electrical and magnetic properties and radioactivity of the biomaterial. With respect to interactions within the host-material system, some factors that should be taken into account also include the extent and method of surgery, subsequent complications, and the duration of contact between the host and biomaterial (Thevenot et al. 2009). When assessing the potential of implanted materials to cause cancer in humans, host characteristics such as sex, age, occupation, lifestyle, genetic factors as well as pharmacological status must be considered.

2.7 Corrosion, Wear, and Fatigue Properties

Biomaterial should be highly corrosion-resistant. This is particularly problematic in the case of metal implants. Corrosion is one of the most important factors impacting the life and performance of metal and alloy orthopedic implants. Metal implants can produce metallic ions that can collect in adjacent tissues or be transferred to other parts of the body when bodily fluids are present. Released ions have been discovered to trigger allergic and harmful reactions (Geetha et al. 2019). The corrosion resistance of an implant material influences its usefulness and endurance, as well as being a key factor in biocompatibility. The body's environment is hostile, posing a number of issues in terms of corrosion control. Some of the major issues associated with medical device implant failure have been identified as fatigue, fracture, and wear. Though fatigue fracture and wear are common in orthopedic applications such as hip joint prostheses, they can also be lethal in mechanical heart valves (Teoh 2000). The prevention of biomaterial corrosion has become essential, especially in order to avoid the inflammation and allergic reactions generated by biomaterial implants in the human body (Asri 2017).

Low wear resistance leads the implant to discharge wear debris into the surrounding tissue, resulting in the development of harmful enzymes, inflammation, infection, osteolysis, pain, and bone resorption (Li et al. 2004). The material's wear resistance is important for the biomaterial's correct functioning, preventing it from unraveling and causing reactions in the tissue where it is deposited, therefore increasing the patient's quality of life (Gobbi and Gobbi 2018). Therefore, the implants should be designed with wear-resistant materials.

2.8 Design and Manufacturability

A biomaterial is a non-viable substance designed to interact with living processes. One of the most important considerations for biomaterials is proper material design. The biomaterials' reliable and efficient qualities allow them to be used in a physiological medium (Yoruc and Sener 2012). In order to create well-known biomaterials, these distinctive features are equipped with an appropriate combination of chemical, biological, mechanical, and physical capabilities (Williams 1987). These biomaterials were created by combining polymers, viz. metals, ceramics, and composite materials. Manufacturability refers to an item's ability to be manufactured with relative ease and at a low cost while maintaining high reliability. In the case of biomaterials, manufacturability refers to the material's ability to be sterilized using a verified sterilization process that is considered adequate for biomedical purposes. Standard sterilization treatments, such as autoclaving, dry heat, ethylene oxide, radiations, etc., are expected not to harm the material. Advances in material technology have helped to broaden the ingredients used to design various types of orthopedic implants. As a result, nanoparticles may mimic the surface characteristics of natural tissues, such as wettability, topography, chemistry, and energy (Chen et al. 2022).

2.9 Absence of Foreign Body Reaction

Foreign body response (FBR) is an inherent reaction that occurs when any foreign material is introduced into the body (Franz et al. 2011). Immune responses to foreign bodies defend the organism from bacteria, fungi, viruses, and traumas, but they also isolate the implant, rendering it non-functional. FBR generally consists of foreign body giant cells and macrophages that may remain at the interface of tissue and implant for the rest of the implant's life (Anderson 2001). In severe cases, the implants must be removed or replaced. The implantation procedure damages the tissue surrounding the foreign object, resulting in an inflammatory response. This inflammatory response transforms into a fibrotic response that surrounds and isolates the implanted material (Carnicer-Lombarte 2021). When a foreign material is implanted with the intention of delivering a therapy, both the acute and chronic stages of FBR pose significant challenges to its integrity and therapeutic function. The body's system activates the rejection mechanism, which causes the foreign body to be eliminated if the chemical signal from implanted material is detected as a threat and cannot be terminated (Bharadwaj 2021). There are some ways by which foreign body reactions can be prevented or decreased. The surface of the artificial implant can be changed with condensed aromatic structures containing free radicals, allowing for covalent attachment of host proteins in their native shape. Immune cells are not stimulated because the entire protein coating precludes direct contact with the implant surface. As a result, the organism's immune response is not elicited, and the artificial implant is not isolated from the surrounding tissue. There is no collagen capsule, macrophage activity is low, cell proliferation is low, and inflammatory activity is minimal. The best implants are ones that are reactive yet do not induce immunological reactions (Sohling et al. 2022). According to the literature review, all artificial implants generate an immunological reaction and 35% of them require a secondary operation. Immunosuppressive drugs can control the immune response to artificial implants (Kastellorizios 2015). However, such a depressed immune system cannot protect the living body from diseases.

3. Conclusion

The increasing demand in the medical field for bio-alternatives that may perform the live activities of biological organs in their absence has perked researchers' interest in developing biomaterials with unique properties. Currently, numerous biomaterials with different properties are accessible for designing medical devices. However, there are several obstacles to overcoming different reactions of the body tissues against them. Henceforth, there is a strong need for novel research to generate biomaterials that are more compatible with living bodily tissues. Because of its properties, such as strong corrosion and wear resistance, high strength, and stiffness compared to other types of biomaterials, composite or biocomposite biomaterials may be able to meet these needs in the future. In order to get the best clinical results, future trends must combine mechanically superior metals with ceramics and polymers' biocompatibility and biofunctionality. Further, biomaterial properties covered here provide the necessary basis for stakeholders, tissue engineers, material scientists, and emerging researchers to design befitting biomaterial devices and implants.

References

Anderson, J. M. 2001. Biological responses to materials. Annu. Rev. Mater. Res. 31: 81–110.

Asri, R. I., W. S. Harun, N. A. Lah et al. 2017. Corrosion and surface modification on biocompatible metals: A review. Materials Science and Engineering: C, 77: 1261–1274.

Bharadwaj, A. 2021. An overview on biomaterials and its applications in medical science. 2021 IOP Conf. Ser.: Mater. Sci. Eng. 1116 012178.

Boretos, J. W. and M. Eden (Eds.). 1984. Contemporary biomaterials, Material and host response, Clinical applications, New Technology and Legal Aspects (pp. 27–88, 128–132, 193–253). Noyes Publications, Park Ridge, NJ.

Bostman, O. and H. Pihlajamaki. 2000. Clinical biocompatibility of biodegradable orthopaedic implants for internal fixation: A review. Biomaterials 21: 2615–2621.

Braune, S., R. A. Latour, M. Reinthaler et al. 2019. *In vitro* thrombogenicity testing of biomaterials. Adv. Healthcare Mater. 8(1900527): 1–17. doi: https://doi.org/10.1002/adhm.201900527.

Carnicer-Lombarte A., S. T. Chen, G. G. Malliaras et al. 2021. Foreign body reaction to implanted biomaterials and its impact in nerve neuroprosthetics. Front. Bioeng. Biotechnol. 9: 622524. doi: 10.3389/fbioe.2021.622524.

Chen, M. Q. 2022. Recent advances and perspectives of nanotechnology-based implants for orthopedic applications. Front. Bioeng. Biotechnol. 10: 878257. doi: 10.3389/fbioe.2022.878257.

Costerton, J. W., P. S. Stewart and E. P. Greenberg. 1999. Bacterial biofilms: a common cause of persistent infections. Science 284(5418): 1318–22. doi: 10.1126/science.284.5418.1318.

Davis, R., Singh, A., Jackson, M. J. et al. 2022. A comprehensive review on metallic implant biomaterials and their subtractive manufacturing. Int. J. Adv. Manuf. Technol. 120(3-4): 1473–1530. doi: 10.1007/s00170-022-08770-8.

Fathi, M. H., M. Salehi, A. Saatchi et al. 2003. *In vitro* corrosion behaviour of bioceramic, metallic, and bioceramic-metallic coated stainless steel dental implants. Dental Materials 19(3): 188–198.

Franz, S., S. Rammelt, D. Scharnweber et al. 2011. Immune responses to implants: A review of the implications for the design of immunomodulatory biomaterials. Biomaterials 32: 6692–6709.

Freire, W. P., M. V. Fook, E. F. Barbosa et al. 2015. Biocompatibility of dental restorative materials. Materials Science Forum. 805: 19–25.

Freire, W. P., M. V. Fook, E. F. Barbosa et al. 2015. Biocompatibility of dental restorative materials. Materials Science Forum 805: 19–25.

Geetha, M., A. K. Singh, R. Asokamani et al. 2009. Ti-based biomaterials, the ultimate choice for orthopaedic implants-A review. Progress in Materials Science 54: 397–425.

Gobbi, S. J. and V. J. Gobbi. 2018. Wear resistance of metallic orthopedic implants—Mini review. Biomedical Journal Scientific & Technical Research 12(3): 1–2.

Gobbi, S. J., G. Reinke, V. J. Gobbi et al. 2020. Biomaterial: concepts and basics properties. European International Journal of Science and Technology 9(2): 23–42.

Hanker, J. S. and B. L. Giammara. 1988. Biomaterials and biomedical devices. Science 242(4880): 885–892.

Hans, M. 2022. Mechanical properties of biomaterial. J. Biotechnol. Biomater. 11(6): 490.

Heimann, Robert B. Chapter 1. Biomaterials- characteristics, history, applications. Materials for Medical Application, edited by Robert B. Heimann, Berlin, Boston: De Gruyter, 2020, pp. 1–74. doi: https://doi.org/10.1515/9783110619249-001.

Helmus, M. N. ed. 2003. Biomaterials in the design and reliability of medical devices. Kluwer Academic/Plenum Publishers and Landes Bioscience, New York and Georgetown, TX.

Helmus, M. N., D. F. Gibbons and D. Cebon. 2008. Biocompatiblity: meeting a key functional requirements of next generation medical devices. Toxicologic Pathology 36: 70–80.

Jenkoa, M., M. G. Sekb, M. Godeca et al. 2018. Surface chemistry and microstructure of metallic biomaterials for hip and knee endoprostheses. Applied Surface Science, 427: 584–593.

Kastellorizios, M., F. Papadimitrakopoulos and D. J. Burgess. 2015. Prevention of foreign body reaction in a pre-clinical large animal model. Journal of Controlled Release 202: 101–107.

Katti, D. S., S. Lakshmi, R. Langer et al. 2002. Toxicity, biodegradation and elimination of polyanhydrides. Adv. Drug. Deliv. Rev. 54: 933–61.

Li, S. 1999. Hydrolytic degradation characteristics of aliphatic polyesters derived from lactic and glycolic acids. J. Biomed. Mater. Res. 48: 342–53.

Li, S. J., R. Yang, S. Li et al. 2004. Wear characteristics of Ti-Nb-Ta-Zr and Ti-6Al-4V alloys for biomedical applications. Wear 257: 869–876.

Lloyd, A.W. 2002. Interfacial bioengineering to enhance surface biocompatibility. Med. Device. Technol. 13: 18–21.

Markhoff, J., M. Krogull, C. Schulze et al. 2017. Biocompatibility and inflammatory potential of titanium alloys cultivated with human osteoblasts, fibroblasts and macrophages. Materials, 10(1): 52. doi: 10.3390/ma10010052.

Miljana, Z., Rubezic, Aleksandra, B., Krstic, Hristina, Z., Stankovic et al. 2020. Different types of biomaterials: structure and application: a short review. Advanced Technologies. 9(1): 69–79.

Mitragotri, S. and J. Lahann. 2009. Physical approaches to biomaterial design. Nature Materials 8(1): 15–23. doi: https://doi.org/10.1038/nmat2344.

Naahidi, S., M. Jafari, F. Edalat et al. 2013. Biocompatibility of engineered nanoparticles for drug delivery. J. Control Release 166(2): 182–94. doi: 10.1016/j.jconrel.2012.12.013.

Naira, L. S. and T. Cato. 2007. Laurenc in biodegradable polymers as biomaterials. Prog. Polym. Sci. 32: 762–798.

Niinomi, M., M. Nakai and J. Hieda. 2012. Development of new metallic alloys for biomedical applications. Acta Biomateriali 8: 3888–3903.

Prasad, K., O. Bazaka, M. Chua et al. 2017. Metallic biomaterials: current challenges and opportunities. Materials (Basel, Switzerland) 10(8): 884. doi: https://doi.org/10.3390/ma10080884.

Raghavendra, G. M., K. Varaprasad and T. Jayaramudu. 2015. Biomaterials: design, development and biomedical applications, Editor(s): Sabu Thomas, Yves Grohens and Neethu Ninan, Nanotechnology Applications for Tissue Engineering, William Andrew Publishing, pp. 21–44. doi: https://doi.org/10.1016/B978-0-323-32889-0.00002-9.

Ratner, B. D., A. S. Hoffman and F. J. Schoen. 2003. Jack lemons biomaterials science: a multidisciplinary endeavour. Biomaterials Science 16(14): 1–19.

Sohling, N., M. Ondreka, K. Kontradowitz et al. 2022. Early immune response in foreign body reaction is implant/material specific. Materials 15: 2195. doi: https://doi.org/ 10.3390/ma15062195.

Suwardi, A., F. Wang, K. Xue et al. 2022. Machine learning-driven biomaterials evolution. Adv. Mater. 34: 2102703. doi: https://doi.org/10.1002/adma.202102703.

Teoh, S. H. 2000. Fatigue of biomaterials: a review. International Journal of Fatigue 22(10): 825–837. doi: https://doi.org/10.1016/S0142-1123(00)00052-9.

Thevenot, P., W. Hu and L. Tang. 2008. Surface chemistry influences implant biocompatibility. Current Topics in Medicinal Chemistry 8(4): 270–280. doi: https://doi.org/10.2174/156802608783790901.

Todros, S., M. Todesco and A. Bagno. 2021. Biomaterials and their biomedical applications: from replacement to regeneration. Processes 9(1949): 1–20. doi: https:// doi.org/10.3390/pr9111949.

Vert, M. 2005. Aliphatic polyesters: great degradable polymers that cannot do everything. Biomacromolecules. 6: 538–546.

Wagner, W. R. 2020. Chapter 1.3.1—The materials side of the biomaterials relationship. pp. 83–84. *In*: Wagner, W. R., S. E. Sakiyama-Elbert, G. Zhang et al. (eds.). Biomaterials Science (Fourth Edition): Academic Press.

Waterhouse, A., S. G. Wise, K. C. Martin et al. 2011. Elastin as a nonthrombogenic biomaterial. Tissue Eng. Part B. Rev. 17(2): 93–9. doi: 10.1089/ten.TEB.2010.0432.

Williams, D. F. 1987. Review: tissue biomaterial interactions. J. Mat. Sci. 22(10): 342145.

Williams, D. F. 2008. On the mechanisms of biocompatibility. Biomaterials 29(20): 2941–53. doi: 10.1016/j.biomaterials.2008.04.023. Epub 2008 Apr 28.

Williams, D. F., J. Black and P. J. Doherty. 1992. Consensus report of second conference on definitions in biomaterials. pp. 525–533. *In*: Doherty, P. J., R. L. Williams, D. F. Williams (Eds.). Biomaterial-Tissue Interfaces. Elsevier: Amsterdam, The Netherlands, 1992; Volume 10.

Yoruç A. B. H. and B. C. Sener. 2012. Biomaterials. pp. 1–242. *In*: Prof. Kara, S. (Eds.). A Roadmap of Biomedical Engineers and Milestones. doi: 0.5772/2494.

3

Designing of Materials for Biodegradation and Bio-Erosion

Poornima Sharma,[1] *Arpana Parihar*[2] and *Raju Khan*[3,*]

1. Introduction

A great deal of research has been employed to examine human well-being and ease of lifestyle, which could thrash health because of serious diseases. This concern has raised the importance of modern methods and techniques to regularly design materials to detect and diagnose various diseases or allied factors as precautionary measures. The contribution of these techniques implies using some basic tools and techniques to detect factors associated with disease easily and selectively (Asapur and Banerjee 2020). The scientific community has diverted the development of technology from conventional techniques, which could dictate the quality of materials (Wu et al. 2018) and resulted in (Kang et al. 2006) (Figure 1) the possibility of using such biomaterials in clinical sample monitoring (Nguyen 2020). Technological advancement and superior scientific understanding in the field of nanotechnology by intersecting with various other branches of science for research and development have delivered a far-reaching impact on all forms of life (Parihar et al. 2023; Dong 2020; Oliveira 2018) (Tables 1, 2 and 3).

1.1 Application-Based Biomaterials

Since chemical synthesis leads to toxic compound production, biomaterials of numerous scientific interests are effectively a bridge between atomic or molecular

[1] Department of Education in Science & Mathematics (DESM), Regional Institute of Education, Bhopal, Madhya Pradesh, India.

[2] Department of Translational Medicine, All India Institute of Medical Sciences, Bhopal, Madhya Pradesh, India.

[3] Industrial Waste Utilization, Nano and Biomaterials, CSIR-Advanced Materials and Processes Research Institute (AMPRI), Bhopal, Madhya Pradesh, India; Academy of Scientific and Innovative Research (AcSIR), Ghaziabad, Uttar Pradesh, India.

* Corresponding author: rajukhan.ampri@gmail.com

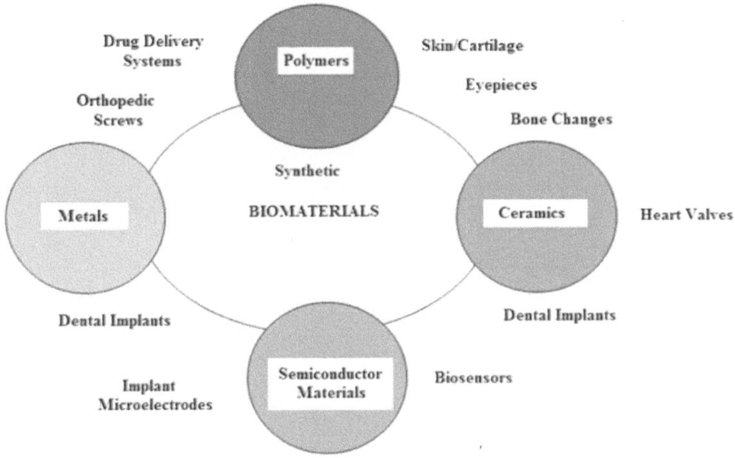

Figure 1: Biomaterials of numerous medical applications (Gezer et al. 2021).

structures (Inshakova 2017). The other concern is that chemical synthesis may lead to toxic compound production, creating fatal responses on the surface and adversely affecting human health. Materials like plant extracts, microbes and other biomaterials are eco-friendly and simultaneously remove the constraints due to toxic material (Ribeiro et al. 2015; Li et al. 2011; Khan et al. 2014). Utilization of correct material becomes the most selected choice because particles driven must be available easily, do not have complex metabolic pathways, can tolerate heavy metals and are cost-effective (Parihar et al. 2024; Skeffington et al. 2018; Yan et al. 2017). Compared to microbial systems, plant systems also require incubation for less time, but the thing is that local and endemic species favor this process more (Gama et al. 2012) (Figure 2). Synthesis of gold nanoparticles has been studied from a wide array of plants where the protective and reducing properties of molecules in plants are accountable for such synthesis because such synthesis generally is biocompatible and nontoxic (Nwodo et al. 2012; Malvankar et al. 2012; Prasad et al. 2016; Dragone et al. 2017; Petros et al. 2010; Parihar et al. 2024; Wang et al. 2010; Du et al. 2007; Ramya et al. 2015; Bao et al. 2010).

The growing interest and recent advances have influenced the response of materials with high levels of complexity and many factors required for applications. The studies that are concerned with response to external inputs and various stimuli have developed a great deal of attention. The performance of unique and excellent synthesis pathways, along with some important factors, adds responsive mechanisms in future applications. Significant modifications in the structure to identify biodegradable polymers will act as a problem-solving element related to pollution through synthetic polymers in dangerous proportions (Wade and Burdick 2014).

1.2 Electrochemical Biomaterials With Bio-Sensing Application

Recently, sophisticated diagnostic methods based on biosensors have demonstrated promise for the early diagnosis of lethal diseases like cancer. Optics, electrochemistry, and piezoelectric are a few of the detection methods used. A lot of interest has been

Table 1: Classification of biodegradable polymers on the basis of origin, example and applications.

S. No.	Natural Polymer Type	Example	Applications
1	Plant origin: polysacharides	Cellulose, starch, and alginate	Chemical transformations
2	Animal Origin: polysaccharides	Chitin (chitosan), hyaluronate	Non-toxicity, heat tolerance, food industry
a	Proteins	Collagen (gelatin), albumin	Wound healing
3	Microbial origin		
a	Protein	Poly (3-hydroxyalkanoate)	Packaging material in orthopedic surgery
b	Polysaccharides	Hyaluronate	Ophthalmic solutions

Table 2. Types of synthetic polymer, examples and applications.

S. No.	Synthetic Polymer Type	Example	Applications
1	Aliphatic polyesters		
1.1	Glycol and dicarbonic acid polycondensates	Poly (ethylene succinate), poly (butylene terephthalate)	In bioplastics
1.2	Polylactides	Polyglycolide and polylactides	In human extraction sockets and bone formation
1.3	Polylactones	Poly-caprolactone	Polyurethane and health care
1.4	Miscellaneous	Poly (butylene terephthalate)	Electronics and civil electrical appliances
2	Polyols	Poly (vinyl alcohol)	Adhesives and paints
3	Polycarbonates	Poly (ester carbonate)	Adhesive or thermoplastic blend
4	Miscellaneous	Polyanhydrides, poly s-cyanoacrylate, polyphosphazenes, poly (orthoesters)	Drug delivery

drawn to the electrochemical-based detection of a number of biomarkers, including EpCAM, CD44, VEGF, Mucin 1, and CEA. Amperometry, impede metric, and potentiometric electrochemical techniques provide a sensitive, quick, and economical platform for the early identification of cancer biomarkers. Electrochemical devices are becoming more portable thanks to their downsizing (Parihar et al. 2022). The electrochemical methods also provide good selectivity and can detect attomole levels. In order to improve the functionality of electrochemical devices for the early detection of cancer, recent research has demonstrated the critical role that nanomaterials play (Khan et al. 2021). These nanomaterials exhibit outstanding mechanical, electrical, electrochemical, and optical characteristics. Since the beginning of the last decade, the outstanding electrical, mechanical, electrochemical,

Table 3: Biopolymers and applications in various domain(s):

S. No.	Application of Polymers	Example	References
1	Medicinal	Developing significant surgical reactions, Biocompatible	Kweon et al. 2001; Salehi et al. 2020; Hennecke et al. 2013
2	Surgical use	Suturing	Huemmerich et al. 2004; Zeplin et al. 2014; Bodin et al. 2007; Nimeskern et al. 2013; Zang et al. 2015
3	Pharmaceutical use	Drug delivery system	Brown et al. 2013; Cherian et al. 2011; Letchford et al. 2011; Kolakovic et al. 2011
4	Tissue engineering	Replacements of defective or lost tissues using cells and cell growth factors	Barbosa et al. 2016; Fu et al. 2013; Fu et al. 2012; Park et al. 2012
5	Ecological applications	Replace biostable plastics	Czaja et al. 2007

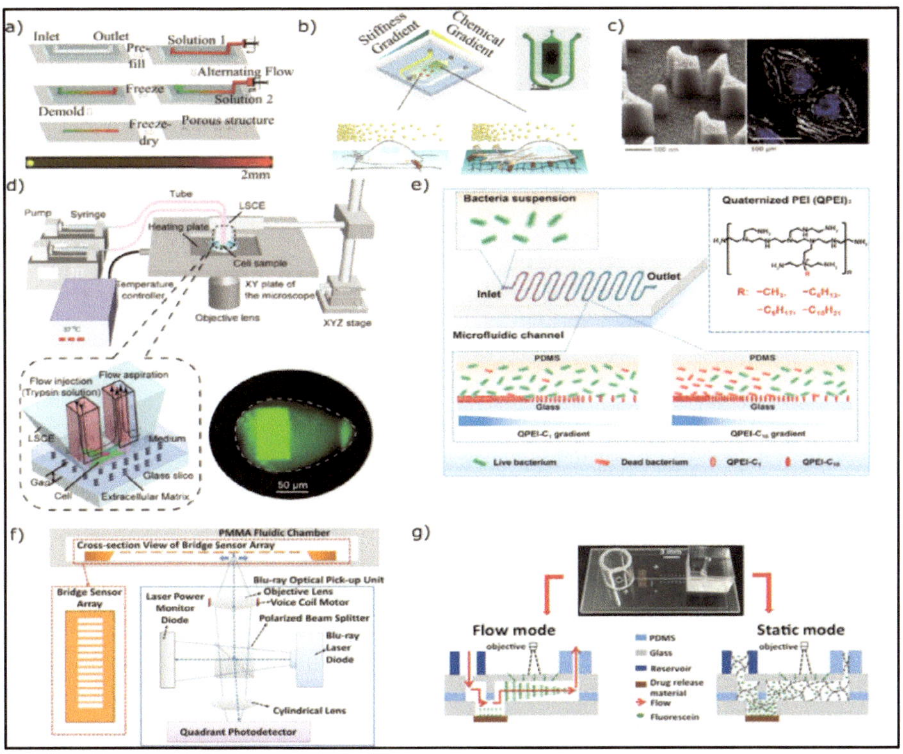

Figure 2 (a–g): Advances in the interface between microfabrication and biomaterials research (Alexander et al. 2021).

and optical properties of 2D materials have attracted considerable interest due to their broad applicability in a variety of sectors, including disease diagnosis and therapies. Numerous 2D materials have been studied for use in various bio-sensing applications, including graphitic carbon nitride transition metal dichalcogenides, black phosphorous, hexagonal boron nitride (borophene), grapheme, metal halides, metal oxides, metal-organic frameworks, and certain polymers. Early transition metal carbides and/or nitrides (MXene) stand out from the others due to their exceptional hydrophilicity, electrochemical, mechanical, and optical characteristics. They are frequently made utilizing the HF etching technique. MXene has recently been successfully exploited for a wide range of bio-sensing applications, targeted drug administration, cancer therapeutics, energy storage, the manufacture of heat-resistant materials, and many other things. Due to their numerous distinctive features, the MXenes are regarded as a potential material in analytical chemistry applications. Some of these characteristics include a wide surface area, high surface functionality, hydrophilicity, large-scale manufacturability, high stability, high conductivity, and non-hazardous nature. Additionally, by varying the size, spacing, and thickness of the layers, the characteristics of the MXenes can be tailored appropriately. Excellent biocompatibility is displayed by the MXenes. Electrochemical devices based on MXene have the ability to detect cancer biomarkers and exhibit extremely high sensitivity in target analyte detection (Parihar et al. 2022).

The major problem that arises here is the overlapping understanding of bio-erosion and biodegradation. Some ecologists define biodegradable polymers as solely related to biological environments, while others compare them to processes such as thermal oxidation, photolysis, radiolysis, hydrolytic breakdown, etc. However, Gliding proposed that biodegradable polymers actually undergo *in vivo* degradation, while Graham and Wood define them as systems that automatically degrade into simple and soluble products that can easily escape from the site of degradation (Christopherson et al. 2009). According to Griffin, only such substances can be called biodegradable polymers, which require bio-substances like enzymes or microorganisms for degradation. However, polylactides cannot fall under this category because they require only a high temperature and neutral pH, which are physical factors for the degradation of substances. A polymer can be further categorized as absorbable, resorbable or bioabsorbable with respect to the time it requires for degradation, regardless of both enzymatic and non-enzymatic hydrolysis.

1.3 Bio-Erosion

Heller described "bio-erosion" as a conversion of water-insoluble to water-soluble material, where the involvement of major chemicals is not mentioned, whereas Langer and Peppas discussed erodibility in context with the ability of a polymer to degrade and, as a result, control rate of drug release (Repanas 2016). Kumar coined the term "biodegradable polymer", referring to polymers that secrete degrading enzymes linked to physiological cleavage in the molecular structure of the polymer. However, bio-erodible or bio-assimilable processes involve hydrolytic cleavage without enzyme participation (Corey 2007) (Figure 3). Vert et al. have elaborated this process as a breakdown of polymer without any proof of exclusion of substances

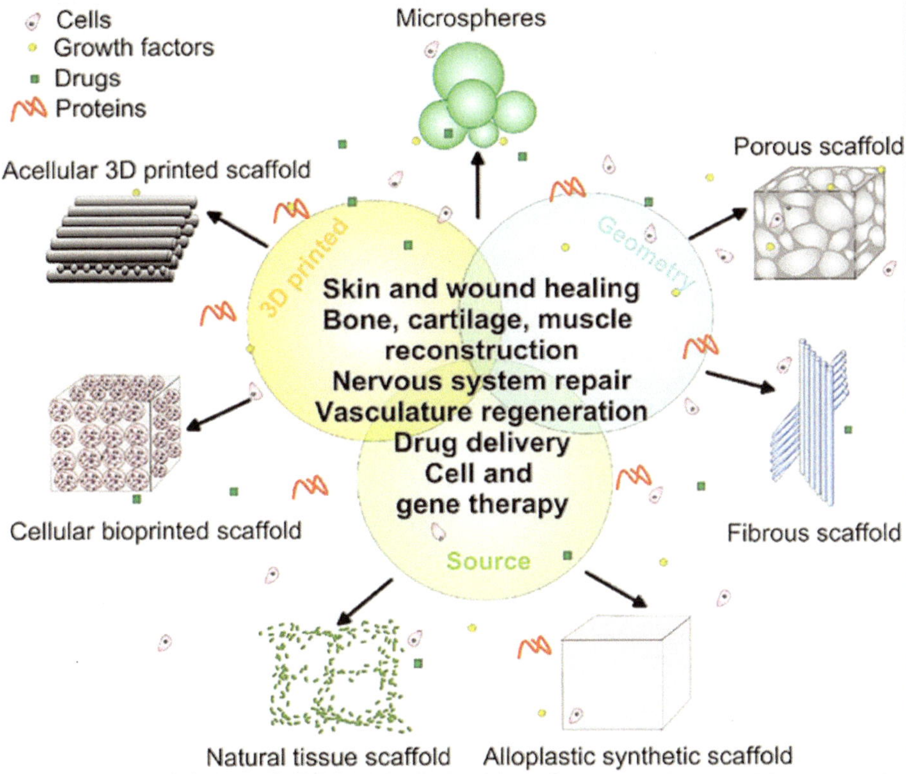

Figure 3: Recent advances in biomaterials for 3D scaffolds (Valent et al. 2012).

from the body, whereas they defined bioresorbable as substances which, after degradation, further resort and elimination of materials is done through natural pathways respectively as by-products of metabolism (Frohbergh 2012).

1.4 Nanomaterials

The biosynthesis of materials has gained increasing attention for the use of novel products, leading to the development of new products and devices with great impacts across different fields, from sensoristics to biomedicine. The biosynthesis of nanomaterials by microorganisms has recently attracted interest as a new, exciting approach toward the development of nanomanufacturing as compared to traditional methods. This chapter highlights the synthesis of nanomaterials from microorganisms such as bacteria, yeast, molds, microalgae, etc., for their applications in devices for therapeutic/diagnostic applications. For this, much literature has been reviewed, focusing on various specific features and taking various aspects from scientific works where the main organisms are bacteria and microalgae principally used in various biomedical applications because they have the ability to form advantageous nanostructures with characteristic properties associated with respective applications. However, variations in intrinsic properties in biological systems along with biosynthetic properties require standardized protocols to obtain nanomaterials with increasingly uniform and engineered breakthrough development for future scaling and industrial developments (Zhao et al. 2018).

1.5 Nanocellulose and Nanocarbons

Investigation of various nanomaterials, such as nanocellulose/nanocarbons, has become an important field since the emergence of nanotechnology in recent decades. Such nanomaterials have become an important field of research and development. In spite of varied applications, they also look toward environmental and sustainability concerns, which also include a wide range of domains that influence our society and our way of life (Lin and Kang 2021). Investigations for improvements in properties and new functionalities are important goals for the discovery and better understanding, characterization, and manufacturing of new nanomaterials for adequate systems (Kim et al. 2019). Investigation of nanocellulose/nanocarbons hybrid materials have demonstrated both academic and technological importance and offered great research opportunities within cross-disciplinary areas.

Nanocellulose, which refers to cellulose with a nanoscale dimension, encompasses cellulose nanocrystals, cellulose nanofibrils, and bacterial cellulose (Hutmacher et al. 2015). It displays outstanding features such as biocompatibility, eco-friendliness, renewability, interesting reinforcing potential, hydrogen-bonding capacity, low thermal expansion coefficient, dimensional stability, high elastic modulus, high specific surface area and low density (Parihar et al. 2024; Schuman et al. 1995; Phull et al. 2016; Wakitani et al. 1989; Wiesmann et al. 2003). Various nanocellulose-based composites, which found application in different fields, have been developed during the last two decades. However, numerous challenges need to be addressed to achieve the full requirement of advanced materials (Yang et al. 2018; Khan et al. 2018; Marks et al. 1991; Lepistö et al. 1994; Wachol-Drewek et al. 1996). Accordingly, several research activities continue to be performed by different research groups worldwide to develop the next generation of nanomaterials and fully explore the potential of nanocellulose. Nanocarbons are another exciting type of potential nanomaterials, which has received tremendous attention since the discovery of fullerenes over 30 years ago (Vaneerdeweg et al. 2003; Mitra et al. 2013; Liu et al. 2020; Bello et al. 2020; Young et al. 2005). They include carbon nanotubes, graphene and carbon dots. They exhibit outstanding thermal, electrical, optical and mechanical features, and they can be employed in a wide range of applications such as composites, microelectronics, biomedical, conductive films, sensors, adsorbents, catalysis, energy storage, and coating, among others.

Based on these premises, the present special issue in nanomaterials aims to further contribute to the momentum of research and development in nanocellulose/nanocarbon hybrids by featuring 10 (Siepmann and Siepmann 2006) original research articles as well as 2 (Kumar 2011) review articles, authored and reviewed by experts in the field. It targets a broad readership of materials scientists, chemists, physicists, and nanotechnologists, among others. Most of the research papers highlight theoretical concepts and practical approaches of interest for real-world applications related to nanocellulose and nanocarbons (Figure 4). Some interesting

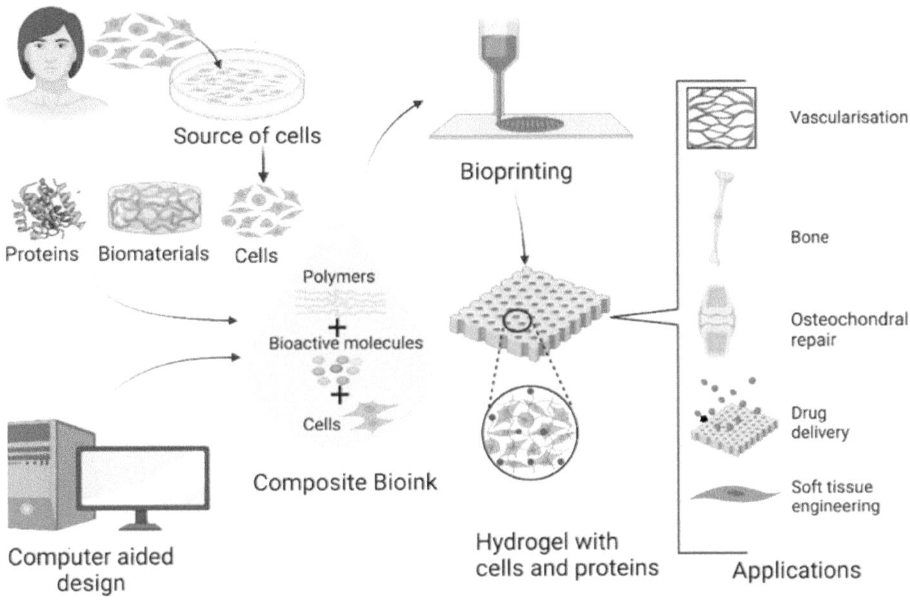

Figure 4: Nature-based biomaterials and their application in biomedicine (Troy et al. 2021)

research works on nanocellulose-based materials have been published in the present special issue. Prepared wood-based cationic cellulose nanocrystals (Case et al. 2021).

1.6 Biodegradable Polymers

In the past few decades, a newly emerging field called biodegradable polymers is also useful for various applications in medicine, agriculture, etc. Hence, various methods have been improved both for the enhancement of the mechanical properties and the final products after biodegradation. The main reason is their price level, which is still competitive, and they could find another way to prepare biodegradable materials with significant characteristics for biodegradation. The various physical properties that provide value to biodegradable polymers are also applied to some advanced technologies. Due to their potential applications, only a few groups of the mentioned biopolymers are of market importance (Madan et al. 2013).

1.7 Need For Biodegradable Polymers

The classifications of biodegradable polymers are based on their origin, either natural or synthetic material, and they are preferred for degradation in many fields like biomedical, agricultural and ecological uses that undergo degradation in varied physiological or physical environments. However, it can be proposed that the ease of conventional thermoplastics, after performing a function as moisture or odor barrier or as a carrier for a payload of an active substance, simply erodes in a harmless way than the one which has recently been proposed for a number of problem areas in packaging, agro, pharmaceutical and other industries. Perhaps the greatest ecological problems have stemmed from recent legislation in Italy and the USA against various

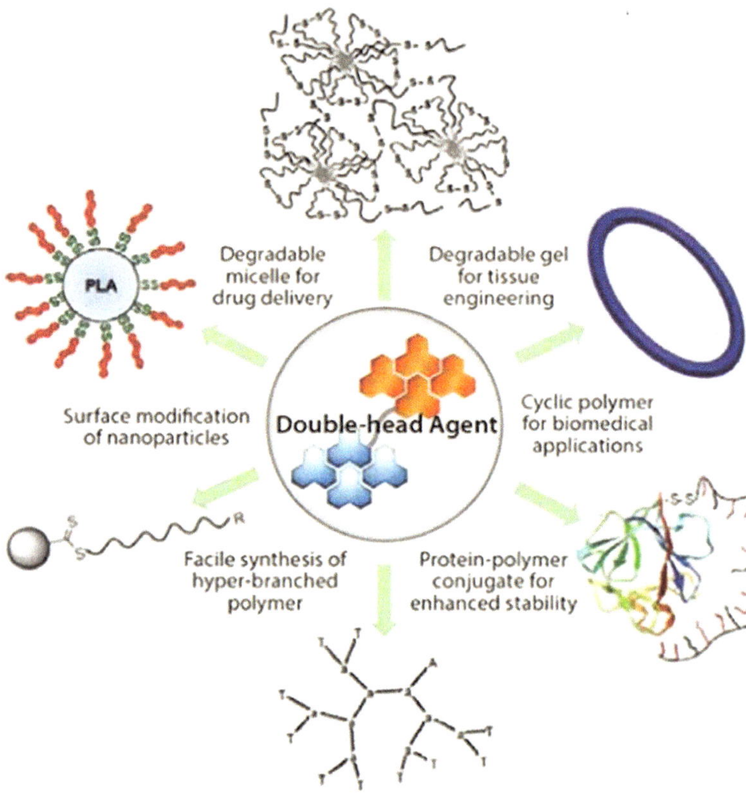

Figure 5: Functional biomaterials synthesized by double-head polymerization agents (Wang et al. 2016).

conventional plastics in use, which result in obvious indigestible litter (Madan et al. 2013; Lee et al. 2013). These biodegradable polymers contribute to two major applications, one in the clinical care of patients and the other as ecological polymers to keep the earth's environments clean. So, the currently available biodegradable polymers are either used for the two purposes or wherever applicable (Lee et al. 2013) (Figure 5). Furthermore, they are used in clinical and surgical treatments, including preventive medicine and clinical inspections (Tielens et al. 2017).

Polymeric biomaterials are among the polymers employed for medical purposes, which are also known as specified groups. However, their typical applications lie in disposable products (e.g., syringes, etc.), materials supporting surgical operation (e.g., sutures, etc.), prostheses for tissue replacements (e.g., intraocular lens, etc.), and artificial organs for temporary or permanent assist (e.g., artificial kidney, etc.). These biomaterials are quite unique in their features and use as they have strict industrial manufacturing permissions and are allowed unless they clear stringent governmental regulatory issues. Hence, biocompatibility needs more grounds to evolve as it is highly desirable but, at the same time, not indispensable since most of the clinically used biomaterials lack excellent biocompatibility; nonetheless,

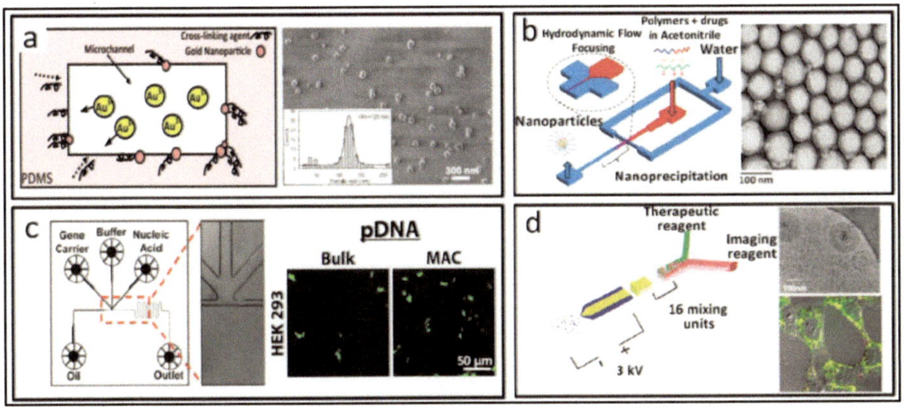

Figure 6 (a-d): Biomaterials meet microfluidics: from synthesis technologies to biological applications (Jingyun et al. 2017).

many efforts have been employed for the development of biocompatible materials (Figure 6). A large unsolved limitation is the lack of biocompatibility cost ineffectiveness especially when they are used not temporarily but permanently as implants in our body (Nguyen et al. 2019).

1.8 Surgical Use

The application of biodegradable polymers to medicine did not start recently and has already a long history. Actual and possible applications of biodegradable polymers in medicine are shown, and most of the applications are for surgery. The largest and longest use of biodegradable polymers is for suturing. Suture materials are required to keep wounds together until they hold sufficiently well by themselves. They provide artificial fiber support until the natural fiber (collagen) is synthesized and woven into a strong scar. Collagen fibers obtained from animal intestines have long been used as absorbable sutures following chromium treatment (Meftahi et al. 2009).

Throughout the history of surgery, absorbable sutures have been derived almost exclusively from various forms of collagen, such as catgut. Catgut, however, produces an allergic reaction in a number of patients, along with other problems such as fast loss of mechanical strength, problems with knot holding, etc. (Figure 7). Baptist presented the idea of synthetic biodegradable sutures based on PHB, and in 1966, Soppimath et al. reported the manufacture of biodegradable poly (L-lactic acid) sutures (Garg et al. 2004)

Huggins (Yang et al. 2018) reported the spinning of poly (glycolic acid in 1954). In 1967, Schmitt and Polistina obtained a patent on PGA sutures, which led to the first commercial synthetic biodegradable suture (Dexon®) manufactured by American Cynamid Co in 1970 (Yang et al. 2018). DuPont considered the polymer of homologous poly (lactide) (PLLA) for the same purpose (Ma et al. 2020). A copolymer of 92 mol% glycolide and 8 mol% lactide led to the development of the second commercial synthetic biodegradable suture (Vicryl®) (Sun et al. 2020; Gomes et al. 2019). There has also been a huge development in antagonist reactions

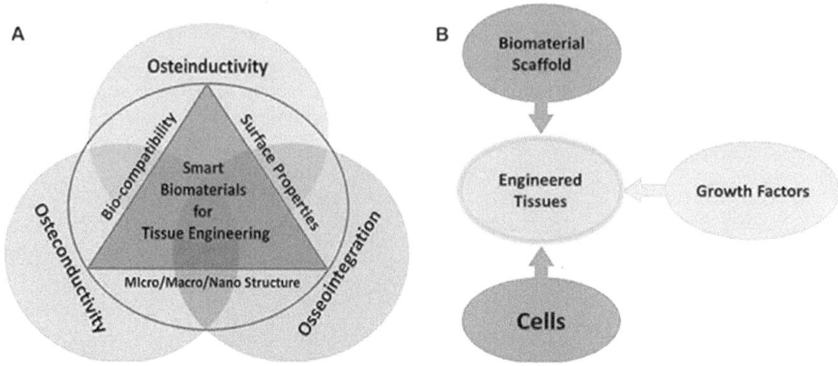

Figure 7: Advanced functional biomaterials for tissue engineering (Terzopoulou 2022).

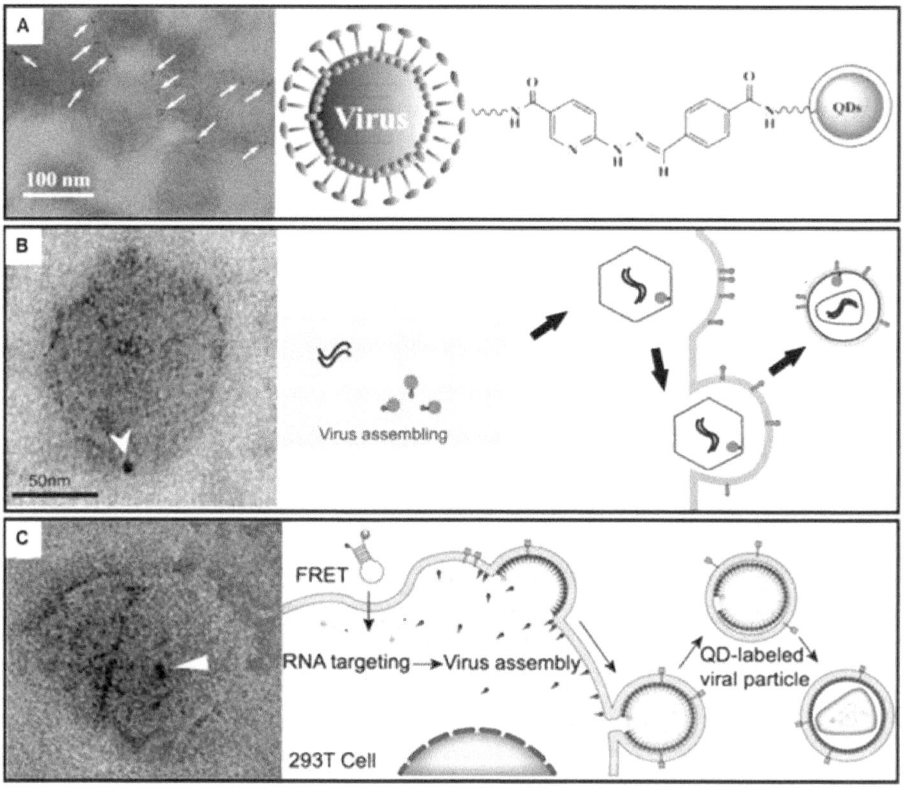

Figure 8 (A-C): Progress on biomaterials fighting against viruses (Xiao et al. 2021).

to viruses (Figure 8). Thick biodegradable mono-filament sutures were developed from poly (p-dioxanone) (PDS) and from PGA/TMC (Maxon®) (Sun et al. 2020; Gomes et al. 2019).

1.9 Advanced Materials

Commercial polymers used for this purpose include polyglycolide, which is still the largest in volume production, together with a glycolide-L-Iactide (90:10) copolymer (Ahmed and Ikram 2016; Luben et al. 2017). The sutures made from these glycolide polymers are of braid type processed from multi -filaments, but synthetic absorbable sutures of mono-filament type also at present are commercially available. Biodegradable sutures have wide applications not only in closing wounds in soft tissues but also in repair of tendons, ligaments, and dislocation of joints (Ahmed and Ikram 2016; Lubben et al. 2003). Also, by constructing gauzes, felt or velour dressings from biodegradable fibers one can manufacture surgical dressing to protect a wound surface, such as burn, traumatic injury or surgical incision and adhesion to tissues (Zhao et al. 2017). Liquid-type products are mostly used for these purposes. Immediately after application of a liquid to a defective tissue where homeostasis, sealing or adhesion is needed, the liquid sets to a gel and covers the defect to stop bleeding, seal a hole, or adhere two separated tissues. As the gelled material is no longer necessary after healing of the treated tissue, it should be biodegradable and finally absorbed into the body.

The biomaterials used to prepare such liquid products include fibrinogen (a serum protein), 2-cyanoacrylates, and a gelatin/resorcinol/formaldehyde mixture. 2-cyanoacrylates solidify upon contact with tissues as a result of polymerization to polymers that are hydrolyzable at room temperature and neutral pH but yield formaldehyde as a hydrolysis by-product (Chenouard et al. 2014). Regenerated collagen is also used as a hemostatic agent in forms of fiber, powder, and assemblies.

Another possible application of biodegradable polymers is the fixation of fractured bones. Currently, metals are widely used for this purpose in orthopedic and oral surgeries in the form of plates, pins, screws, and wires, but they need removal after re-union of fractured bones by further surgery. It would be beneficial to patients if these fixation devices could be fabricated using biodegradable polymers because there would be no need for a re-operation.

1.10 Pharmaceutical Use

In order to deliver drugs to diseased sites in the body in a more effective and less invasive way, a new dosage form technology, called drug delivery systems (DDS), was started using polymers. The objectives of DDS include the sustained release of drugs for a desired duration at an optimal dose, targeting drugs to diseased sites without affecting healthy sites, controlled release of drugs by external stimuli and simple delivery of drugs mostly through skin and mucous membranes. Polymers are powerful for this new pharmaceutical technology. If a drug is administered through a parenteral route like injection, the polymer used as a drug carrier should be preferably absorbable because the polymer is no longer required when the drug delivery has been accomplished. Therefore, biodegradable polymers are widely used, especially for the sustained release of drugs through administration by injection or implantation into the body. For this purpose, absorbable nanospheres, microspheres, beads, cylinders and disks are prepared using biodegradable polymers (Lee and Mooney 2012). The shape of the most widely used drug carriers is a microsphere,

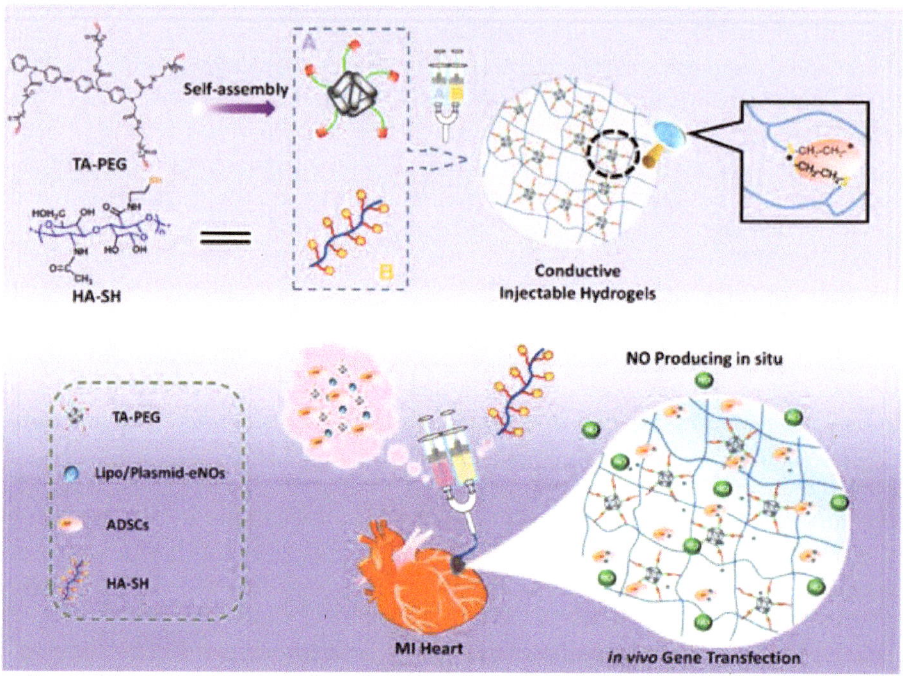

Figure 9: Immunomodulatory biomaterials and their application in therapies for chronic inflammation-related diseases (Nakkala et al. 2021).

which incorporates drugs and releases them through physical diffusion, followed by resorption of the microsphere material (Figure 9). If the drug carrier is soluble in water, the polymer need not to be biodegradable because this polymer will be excreted from the body, associated with urine or feces, although excretion will take a long time if the molecular weight of the polymer is extremely high.

1.11 Use for Tissue Engineering

Tissue engineering is an emerging technology that creates biological tissues to replace defective or lost tissues using cells and cell growth factors. Also, scaffolds are required for tissue construction if the lost part of the tissue is so large that it cannot be cured by conventional drug administration. At present, such largely diseased tissues and organs are replaced either with artificial organs or transplanted organs, but both therapeutic methods involve some problems. As mentioned earlier, the biocompatibility of clinically used artificial organs is mostly not satisfactory enough to prevent severe foreign-body reactions and fully perform the objective of the artificial organs aimed at patients. The biofunctionality of current artificial organs is still poor. On the contrary, the biofunctionality of transplanted organs is as excellent as that of healthy human organs, but the patients with transplanted organs are suffering from side effects induced by the immunosuppressive drugs administered. Another major problem of organ transplantation is the shortage of organ donors. The final objective of tissue engineering is to solve these problems by providing biological tissues and organs that are more excellent in both biofunctionality and biocompatibility than conventional artificial organs. Biodegradable polymers are required to fabricate scaffolds for cell proliferation and differentiation, which result in tissue regeneration or construction (Queen et al. 2004). Biodegradable polymers are also necessary for a sustained release of growth factors at the location of tissue regeneration. Generally, scaffolds used in tissue engineering are porous and three-dimensional to encourage the infiltration of a large number of cells into the scaffolds. Currently, the polymers used for scaffolding include collagen, glycolide-lactide copolymers, other copolymers of lactide, and cross-linked polysaccharides.

1.12 Ecological Applications (Processing of Plastic Wastes)

The other major application of biodegradable polymers is in plastic industries to replace biostable plastics for keeping our earth's environment clean. The first choice for processing plastic waste is reuse. However, only some plastic products can be reused after adequate processing, and many of them are difficult to recycle. In these cases, wastes are processed by landfill or incineration, but these processes often pollute the environment. If biodegradation by-products do not exert adverse effects on animals and plants on the earth, biodegradable plastics can be regarded as environment-friendly or ecological materials. Therefore, much attention has been focused on manufacturing biodegradable plastics, which, however, should address several requirements. They are to be low in product cost, satisfactory in mechanical properties, and not harmful to animals and plants when biodegraded. Biodegradation kinetics is also an important issue of biodegradable plastics.

Applications of biodegradable polymers in plastic industries can be seen in a wide range of industries, including agriculture, fishery, civil engineering, construction, outdoor leisure, food, toiletry, cosmetics and other consumer products. It is possible that the waste left as a result of outdoor activity and sports will stay for a long time in natural environments, possibly damaging them (Figure 10). On the other hand, when plastics are used indoors as food containers that are difficult to separate from the food remaining after use, the waste can be utilized as compost if it is biodegradable (Balakrishnan et al. 2006).

1.13 Modes of Biodegradation

Polymer degradation occurs mostly through the scission of the main chains or side chains of polymer molecules, induced by their thermal activation, oxidation, photolysis, radiolysis, or hydrolysis. Some polymers undergo degradation in biological environments when living cells or microorganisms are present around the polymers (Rabbany et al. 2010) (Figure 11).

Type A: In these systems, water-soluble polymers are insolubilized using hydrolytically or enzymatically unstable cross-links. As a result, the polymeric matrix is highly hydrophilic and completely permeated by water. Since the active agent is located in an aqueous environment, compounds with high water solubility will be rapidly leached out independent of the matrix erosion rate. A useful application of these systems is for the slow release of water-soluble macromolecules that can be

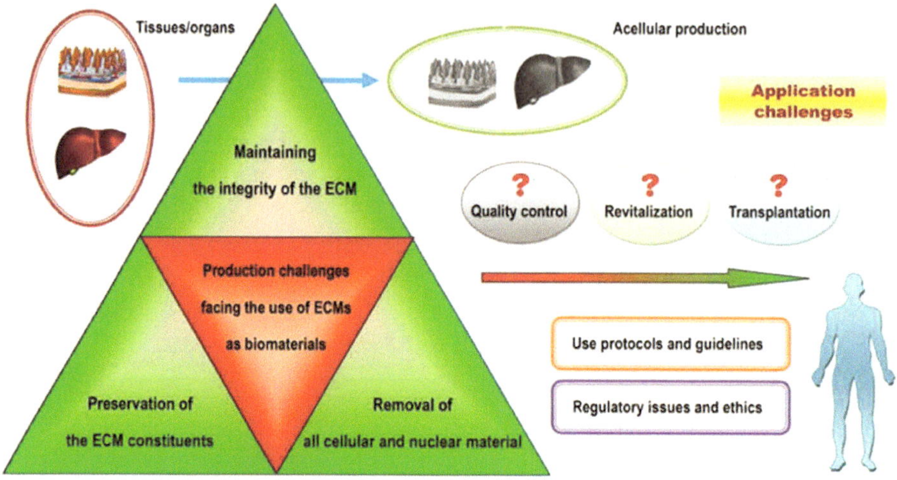

Figure 10: Advancing biomaterials of human origin for tissue engineering (Chen and Liu 2016).

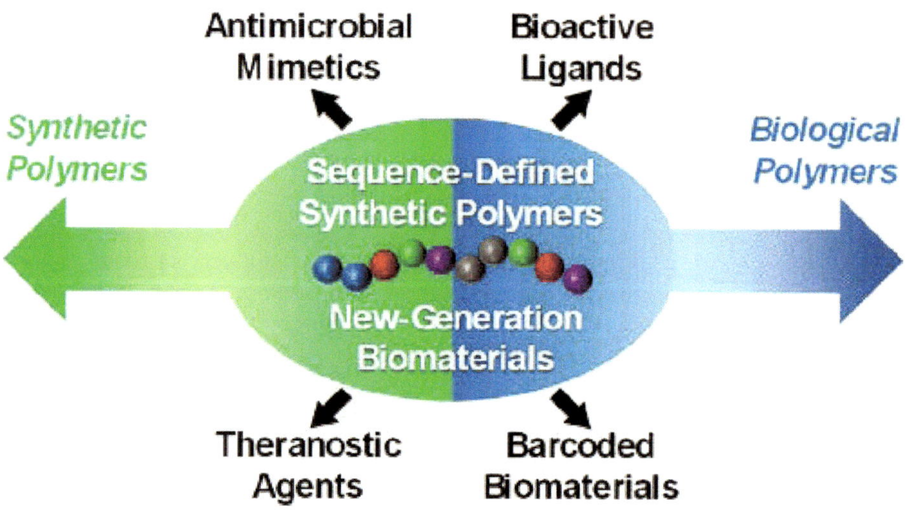

Figure 11: Sequence-defined synthetic polymers for new-generation functional biomaterials (Deng et al. 2021).

entangled in these systems and consequently cannot escape until a sufficient number of cross-links have cleaved and the matrix cross-link density has been reduced.

Type B: These systems include polymers that are initially water insoluble and are solubilized by hydrolysis, ionization or protonation of a pedant group. Since no backbone cleavage takes place, the solubilization does not result in any significant changes in MW. Polymers in this category are only useful in topical applications where the elimination of high molecular weight, water-soluble macromolecules proceeds with no difficulty, e.g., partial esters of methyl vinyl ether and maleic anhydride copolymers.

Type C: This category includes hydrophobic polymers that are converted to low molecular weight water-soluble molecules by backbone cleavage, e.g., the most commonly used biodegradable polymers of this category are PLA, PLGA copolymers, polycaprolactones, poly orthoesters, poly amino acids and poly anhydrides (Wiegand et al. 2009).

1.14 Quantitative Aspect of Polymer Degradation

In order to successfully design chemically controlled drug release systems, it is necessary to understand the main factors that determine the overall rate of the erosion or the degradation process. Zaikov has reported the *in vivo* breakdown of a polymer in terms of three degradative components, these being water (diffusivity), salt content/ pH of the physiological environment, and lastly, enzymatic attack (Boontheekul et al. 2005).

1.15 Designing of Biodegradable Polymers

The synthesis and development of biodegradable polymers are among the leading frontiers of research in polymer science at present. More recently, with the emphasis on reducing environmental pollution, attention has been directed toward biodegradable polymers (Boontheekul et al. 2005). In addition, there is now a wide variety of medical applications of synthetic and natural polymers that are biodegradable.

Rodriq has discussed a number of possible approaches to biodegradable systems (Lee et al. 2011). He suggested a search for new organisms for conventional polymers, the introduction of vulnerable units in the polymer chain, or adding nutrients, e.g., cellulose or amylose, on reinforcement in polymer composites. Gilbert have reviewed two approaches to the designing of biodegradable systems (Madan et al. 2013). The first is concerned with the synthesis and *in vivo* caprolactone) and its copolymers. The second approach was based on synthesizing block and graft copolymers containing naturally occurring segments, such as cellulose and amylose, which are known to be biodegradable. Sudesh Kumar has classified the possible approaches to developing new biodegradable polymers in the following four areas: studies of a series of polyesters, such as PLA, PGA, and poly (Lee et al. 2013).

- Modification of natural polymer
- Biodegradable polymer composites
- Synthesis of new biodegradable polymers
- Modification of synthetic polymers

1.16 Modification of Natural Polymers

Chemical modification of the natural polymers serves the two-fold purpose of utilizing renewable, naturally derived products such as polysaccharides and proteins as replacements for synthetic, petroleum-based products and as biodegradable compositions, which can be tailored for a slower or faster rate of degradation. Several new polymeric materials have been prepared through the chemical modification of natural polymers.

Among the polysaccharides, the commonly employed polymers include cellulose, starch, dextran and amylose. Various cellulose ethers have been reported to be biodegradable by various microorganisms (Zulkifli et al. 2021). 1,4 linkage between the glucose units in the cellulose could be made susceptible to hydrolysis by introducing chemical modifications. One such product is periodate-oxidized cellulose, which is formed by oxidizing cellulose with sodium periodate, leading to the conversion of anhydro-D-glucose units to dialdehyde units. Also (Wu et al. 2019), Singh et al. have reported that this product is degradable in the physiological environment, and the degradation products are glycolic acid and 2,4-dihydroxy butyric acid (Nam et al. 2019; Ravichandran et al. 2015). Another approach reported by Gilbert et al. was the biodegradable graft and copolymers of cellulose and amylose (Parihar et al. 2023). Grafting reaction leaves the main chain of a polymer intact, and it can be expected to add new properties associated with the side chains without drastically changing the main chain properties. Regenerated cellulose, cellulose acetate, and ethyl cellulose are commercially available and find application in packaging. Ferruti et al. reported succinic esters of starch and dextran as biodegradable drug-release matrixes. Dextran, as such, is used clinically as a plasma expander (Salehi et al. 2020).

Schacht has many questions concerning the immunological properties of polysaccharide derivatives (Hennecke et al. 2013). Also, the strength of polysaccharides is not outstanding when compared to other classes of materials. Bishop reported that fibers useful in surgery could be prepared by extruding the ethyl ether of dextran (Huemmerich et al. 2004).

Among the proteins, collagen and gelatin-based polymers are commonly employed. Collagen was used for a long time as a bioabsorbable suture. Gelatin has found many applications in the biomedical field as a bioabsorbable agent. For surgical use, gelatin is prepared as a foam or film and cross-linked with formaldehyde to decrease solubility. It is used to fill voids after tissue resection. A new surgical procedure was developed using a cross-linked gelatin cup prosthesis, which participated in the temporary substitution of tissues. Kumar et al. investigated the biodegradation of gelatin graft copolymers with poly (methylacrylate) using pure cultures and mixed bacterial inoculums (Huemmerich et al. 2004). As they are proteinaceous in nature, both collagen and gelatin also have the potential to incite an immune response.

1.17 Polymer Composites

Several polysaccharide plastic blends have been developed as degradable plastics, with the majority of them containing starch, cellulose or their derivatives as polysaccharides (Barbosa et al. 2016). The primary goal of that work was to replace plastic with inexpensive renewable resource materials derived from plants. These blends have been suggested as desirable alternatives to non-degradable plastics for which a number of solid waste disposal and environmental problems exist (Table 4). The principal degradable materials are mixtures of starch and polyethylene or starch, polyethylene and ethylene-co-acrylic acid. Poly (vinyl chloride) compositions filled with starch have also been reported (Park et al. 2012).

Table 4. Synthetic biodegradable polymers, features and applications.

S. No.	Biodegradable Synthetic Polymers	Features	Applications	References
1.	Polyamide	Amide linkage is subject to attack by proteolytic enzymes	Biodegradable matrixes enzymatic degradation	Yang et al. 2018; Ma et al. 2020; Sun et al. 2020; Latifi et al. 2018; Gomes et al. 2019
2.	Polyester	Ensures polycondensation results in low molecular weight compound with optimum properties	Surgery and drug delivery	Wu et al. 2019; Lee et al. 2018; Chatterjee et al. 2021; Chatterjee et al. 2019; Pasparakis and Tsitsilianis 2020
3.	Polyanhydride	Bulk metal condensation, ring-opening polymerization and interfacial polymerization	biodegradable materials for drug delivery and medical applications	Silva et al. 2015, Nam et al. 2019, Ravichandran et al. 2015; Kong et al. 2020 Bai et al. 2020
4.	Polyorthoesters	Prepared by transesterification	Biodegradable polymer	Chatterjee and Hui et al. 2021; Chatterjee et al. 2020; Pham et al. 2019
5.	Polyamino Carbonates	Dramatically increase the hydrolytic liability of the backbone without appreciably affecting the physicochemical properties of the polymer	Easily accessible with excellent engineering techniques	Wang et al. 2018; Hu et al. 2020; Mu and Ebara 2020
6.	Polyphosphazenes	Chemical link drug to polymer	Degrade non-toxic products	Wu et al. 2019; Lee et al. 2018; Chatterjee and Hui 2021; Pasparakis and Tsitsilianis 2020
7.	Polyphosphoesters	Biodegradability and mechanical strength	Degrade polymer to non-toxic products such as phosphate and other side constituents	Yang et al. 2018; Ma et al. 2020; Latifi et al. 2018; Gomes et al. 2019

A significant development in this area is the development of a biodegradable ceramic plastic composite for use as a biomaterial for hard tissue or bone replacement therapy (Czaja et al. 2019). When a biodegradable additive is used, the additive undergoes rapid biodegradation, leaving a porous and mechanically weakened

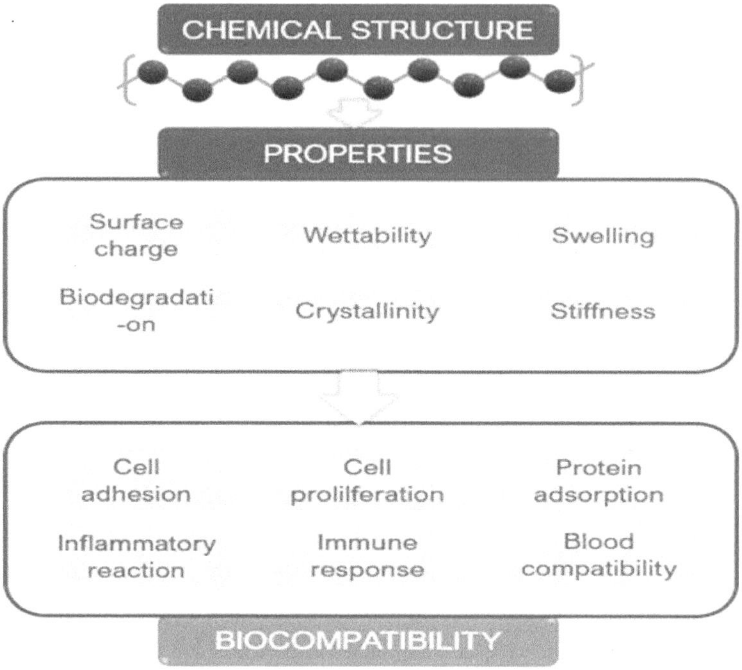

Figure 12: Biocompatible synthetic polymers for tissue engineering purposes (Terzopoulou et al. 2022).

ungraded polymer behind (Figure 12). The microorganisms, once having gained a foothold on the residual polymer, start degrading it slowly.

1.18 Synthetic Polymers

The various classes of synthetic polymers explored are polyesters, polyamides, polyanhydrides, polyorthoesters, polyimino carbonate, polyphosphazenes and polyphosphoesters.

1.19 Polyamide

Polyamides $(\text{-RCONH-})_n$ have received considerable attention as biodegradable polymers because of the general assumption that amide linkage is subject to attack by proteolytic enzymes. The utilization of polyamides in preparing biodegradable matrixes began following the observation that nylon implants deteriorated with time *in vivo* (Yang et al. 2018). Although nylon microcapsules have been used to study the release of several drugs, their slow rate of biodegradation has focused attention on more hydrolytically labile, hydrophilic polyamides (Ma et al. 2020).

The naturally occurring proteins, e.g., collagen, gelatin and albumin, have been used for drug delivery applications (Sun et al. 2020; Latifi et al. 2020). However, problems can arise due to their potential antigenicity (Gomes et al. 2019). The hydrolytic instability of the amide bond in synthetic amino acid polymers has been viewed in the formation of intentionally biodegradable polymers. As reviewed by Kopecek and Rejmanova, these polymers may be designed to give controlled biodegradation by introducing segments susceptible to specific enzymes. For, e.g., polymers of N-(2-hydroxypropyl)-methacrylamide and p-nitrophenyl esters of N-methyl acrylate amino acids are reacted with compounds containing an aliphatic amino group, with the formation of an amide bond. If this bond originates in an amino acid specific for a certain enzyme, an enzymatically cleavable bond is formed, e.g., L-phenylalanine, L-tyrosine and L-leucine are degraded by chymotrypsin. A wide variety of biodegradable polymers have been prepared in this way.

Poly (amino acids) based on synthetic amino acids such as poly (glutamic acid) have also been increasingly reported. Poly (glutamates) with increasing ester contents and chemically or enzymatically degradable bonds have been evaluated as drug carriers. The poly (t-butyloxy carbonyl methyl) glutamates are obtained by partially esterifying polyglutamic acid with t-butyl bromoacetate. Other biodegradable nonpeptidic polyamides have also been reported. Vanderbilt et al. have reported the polyamides obtained by polymerization of 4, 4-spirorbis butyrolactone with amines (Sun et al. 2020). Harris and Eury have obtained polyamides by the ring opening polymerization of 2, 2'-bis (5 (4h)-oxazolones) with diamines (Latifi et al. 2020).

1.20 Modification of Synthetic Polymers

One approach involves the copolymerization of a desired monomer with a monomer possessing a polar functional group, which might serve as a point of attack for microbial enzymes. The polymers can also be modified by substitution to enhance their degradability. It is reported that poly (vinyl butyrate) was rapidly degradable and at a rate comparable to catgut. Also, it was shown that amide bonds in side chains of synthetic polymers are cleaved by enzymes if they originate from the L-amino acids specific to the given enzymes (Wu et al. 2019). Kopecek et al. have reported the water-soluble copolymer based on N-2-hydroxypropyl methacrylamide and p-nitrophenyl esters of N-methacrylatedoligopeptides, bearing in their side chains a chromogenic substrate for chymotrypsin (Nam et al. 2019; Ravichandran et al. 2015). The substrate was L-phenylalanine-4'-nitro aniline, linked by its amino group to the terminal carboxyl group of the side chain. They have critically investigated the influence of the detailed oligopeptide structures and structural and steric factors on the degradability of the polymeric substrates both *in vivo* and *in vitro* (Kreller et al. 2021; Tripathi et al. 2015). Kumar et al. have reported the approach to rendering the polybutadiene moiety bio-susceptible by incorporating ester linkages by the condensation of the carboxy terminated polybutadiene with glycerol (Law et al. 2017). It was thought that the enzyme would get a foothold on the polymer surface (Figure 13). Thus, leading to progressive scissoring of C_2-oxidation of fatty acids b-units form the carbon chain, similar to well-known.

2. Conclusion and Future Prospects

The development of natural and synthetic biodegradable polymers is apparent and has progressed in response to a clearer definition of the need for materials with special characteristics. For instance, fibers that are strong and required for sutures, bulky, and perhaps porous masses are indicated for tissue void filling or scaffolding, and flexible but rapidly absorbing forms are suitable for drug release vehicles. The

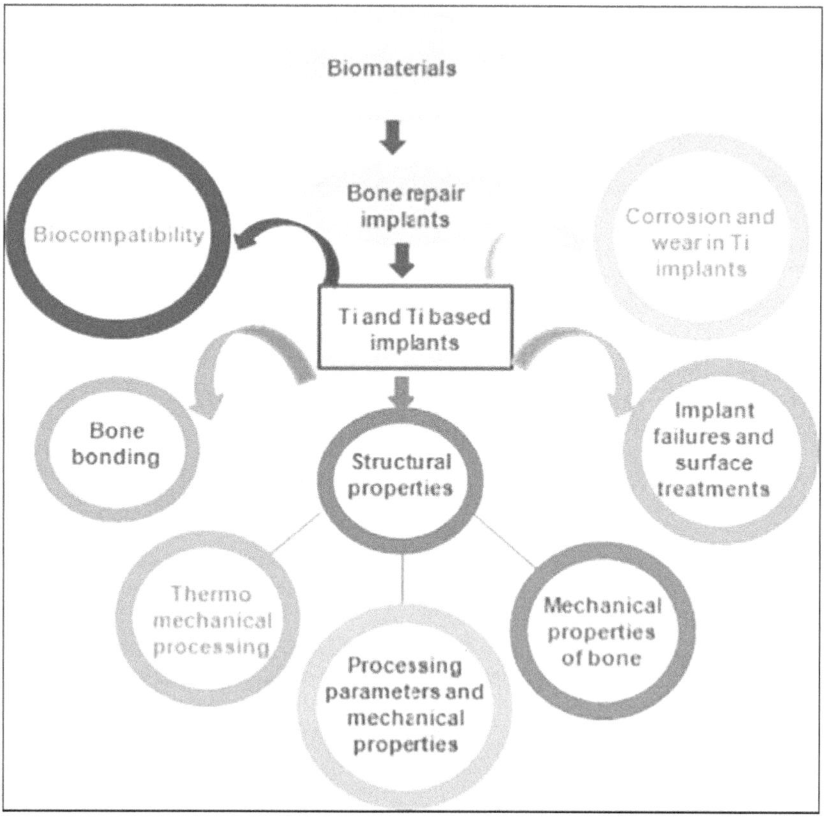

Figure 13: Polymeric biomaterials for biomedical applications (Kaliranjan et al. 2021).

wide variety of polymer systems expected with varied and special characteristics will allow an increase in successful applications to many problems in medicine and surgery requiring such materials. Future challenges include meeting the rising demand for improved sensitivity and selectivity, enabling real-time, low-cost monitoring of molecules in analyzing challenging clinical and environmental samples. The impact of nanotechnology on biosensors is also extensively covered in this study, along with the past, present, and future of these rapidly developing and burgeoning research endeavors. It is supported by the quick advancements in nanotechnology; these works of multidisciplinary research are progressing and starting to find their way to the market. The study emphasizes the main milestones accomplished and clarifies the emerging trends further by concentrating on the key advances from the views of novel smart materials, device topologies, and functionalities.

References

Aḥmed, S. and S. Ikram. 2016. Biosynthesis of gold nanoparticles: a green approach. Journal of Photochemistry and Photobiology B: Biology 161: 141–153.

Alexander, S.P., A. Mathie, J.A. Peters et al. 2021. The concise guide to pharmacology 2021/22: Ion channels. British Journal of Pharmacology 178: S157–S245.

Ankamwar, B., C. Damle, A. Ahmad et al. 2005. Biosynthesis of gold and silver nanoparticles using *Emblica officinalis* fruit extract, their phase transfer and transmetallation in an organic solution. J. Nanosci. Nanotechnol. 5(10): 1665–71.

Aromal, S. A. and D. Philip. 2012. Green synthesis of gold nanoparticles using *Trigonella foenum-graecum* and its size-dependent catalytic activity. Spectrochimica acta Part A: Molecular and Biomolecular Spectroscopy 97: 1–5.

Arunachalam, K. D., S. K. Annamalai and S. Hari. 2013. One-step green synthesis and characterization of leaf extract-mediated biocompatible silver and gold nanoparticles from *Memecylon umbellatum*. Int. J. Nanomedicine 8: 1307.

Arya, D. and R. K. Gupta. 2014. Simulation studies on variable cluster size in FCC riser reactor. Part. Sci. Technol. 32(2): 144–50.11.

Asano, M., H. Fukuzaki, M. Yoshida et al. 1990. Application of poly DL-lactic acids of varying molecular weight in drug delivery systems. Drug Des Deliv. 5: 301–20.

Asano, M., M. Yoshida, I. Kaetsu et al. 1985. Biodegradability of a hot-pressed poly (lactic acid) formulation with controlled release of LH-RH agonist and its pharmacological influence on rat prostate. Die Makromolekulare Chemie, Rapid Communications 7: 509–13.54.

Asano, M., M. Yoshida, I. Kaetsu et al. 1983. Biodegradable random copolypeptides of β-benzyl l-aspartate and γ-methyl l-glutamate for the controlled release of testosterone. Die Makromolekulare Chemie. 184(9): 1761–70.51.

Asapur, P. C. and I. Banerjee. 2020. Smart materials for medical applications. In *POLYM-PLAST TECH MAT*. 177–196). Woodhead Publishing.

Assi, M. A., M. N. Hezmee, A. W. Haron et al. 2016. The detrimental effects of lead on human and animal health. Vet World. 9(6): 660–71.28.

Awual, M. R. 2019. Innovative composite material for efficient and highly selective Pb (II) ion capturing from wastewater. J. Mol. Liq. 284: 502–10.

Bai, Z., W. Dan, G. Yu et al. 2018. Tough and tissue-adhesive polyacrylamide/collagen hydrogel with dopamine-grafted oxidized sodium alginate as crosslinker for cutaneous wound healing. RSC Adv. 8: 42123–42132.

Balakrishnan, K., A. Datar, W. Zhang et al. 2006. Nanofibril self-assembly of an arylene ethynylene macrocycle. Journal of the American Chemical Society 128(20): 6576–6577.

Bao, H., Z. Lu, X. Cui et al. 2010. Extracellular microbial synthesis of biocompatible CdTe quantum dots. Acta Biomaterialia 6(9): 3534–3541.

Barbosa, A.M., E. Robles, J. S. Ribeiro et al. 2016. Cellulose nanocrystal membranes as excipients for drug delivery systems. Materials 9.1002.

Belen'kii, L. I. 1964. Present-day methods of synthesising macrocyclic compounds. Russ. Chem. Rev. 33(11): 551.71.

Bello, A. B., D. Kim, D. Kim et al. 2020. Engineering and functionalization of gelatin biomaterials: from cell culture to medical applications. Tissue Eng. Part B Rev. 26: 164–180.

Benicewicz, B. C. and P. K. Hopper. 1991. Polymers for absorbable surgical sutures-Part II. J. Bioact. Compat. Polym. 1: 64–94.

Bodin, A., S. Concaro, M. Brittberg et al. 2007. Bacterial cellulose as a potential meniscus implant. J. Tissue Eng. Regen. Med. 1: 406–408.

Boontheekul, T., H.J. Kong and D.J. Mooney. 2005. Controlling alginate gel degradation utilizing partial oxidation and bimodal molecular weight distribution. Biomaterials 26(15): 2455–2465.

Braden, M. and R. L. Clarke. 1984. Water absorption characteristics of dental microfine composite filling materials: I. Proprietary materials. Biomaterials 5(6): 369–72.

Brosse, J. C., D. Derouet, L. Fontaine et al. 1989. Fixation of pharmacologically active amines on polyphosphonates, 2. Application to benzocaïne and phenethylamine. Die Makromolekulare Chemie: Macromol. Chem. Phys. 190(9): 2339–45.93.

Brown, E. E., D. Hu, N. Abu Lail et al. 2013. Potential of nanocrystalline cellulose–fibrin nanocomposites for artificial vascular graft applications. Biomacromolecules 14: 1063–1071.

Bruck, S. D. 1983. Controlled drug delivery. Boca Raton, Florida: CRC Press.

Case, D. A., H. M. Aktulga, K. Belfon et al. 2021. Amber 2021. University of California, San Francisco.

Chandran, S. P., M. Chaudhary, R. Pasricha et al. 2006. Synthesis of gold nanotriangles and silver nanoparticles using *Aloe vera* plant extract. Biotechnol. Prog. 22(2): 577–83.

Chasin, M. 1990. Biodegradable polymers as drug delivery systems. Informa Health Care.
Chatterjee, S. and P. C. Hui. 2019. Review of stimuli-responsive polymers in drug delivery and textile application. Molecules. 24: 2547.
Chatterjee, S. and P. C. Hui. 2021. Review of applications and future prospects of stimuli-responsive hydrogel based on thermo-responsive biopolymers in drug delivery systems. Polymers 13: 2086.
Chen, F. M. and X. Liu. 2016. Advancing biomaterials of human origin for tissue engineering. Progress in Polymer Science 53: 86–168.
Chenouard, N., I. Smal, F. De Chaumont et al. 2014. Objective comparison of particle tracking methods. Nature methods 11(3): 281–289.
Cherian, B. M., A. Leao, S. F. de Souza et al. 2011. Cellulose nanocomposites with nanofibres isolated from pineapple leaf fibers for medical applications. Carbohydr. Polym. 86: 1790–1798.
Christopherson, G. T., H. Song and H. Q. Mao. 2009. The influence of fibre diameter of electrospun substrates on neural stem cell differentiation and proliferation. Biomaterials 30(4): 556–564.
Cohen, S., M. C. Bano, K. B. Visscher et al. 1990 Ionically crosslinkable polyphosphazene: a novel polymer for microencapsulation. J. Am. Chem. Soc. 112(21): 7832–3.90.
Corey, J. M., D. Y. Lin, K. B. Mycek et al. 2007. Aligned electrospun nanofibers specify the direction of dorsal root ganglia neurite growth. J. Biomed. Mater. Res. Part A: An Official Journal of The Society for Biomaterials, The Japanese Society for Biomaterials, and The Australian Society for Biomaterials and the Korean Society for Biomaterials 83(3): 636–645.
Czaja, W., A. Krystynowicz, M. Kawecki et al. 2007. Biomedical applications of microbial cellulose in burn wound recovery. *In*: Brown, R. M. and I. M. Saxena (Eds.). Cellulose: Molecular and Structural Biology. Springer: Dordrecht, Germany.
Deng, Z., Q. Shi, J. Tan et al. 2021. Sequence-defined synthetic polymers for new-generation functional biomaterials. ACS Mat. Lett. 3(9): 1339–56.
Diba, M., S. Spaans, K. Ning et al. 2018. Self-healing biomaterials: from molecular concepts to clinical applications. Adv. Mater. Interfaces 5(17): 1800118.
Domb, A. J. and R. Langer. 1989. Solid-state and solution stability of poly (anhydrides) and poly (esters). Macromolecules 22(5): 2117–22.
Dong, J., J. Tong, J. Luo et al. 2020. Gold nanoparticles for smart and recoverable catalyst using thermo-responsive core-crosslinked star polymer as the nanoreactor. Appl. Surf. Sci. 507: 144950.
Dragone, R., G. Grasso, M. Muccini et al. 2017. Portable bio/chemosensoristic devices: innovative systems for environmental health and food safety diagnostics. Front. Public Health 5: 80.
Du, L., H. Jiang, X. Liu et al. 2007. Biosynthesis of gold nanoparticles assisted by *Escherichia coli* DH5α and its application on direct electrochemistry of haemoglobin. Electrochem. Commun. 9(5): 1165–1170.
Dunn, R. L. 1991. Polymeric drugs and drug delivery systems. *In*: Dunn, R. L. and R. M. Ottembrite (Eds.). ACS Symp. Series 545: 11–23.
Engelberg, I. and J. Kohn. 1991. Physico-mechanical properties of degradable polymers used in medical applications: a comparative study. Biomaterials 12(3): 292–304.65.
Feng, X. D., C. X. Song and W. Y. Chen. 1983. Synthesis and evaluation of biodegradable block copolymers of ε-caprolactone and dl-lactide. Journal of Polymer Science: Polymer Letters Edition. 21(8): 593–600.
Frohbergh, M. E., A. Katsman, G. P. Botta et al. 2012. Electrospun hydroxyapatite-containing chitosan nanofibers crosslinked with genipin for bone tissue engineering. Biomaterials 33(36): 9167–9178.
Fu, L., Y. Zhang, C. Li et al. 2012. Skin tissue repair materials from bacterial cellulose by a multilayer fermentation method. J. Mater. Chem. 22: 12349–12357.
Fu, L., P. Zhou, S. Zhang et al. 2013. Evaluation of bacterial nanocellulose-based uniform wound dressing for large area skin transplantation. Mater. Sci. Eng. C. 33: 2995–3000.
Fukuzaki, H., M. Yoshida, M. Asano et al. 1990. Synthesis of low-molecular-weight copoly (l-lactic acid/ε-caprolactone) by direct copolycondensation in the absence of catalysts, and enzymatic degradation of the polymers. Polymer. 31(10): 2006–14.57.
Ganaie, S. U., T. Abbasi, J. Anuradha et al. 2014. Biomimetic synthesis of silver nanoparticles using the amphibious weed ipomoea and their application in pollution control. JKSUES. 26(3): 222–229.
Garg, M., A. Parihar and M.S. Rahman. 2024. Advanced and personalized healthcare through integrated wearable sensors (versatile). Materials Advances. 5: 432–452.

Gezer, F., R.G. Aykroyd and S. Barber. 2021. Statistical properties of Poisson-Voronoi tessellation cells in bounded regions. Journal of Statistical Computation and Simulation 91(5): 915–933.

Gilbert, T.L. 2004. A phenomenological theory of damping in ferromagnetic materials. IEEE Transactions on Magnetics 40(6): 3443–3449.

Gomes Neto, R. J., G. M. Genevro, L. d. A. Paulo et al. 2019.Characterization and *in vitro* evaluation of chitosan/konjac glucomannan bilayer film as a wound dressing. Carbohydr. Polym. 212: 59–66.

Grasso, G., D. Zane and R. Dragone. 2019. Microbial nanotechnology: challenges and prospects for green biocatalytic synthesis of nanoscale materials for sensoristic and biomedical applications. J. Nanomater. 10(1): 11.

Gumargalieva, K. Z., Y. V. Moiseev, T. T. Daurova et al. 1980. Polycaproamide degradation in rabbits and in several model media. Biomaterials 1(4): 214–6.

Guttenplan, A. P., Z. Tahmasebi Birgani, S. Giselbrecht et al. 2021. Chips for biomaterials and biomaterials for chips: recent advances at the interface between microfabrication and biomaterials research. Adv. Healthc. Mater. 10(14): 2100371.

Heller, J., A. C. Chang, G. Rood et al. 1984. Release of insulin from pH-sensitive poly (ortho esters). J. Controlled Release. 13(2-3): 295–302.

Heller, J. 1984. Biodegradable polymers in controlled drug delivery. Crit. Rev. Ther. Drug Carrier Syst. 1(1): 39–90.78.

Hennecke, K., J. Redeker, J. W. Kuhbier et al. 2013. Bundles of spider silk, braided into sutures, resist basic cyclic tests: potential use for flexor tendon repair. PLoS ONE, 8: e61100.

Holland, S. J., B. J. Tighe and P. L. Gould. 1986. Polymers for biodegradable medical devices. 1. The potential of polyesters as controlled macromolecular release systems. J. Controlled Release. 4(3): 155–80.60.

Hu, J., L. Yang, P. Yang et al. 2020. Polydopamine free radical scavengers. Biomater. Sci. 8: 4940–4950.

Huemmerich, D., C. W. Helsen, S. Quedzuweit et al. 2004. Primary structure elements of spider dragline silks and their contribution to protein solubility. Biochemistry 43: 13604–13612.

Hulkoti, N. I. and T. C. Taranath. 2014. Biosynthesis of nanoparticles using microbes-a review. Colloids surf. B: Biointerfaces 121: 474–483.

Hutmacher, D. W., T. B. F. Woodfield and P. D. Dalton. 2015. Scaffold design and fabrication. *In*: van Blitterswijk, C. A. and J. de Boer (Eds.). Tissue Engineering, 2nd ed.; Academic Press: London. UK. 311–346.

Ikada, Y. and H. Tsuji. 2000. Biodegradable polyesters for medical and ecological applications. Macromolecular Rapid Communications 21(3): 117–32.

Ikada. Y. 1998. Tissue engineering research trends at Kyoto University. Tissue Engineering for Therapeutic Use. 1: 1–4.

Inshakova, E. and O. Inshakov. 2017. World market for nanomaterials: Structure and trends. In *MATEC* web of conferences (129). EDP Sciences.

Iravani, S. 2011. Green synthesis of metal nanoparticles using plants. Green Chem. 13(10): 2638–50.15.

Jianrong, C., M. Yuqing, H. Nongyue et al. 2004. Nanotechnology and biosensors. Biotechnol. Adv. 22(7): 505–518.

Kalirajan, C., A. Dukle, A. J. Nathanael et al. 2021. A critical review on polymeric biomaterials for biomedical applications. Polymers 13(17): 3015.

Kang, I., Y. Y. Heung, J. H. Kim et al. 2006. Introduction to carbon nanotube and nanofiber smart materials. Compos. B Eng. 37(6): 382–394.

Karp, J. M. and R. Langer. 2007. Development and therapeutic applications of advanced biomaterials. Curr. Opin. Biotechnol. 1: 18(5): 454–9.

Khan, F., M. Tanaka and S. R. Ahmad. 2015. Fabrication of polymeric biomaterials: a strategy for tissue engineering and medical devices. J Mater Chem B. 3(42): 8224–49.

Khan, R. and M. H. Khan. 2013. Use of collagen as a biomaterial: An update. J. Indian Soc. Periodontol. 17: 539–542.

Khan, R., A. Parihar and S. K. Sanghi. 2021. Biosensor Based Advanced Cancer Diagnostics: From Lab to Clinics. Academic Press.

Khan, T., S. Abbas, A. Fariq. 2018. Microbes: Nature's cell factories of nanoparticles synthesis. In exploring the realms of nature for nanosynthesis. Springer, Cham. 25–50.

Kim, T.-R., M.-S. Kim, T. S. Goh et al. 2019. Evaluation of structural and mechanical properties of Porous artificial bone scaffolds fabricated via advanced TBA-based freeze-gel casting technique. Appl. Sci. 9: 1965.

Kolakovic, R., L. Peltonen, T. Laaksonen et al. 2011. Spray-dried cellulose nanofibers as novel tablet excipient. AAPS Pharm. Sci. Tech. 12: 1366–1373.

Kong, B., Y. Chen, R. Liu et al. 2020. Fibre reinforced GelMA hydrogel to induce the regeneration of corneal stroma. Nat. Commun. 11: 1–12.

Kumar, R., N. Singh, R. Singh et al. 2017. An Introduction of Biodegradable Polymers, Modes of Biodegradation and Designing of Biodegradable Polymers. Res. J. Pharm. Technol. 10(2): 625–40.

Kumar, R., R. C. Nagarwal, M. Dhanawat et al. 2011. *In-vitro* and *in-vivo* study of indomethacin loaded gelatin nanoparticles. J. Biomed. Nanotechnol. 7: 325–333.

Kweon, H., H. C. Ha, I. C. Um et al. 2001. Physical properties of silk fibroin/chitosan blend films. J. Appl. Polym. Sci. 80: 928–934.

Landry, F. B., D. V. Bazile, G. Spenlehauer et al. 1996. Degradation of poly (D, L-lactic acid) nanoparticles coated with albumin in model digestive fluids (USP XXII). Biomaterials 17(7): 715–23.

Latifi, N., M. Asgari, H. Vali et al. 2018. A tissue-mimetic nano-fibrillar hybrid injectable hydrogel for potential soft tissue engineering applications. Sci. Rep. 8: 1–18.

Lee, D., D. N. Heo, H. R. Nah et al. 2018. Injectable hydrogel composite containing modified gold nanoparticles: Implication in bone tissue regeneration. Int. J. Nanomed. 13: 7019.

Lee, E. J., S. A. Khan, J. K. Park et al. 2011. Studies on the characteristics of drug-loaded gelatin nanoparticles prepared by nanoprecipitation. Bioprocess Biosyst. Eng. 35: 297–307.

Lee, K. Y. and Mooney, D.J. 2012. Alginate: properties and biomedical applications. Progress in Polymer Science 37(1): 106–126.

Lee, S. J., J. Y. Yhee, S. H. Kim et al. 2013. Biocompatible gelatin nanoparticles for tumor-targeted delivery of polymerized siRNA in tumor-bearing mice. J. Control. Release. 13(172): 358–366.

Leong, K. W., V. Simonte and R. Langer. 1987. Synthesis of polyanhydrides: melt-polycondensation, dehydrochlorination, and dehydrative coupling. Macromolecules 20(4): 705–12.74.

Lepistö, J., H. Kujari, J. Niinikoski et al. 1994. Effects of heterodimeric isoform of platelet-derived growth factor PDGF-AB on wound healing in the rat. Eur. Surg. Res. 26: 267–272.

Letchford, K., J. K. Jackson, B. Wasserman et al. 2011. The use of nanocrystalline cellulose for the binding and controlled release of drugs. Int. J. Nanomed. 6: 321–330.

Li, N. H., M. Richards, K. Brandt et al. 1989. Poly (phosphate esters) as drug-carriers. In Abstracts of Papers of the American and Chemical society. (Vol. 197, pp. 216-POLY). 1155 16TH ST, NW, Washignton, DC 20036: J. Am. Chem. Soc. 94.

Li, S., X. Guo, R. Gao et al. 2021. Recent progress on biomaterials fighting against viruses. Adv. Mat. 33(14): 2005424.

Lin, C.-Y. and J.-H. Kang. 2021. Mechanical properties of compact bone defined by the stress-strain curve measured using uniaxial tensile test: a concise review and practical guide. Materials 14: 4224.

Liu, S., D. Huang, Y. Hu et al. 2020. Sodium alginate/collagen composite multiscale porous scaffolds containing poly(ε-caprolactone) microspheres fabricated based on additive manufacturing technology. RSC Adv. 10: 39241–39250.

Loo, Y. Y., B. W. Chieng, M. Nishibuchi et al. 2012. Synthesis of silver nanoparticles by using tea leaf extract from *Camellia sinensis.* Int. J. Nanomedicine. 7: 4263.

Lovene, A., Y. Zhao, S. Wang et al. 2021. Bioactive polymeric materials for the advancement of regenerative medicine. J. Funct. Biomater. 12(1): 14.

Luben, T. J., J. L. Nichols, S. J. Dutton et al. 2017. A systematic review of cardiovascular emergency department visits, hospital admissions and mortality associated with ambient black carbon. Environment International 107: 154–162.

Ma, J., Y. Wang and J. Liu. 2017. Biomaterials meet microfluidics: from synthesis technologies to biological applications. Micromachines 18(8): 255.

Ma, Z., W. Song, Y. He et al. 2020. Multilayer injectable hydrogel system sequentially delivers bioactive substances for each wound healing stage. ACS Appl. Mater. Interfaces. 12: 29787–29806.

Madan, J., R. S. Pandey, U. K. Jain et al. 2013. Sterically stabilized gelatin microassemblies of noscapine enhance cytotoxicity, apoptosis and drug delivery in lung cancer cells. Colloids Surf. B. Biointerfaces 107: 235–244.

Malvankar, N. S. and D. R. Lovley. 2012. Microbial nanowires: a new paradigm for biological electron transfer and bioelectronics. Chem. Sus. Chem. 5(6): 1039–1046.

Marks, M. G., C. Doillon and F. H. Silvert. 1991. Effects of fibroblasts and basic fibroblast growth factor on facilitation of dermal wound healing by type I collagen matrices. J. Biomed. Mater. Res. 25: 683–696.

Meftahi, A., R. Khajavi, A. Rashidi et al. 2009. The effects of cotton gauze coating with microbial cellulose. Cellulose 17: 199–204.

Miller, N. D. and D. F. Williams. 1987. On the biodegradation of poly-β-hydroxybutyrate (PHB) homopolymer and poly-β-hydroxybutyrate-hydroxyvalerate copolymers. Biomaterials 8(2): 129–37.

Miller, R. A., J. M. Brady and D. E. Cutright. 1977. Degradation rates of oral resorbable implants (polylactates and polyglycolates): rate modification with changes in PLA/PGA copolymer ratios. J. Biomed. Mater. Res. 11(5): 711–9.66.

Mitra, A., C. H. Lee and K. Cheng. 2013. Adv. Drug Deliv. Wiley: Hoboken, NJ, USA.

Mu, M. and M. Ebara. 2020. Smart polymers. In Polymer Science and Nanotechnology; Elsevier: Amsterdam, The Netherlands, 257–279.

Nair, L. S. and C. T. Laurencin. 2005. Polymers as biomaterials for tissue engineering and controlled drug delivery. Tissue Eng. 1(1): 47–90.

Nakkala, J. R., Z. Li, W. Ahmad et al. 2021. Immunomodulatory biomaterials and their application in therapies for chronic inflammation-related diseases. Acta Biomater. 123: 1–30.

Nam, S., R. Stowers, J. Lou et al. 2019. Varying PEG density to control stress relaxation in alginate-PEG hydrogels for 3D cell culture studies. Biomaterials 200: 15–24.

Narayanan, K. B. and N. Sakthivel. 2008. Coriander leaf mediated biosynthesis of gold nanoparticles. Mater. Lett. 62(30): 4588–90.

Nguyen, T. D., L. P. Yeo, K. S. Yang et al. 2020. Fabrication and characterization of graphene quantum dots thin film for reducing cross-sectional heat transfer through smart window. Mat. Res. Bull. 127: 110861.

Nguyen, T. P., Q. V. Nguyen, V.-H. Nguyen et al. 2019. Silk fibroin-based biomaterials for biomedical applications: A review. Polymers. 11. 1933.

Nikolova, M. P. and M. S. Chavali. 2019. Recent advances in biomaterials for 3D scaffolds: A review. Bioact. Mater. 4: 271–292.

Nimeskern, L., H. M. Ávila, J. Sundberg et al. 2013. Mechanical evaluation of bacterial nanocellulose as an implant material for ear cartilage replacement. J. Mech. Behav. Biomed. Mater. 22: 12–21.

Niraimathi, K. L., V. Sudha, R. Lavanya et al. 2013. Biosynthesis of silver nanoparticles using *Alternanthera sessilis* (Linn.) extract and their antimicrobial, antioxidant activities. Colloids and surfaces B: Biointerfaces 102: 288–91.

Nwodo, U. U., E. Green and A. I. Okoh. 2012. Bacterial exopolysaccharides: functionality and prospects. Int. J. Mol. Sci. 13(11): 14002–14015.

Oliveira, J., V. Correia, H. Castro et al. 2018. Polymer-based smart materials by printing technologies: Improving application and integration. Addit. Manuf. 21: 269–283.

Ottenbrite, R. M. and S. W. Kim (eds.). 2019. Polymeric drugs and drug delivery systems. CRC Press.

Palit, S. R. and B. M. Mandal. 1968. End-group studies using dye techniques. J. Macromol. Sci-Reviews in Macromolecular Chemistry 2(2): 225–77.

Pannu, R. K. 1997. Synthesis and characterisation of block copoly (DL-lactide-co-ethylene oxide-co-DL-lactide) polymers and evaluation of their use in drug delivery. The University of Manchester (United Kingdom).

Parihar, A., A. Singhal, N. Kumar et al. 2022. Next-generation intelligent MXene-based electrochemical aptasensors for point-of-care cancer diagnostics. Nano-Micro Lett. 14(1): 1–34.

Parihar, A., S. Jain, D. S. Parihar et al. 2022. Biomarkers associated with different types of cancer as a potential candidate for early diagnosis of oncological disorders. In Biosensor Based Advanced Cancer Diagnostics. Academic Press 47–57 pp.

Parihar, A. and R. Khan. 2023. Yttrium functionalized reduced graphene oxide nanocomposite-based aptasensor for ultrasensitive detection of a breast cancer biomarker. ACS Applied Nano Materials https://doi.org/10.1021/acsanm.3c03234.

Parihar, A., S. S Ahmed, P. Sharma et al. (2023). Plant-based bioactive molecules for targeting of endoribonuclease using steered molecular dynamic simulation approach: a highly conserved therapeutic target against variants of SARS-CoV-2. Molecular Simulation 49(12): 1267–1279

Parihar, A., N. K. Choudhary, P. Sharma, R. Khan and A. Gupta. 2024. Evaluation of electrochemical and antimicrobial efficacy of green synthesized silver functionalized reduced graphene oxide nanocomposites. Biomedical Materials & Devices, 1–15.

Parihar, A., P. Vishwakarma and R. Khan. 2024. Miniaturized MXene-based electrochemical biosensors for virus detection. Bioelectrochemistry 158: 108700

Park, S. H., H. S. Shin and S. N. Park. 2018. A novel pH-responsive hydrogel based on carboxymethyl cellulose/2-hydroxyethyl acrylate for transdermal delivery of naringenin. Carbohydr. Polym. 200: 341–352.

Park, S. U., B. K. Lee, M. S. Kim et al. 2012. The possibility of microbial cellulose for dressing and scaffold materials. Int. Wound J. 11: 35–45.

Pasparakis, G. and C. Tsitsilianis. 2020. LCST polymers: Thermoresponsive nanostructured assemblies towards bioapplications. Polymer. 211: 123146.

Paul, B., S. Sanusei, G. Barber et al. 2017. Degradation of polymeric material used for osteosynthesis and comparison of fracture toughness between test and FEA. Int. Orthop. 4(4): 788–95.

Pazmiño Viteri, K., K. S. Vizuete, F. Alexis et al. 2021. A closer look to polyesters: properties, synthesis, characterization, and particle drug delivery applications. Nanosci. & Nanotechnol.-Asia. 11(5): 24–58.

Peppas, N. A. and R. Langer. 1994. New challenges in biomaterials. Science 263(5154): 1715–20.76.

Petros, R. A. and J. M. DeSimone. 2010. Strategies in the design of nanoparticles for therapeutic applications. Nat. Rev. Drug Discov. 9(8): 615–627.

Pham, L., L. H. Dang, M. D. Truong et al. 2019. A dual synergistic of curcumin and gelatin on thermal-responsive hydrogel based on Chitosan-P123 in wound healing application. Biomed. Pharmacother. 117: 109183.

Phisalaphong, M., N. Chiaoprakobkij, M. Gama et al. 2012. Applications and products-Nata de Coco. Bacterial cellulose: Multifunct. Mater. 9(143): 144–154.

Phull A.-R., S. H. Eo, Q. Abbas et al. 2016. Applications of chondrocyte-based cartilage engineering: an overview. BioMed. Res. Int. 1–17.

Prasad, R., R. Pandey and I. Barman. 2016. Engineering tailored nanoparticles with microbes: Quo vadis? Wiley Interdiscip. Rev. Nanomed. Nanobiotechnol. 8: 316–330.

Pretula, J., K. Kaluzynski and S. Penczek. 1986. Synthesis of poly (alkylene phosphates) with nitrogen-containing bases in the side chains. Nitrogen-and carbon-substituted imidazoles. Macromolecules. 19(7): 1797–9.

Pulapura, S., C. Li and J. Kohn. 1990. Structure-property relationships for the design of polyiminocarbonates. Biomaterials 11(9): 666–78.

Queen, R. A., B. M. Gribbon, C. James et al. 2004. Retrotransposon-based molecular markers for linkage and genetic diversity analysis in wheat. Molecular Genetics and Genomics 271: 91–97.

Rabbany, S. Y., J. Pastore, M. Yamamoto et al. 2010. Continuous delivery of stromal cell-derived factor-1 from alginate scaffolds accelerates wound healing. Cell transplantation 19(4): 399–408.

Ramakrishna, S. 2001. Textile scaffolds in tissue engineering. CRC Press: Boca Raton, FL, USA; 50.

Ramteke, C., T. Chakrabarti, B. K. Sarangi et al. 2013. Synthesis of silver nanoparticles from the aqueous extract of leaves of *Ocimum sanctum* for enhanced antibacterial activity. J. chemistry.

Ramya, S., T. Shanmugasundaram and R. Balagurunathan. 2015. Biomedical potential of actinobacterially synthesized selenium nanoparticles with special reference to anti-biofilm, anti-oxidant, wound healing, cytotoxic and anti-viral activities. J. Trace Elem. Med. Biol. 32: 30–39.

Ravichandran, R., M. M. Islam, E. I. Alarcon et al. 2015. Functionalised type-I collagen as a hydrogel building block for bio-orthogonal tissue engineering applications. J. Mater. Chem. B. 4: 318–326.

Repanas, A., S. Andriopoulou and B. Glasmacher. 2016. The significance of electrospinning as a method to create fibrous scaffolds for biomedical engineering and drug delivery applications. J. Drug Deliv. Sci. Technol. 31: 137–146.

Ribeiro, B. D., M. A. Z. Coelho and A. M. Castro de. 2015. Principles of Green Chemistry and White Biotechnology.

Rose, J. C. and L. De Laporte. 2018. Hierarchical design of tissue regenerative constructs. Adv. Healthc. Mate. 7(6): 1701067.
Rosen, H. B., J. Chang, G. E. Wnek et al. 1983. Bioerodible polyanhydrides for controlled drug delivery. Biomaterials 4(2): 131–133.
Salata, O. V. 2004. Applications of nanoparticles in biology and medicine. J. Nanobiotechnol. 2(1): 1–6.
Saleh, T. A., G. Fadillah and E. Ciptawati. 2021. Smart advanced responsive materials, synthesis methods and classifications: from Lab to applications. J. Polym. Res. 28(6): 1–5.
Salehi, S., K. Koeck and T. Scheibel. 2020. Spider silk for tissue engineering applications. Molecules. 25: 737.
Salem, W. M., M. Haridy, W. F. Sayed et al. 2014. Antibacterial activity of silver nanoparticles synthesized from latex and leaf extract of *Ficus sycomorus*. Ind. Crops Prod. 62: 228–234.
Schuman, L., P. Buma, D. Versleyen et al. 1995. Chondrocyte behaviour within different types of collagen gel *in vitro*. Biomaterials 16: 809–814.
Sharma, K., M. A. Mujawar and A. Kaushik. 2019. State-of-art functional biomaterials for tissue engineering. Front. Mater. Sci. 6: 172.
Shukla, M. K., A. Parihar, C. Karthikeyan et al. 2023. Multifunctional GQDs for receptor targeting, drug delivery, and bioimaging in pancreatic cancer. Nanoscale 15(36): 14698–14716.
Sidman, K. R., A. D. Schwope, W. D. Steber et al. 1980. Biodegradable, implantable sustained release systems based on glutamic acid copolymers. J. Membr. Sci. 7(3): 277–91.53.
Siepmann, J. and F. Siepmann. 2006. Microparticles used as drug delivery systems. Prog. Colloid Polym. Sci. 133: 15–21.
Silva, D., A. C. Fernandes, T. G. Nunes et al. 2015. The effect of albumin and cholesterol on the biotribological behavior of hydrogels for contact lenses. Acta Biomater. 26: 184–194.
Sivaraj, R., P. Rahman, P. Rajiv et al. 2014, October. Biogenic zinc oxide nanoparticles synthesis using *Tabernaemontana Divaricate* leaf extract and its anticancer activity against MCF-7 breast cancer cell Lines. In International Conference on Advances in Agricultural, Biological & Environmental Sciences.
Sizentsov, A., G. Karpova, T. Klimova et al. 2019. Evaluation of anionic components of lead on biotoxicity and bioaccumulation ability in respect of probiotic stamps. GEOMATE. 16(55): 8–13.
Skeffington, A. W. and A. Scheffel. 2018. Exploiting algal mineralization for nanotechnology: bringing coccoliths to the fore. Curr. Opin. Biotechnol. 49: 57–63.
Song, R., M. Murphy, C. Li et al. 2018. Current development of biodegradable polymeric materials for biomedical applications. Drug Des. Devel. Ther. 12: 3117.
Soppimath, K. S., T. M. Aminabhavi, A. R. Kulkarni et al. 2001. Biodegradable polymeric nanoparticles as drug delivery devices. Journal of Controlled Release 70(1-2): 1–20
Suggs, L. J., S. A. Moore and A.G. Mikos. 2007. Synthetic biodegradable polymers for medical applications. pp. 939–950. *In*: Physical properties of polymers handbook. Springer, New York, NY.
Sun, Y., D. Nan, H. Jin et al. 2020. Recent advances of injectable hydrogels for drug delivery and tissue engineering applications. Polym. Test. 81: 106283.
Terzopoulou, Z., A. Zamboulis, I. Koumentakou et al. 2022. Biocompatible synthetic polymers for tissue engineering purposes. Biomacromolecules 23(5): 1841–63.
Tielens, S., H. Declercq, T. Górski et al. 2007. Gelatin-based microcarriers as embryonic stem cell delivery system in bone tissue engineering: An *In-Vitro* Study. Biomacromolecules. 8: 825–832.
Trache, D. and V. K. Thakur. 2020. Nanocellulose and nanocarbons based hybrid materials: synthesis, characterization and applications. Nanomaterials 10(9): 1800.
Troy, E., M. A., A. M. Power and J. G. Wall. 2021. Nature-based biomaterials and their application in biomedicine. Polymers 13(19): 3321.
Troy, E., M. A. Tilbury, A. M. Power et al. 2021. Nature-based biomaterials and their application in biomedicine. Polymers 13(19): 3321.
Valent, P., D. Bonnet, R. De Maria et al. 2012. Cancer stem cell definitions and terminology: the devil is in the details. Nature Reviews Cancer 12(11): 767–775.
Vandorpe, J. and E. Schacht. 1996. Synthesis and evaluation of polyphosphazene derivatives with ω-methylpoly (ethylene oxide) side-groups. Polymer. 37(14): 3141–5.
Vaneerdeweg, W., T. Bresseleers, P. Du Jardin et al. 2003. Comparison between plain and gentamicin containing collagen sponges in infected peritoneal cavity in rats. Eur. J. Surg. 164: 617–621.

Vert, M. 2007. Polymeric biomaterials: strategies of the past vs. strategies of the future. Progress in Polymer Science 32(8-9): 755–761.

Wachol-Drewek, Z., M. Pfeiffer and E. Scholl. 1996. Comparative investigation of drug delivery of collagen implants saturated in antibiotic solutions and a sponge containing gentamicin. Biomaterials. 17: 1733–1738.

Wade, R. J. and J. A. Burdick. 2014. Advances in nanofibrous scaffolds for biomedical applications: From electrospinning to self-assembly. Nano Today 9(6): 722–742.

Wakitani, S., T. Kimura, A. Hirooka et al. 1989. Repair of rabbit articular surfaces with allograft chondrocytes embedded in collagen gel. J. Bone Jt. Surgery. Br. 71: 74–80.

Wang, C. E., H. Wei, N. Tan et al. 2016. Sunflower polymers for folate-mediated drug delivery. Biomacromolecules 17(1): 69–75.

Wang, J., G. Chen, Z. Zhao et al. 2018. Responsive graphene oxide hydrogel microcarriers for controllable cell capture and release. Sci. China Mater. 61: 1314–1324.

Wang, T., L. Yang, B. Zhang et al. 2010. Extracellular biosynthesis and transformation of selenium nanoparticles and application in H_2O_2 biosensor. Colloids Surf. B. 80(1): 94–102.

Wani, AL. A. A. and J. A. Usmani. 2015. Lead toxicity: a review. Interdiscip. Toxicol. 8(2): 55.

Weigand, J. J., N. Burford, R. J. Davidson et al. 2009. New synthetic procedures to catena-phosphorus cations: Preparation and dissociation of the first cyclo-phosphino-halophosphonium salts. Journal of the American Chemical Society 131(49): 17943–17953.

Wiesmann, H. P., N. Nazer, C. Klatt et al. 2003. Bone tissue engineering by primary osteoblast-like cells in a monolayer system and 3-dimensional collagen gel. J. Oral Maxillofac. Surg. 61: 1455–1462.

Wizemann, A., S. D. Nandini, I. Stuhldreier et al. 2018. Rapid bioerosion in a tropical upwelling coral reef. PloS one, 13(9): e0202887.

Wu, D., X. Xie, A. A. Kadi et al. 2018. Photosensitive peptide hydrogels as smart materials for applications. Chin Chem Lett. 29(7): 1098–1104.

Wu, T., C. Cui, Y. Huang et al. 2019. Coadministration of an adhesive conductive hydrogel patch and an injectable hydrogel to treat myocardial infarction. ACS Appl. Mater. Interfaces. 12: 2039–2048.

Yan, L., H. Da, S. Zhang et al. 2017. Bacterial magnetosome and its potential application. Microbiol. Res. 203: 19–28.

Yang, N., L. Wei Hong and L. Hao. 2014. Biosynthesis of Au nanoparticles using agricultural waste mango peel extract and it's *in vitro* cytotoxic effect on two normal cells. Mater. Lett. 134: 67–70.21.

Yang, J., Y. Liu, L. He et al. 2018. Icariin conjugated hyaluronic acid/collagen hydrogel for osteochondral interface restoration. Acta Biomater. 74: 156–167.

Yang, P., F. Zhu, Z. Zhang et al. 2021. Stimuli-responsive polydopamine-based smart materials. Chem. Soc. Rev. 50: 8319–8343.

Yang, X., Z. Lu, H. Wu et al. 2018. Collagen-alginate as bioink for three-dimensional (3D) cell printing based cartilage tissue engineering. Mater. Sci. Eng. C. 83: 195–201.

Yannas, I. V. and J. F. Burke. 1980. Design of an artificial skin. I. Basic design principles. J. Biomed. Mater. Res. (1): 65–81.48.

Yao, Q., J. G. Cosme, T. Xu et al. 2017. Three dimensional electrospun PCL/PLA blend nanofibrous scaffolds with significantly improved stem cells osteogenic differentiation and cranial bone formation. Biomaterials 115: 115–127.

Young, S., M. Wong, Y. Tabata et al. 2005. Gelatin as a delivery vehicle for the controlled release of bioactive molecules. J. Control. Release. 109: 256–274.

Yuan, C. and T. Zhang 2013. Environmental implications of nano-manufacturing. In Green Mater. 179–202. Springer, Boston, MA.

Zaikov, G. E. 1985. Quantitative aspects of polymer degradation in the living body. J. Macromol. Sci.-Reviews in Macromolecular Chemistry and Physics 25(4): 551–97.

Zang, S., R. Zhang, H. Chen et al. 2015. Investigation on artificial blood vessels prepared from bacterial cellulose. Mater. Sci. Eng. C. 46: 111–117.

Zeplin, P. H., N. C. Maksimovikj, M. C. Jordan et al. 2014. Spider Silk Coatings as a Bioshield to Reduce Periprosthetic Fibrous Capsule Formation. Adv. Funct. Mater. 24: 2658–2666.

Zhao, P., H. Gu, H. Mi et al. 2018. Fabrication of scaffolds in tissue engineering: A review. Front. Mech. Eng. 13: 107–119.

Zhao, W., S. Li, H. Yao et al. 2017. Molecular optimization enables over 13% efficiency in organic solar cells. Journal of the American Chemical Society 139(21): 7148–7151.

4

Natural Biopolymer

Bhupendra G. Prajapati,[1] *Sunny R. Shah*,[2] *Sudarshan Singh*,[3,*] *Ozioma F. Nwabor*,[4] *Devesh Kapoor*[5] and *Sankha Bhattacharya*[6]

1. Introduction

Polymers are extensive molecules consisting of repeating structural units. They are compounds formed by covalently linking monomeric units into substantial molecular chains. Natural polymers are vast resources with extensive applications in diverse processing and production sectors. A wide range of natural polymers find application in pharmaceuticals for drug delivery, serve as biocompatible composite scaffolds in bone tissue regeneration, and contribute to the preparation of wound healing composites. These versatile materials offer biocompatibility and unique properties that enhance their role in improving healthcare and regenerative medicine. Compared with synthetic polymers, natural polymers are non-toxic and biocompatible with human cells and tissues and hence are preferred in biomedical applications. Several natural polymers are employed in food science as food hydrocolloids, used as stabilisers, gelling agents, and for fabricating packaging composites. Natural polymers are categorised into three groups: polysaccharides, including cellulose and chitin; polypeptides, such as collagen and keratin; and polynucleotides, exemplified by DNA and RNA. These classifications are based on

1 Shree S. K. Patel College of Pharmaceutical Education and Research, Ganpat University, Kherva, 384012, India.

2 B. K. Mody Government Pharmacy College, Gujarat Technological University, Rajkot - 360003, India

3 Department of Pharmaceutical Sciences, Faculty of Pharmacy, Chiang Mai University, Chiang Mai 50200, Thailand.

4 Division of Infectious Diseases, Department of Internal Medicine, Faculty of Medicine, Prince of Songkla University, Hat Yai, Songkhla, 90112, Thailand.

5 Dr. Dayaram Patel Pharmacy College, Bardoli – 394601, Gujarat, India.

6 Department of Pharmaceutics School of Pharmacy & Technology Management, SVKM'S NMIMS Deemed-to-be University, Shirpur, Maharashtra – 425405, India.

* Corresponding author: sudarshansingh83@hotmail.com

distinct structural and compositional properties, forming the foundation for various applications in fields like biomedicine and materials science. Natural polymers are highly reactive, biodegradable, biocompatible, and renewable. Natural polymers are sourced from diverse sources, including plants, animals, bacteria, fungi, and algae.

The demand surge for natural polymers is expected to translate into increased market value. The current projection suggests that the worldwide natural polymer market is expected to grow at a CAGR of approximately 5.6% between 2018 and 2027. This growth reflects the increasing demand for sustainable, bio-based materials across various industries. The economical, non-toxic, and biodegradable nature of natural polymers is anticipated to drive the global natural polymer market growth during this period. Several natural polymers have demonstrated the desired qualities for various industrial, medical, agricultural, and environmental applications. Deploying natural polymers as a replacement for synthetic counterparts is considered a sustainable option. In addition, due to the biodegradability of natural polymers, they pose no hazard to the environment, as do synthetic polymers like nylon, polyethylene, polyester, Teflon, and epoxy. The valorisation of waste from natural resources is a rich source of natural polymer. Chitin and its deacetylated form, chitosan, are extensively obtained from waste materials, including bones, pens, scales, and shells of sea animals and other terrestrial Mollusca (Mohite et al. 2023a). Microorganisms find extensive applications across diverse sectors, including food, pharmaceuticals, cosmetics, agriculture, textiles, biotechnology, and environmental chemistry, playing pivotal roles in product development, sustainability, and scientific advancement (Morin-Crini et al. 2019; Kumar et al. 2023b). Several studies have investigated the biomedical applications of chitosan for a variety of uses, including in wound dressings, antimicrobial agents, anti-cancer, drug delivery systems, gene therapy, tissue engineering, dental repair/replacement, bone regeneration, etc. (Patel et al. 2007; Mohite et al. 2023a; Prajapati and Sawant 2009; Mohite et al. 2023a).

2. Natural Polymer in Biomaterial Fabrication

Natural polymers are increasingly being used in biomaterial fabrication, with various studies showcasing their versatile applications. These bio-friendly materials hold promise in fields like tissue engineering, drug delivery, and wound healing, contributing to innovative and sustainable solutions. The fabrication of hydrogels using natural polymers has been experimentally demonstrated as biomaterials for human use. Research is ongoing, and efforts are being made to improve the medical applications of natural polymers for more efficient and effective processing of biomaterials. Naturally derived polymers such as cellulose, chitosan, alginate, agarose, and composites have undergone extensive research to determine their potential as tissue engineering scaffolds. These materials offer biocompatibility and structural support, making them valuable candidates in the field of regenerative medicine (Ko et al. 2010). Hydrogels are also seen as carrier matrix for drug and bioactive compound delivery. A cellulose-based dual stimuli-responsive hydrogel with an intelligent switch for controlled drug delivery exhibited precise control of the release process, with considerable biocompatibility and antioxidant properties (Song et al. 2021). Natural polysaccharide-based transdermal drug delivery is a promising device with prospective future uses (Li et al. 2022). A novel dual-responsive hydrogel, utilising tamarind gum in conjunction with poly(N-isopropyl acrylamide-co-ethylene glycol vinyl ether), demonstrates significant potential as an innovative platform for precise colon-specific anti-cancer drug delivery (Nagaraja et al. 2022).

3. Source With Derivatives, Characterisation, and Application in the Fabrication of Biomaterials

Natural biomaterials are found in abundance in nature or derived from plants, animals, or fungi sources like cellulose, chitosan, gelatine, chitosan, alginate, etc. In most instances, these biomaterials are biodegradable; however, not all biomaterials are biodegradable, which may limit their application. Natural biopolymers have recently received increased attention from the market, as well as the scientific and industrial communities.

3.1 Source/Extraction/Isolation and Chemical Modification

In the current scenario for natural polymers, there has been a gradual transition in recent years. The increased attention is primarily owing to recent environmental concerns, as well as the inexpensive cost of obtaining natural polymers, which are an excellent option for synthetic polymers that may be harmful. Natural polymers derived from natural sources had their own set of benefits, including no seasonal restrictions on raw material availability, low inorganic salt content, and no regional restrictions on industrial manufacturing. Most of these natural polymers are used in the pharmaceutical industry as well as a possible food additive.

Polymers can be found in abundance in nature. Proteins, cellulose, gelatine, starch, chitin, and chitosan are among the natural polymers found in the human body, plants, and animals (Singh 2022). For increased functionality in diverse applications, these polymers can be isolated, purified, and changed. Polysaccharides, like other naturally generated biopolymers, are harmless and biodegradable, which expands their potential application in a variety of sectors. Polysaccharides are intensively studied, with cellulose being the most important representation due to its abundance and availability. Although the most used biopolysaccharides come obtained from natural resources (e.g., cellulose), they are also obtained from animals (e.g., chitin/chitosan) and microbial sources (e.g., bacterial cellulose). Biopolymer sources and their extraction methods are summarised in Table 1.

Using microorganisms to produce bio-based polymers has received significant interest in the recent decade as a sustainable production method. New blends based on biomass-derived biopolysaccharides, softwood cellulose fibres like hemp fibres, and fungal fibres have been prepared and characterised for strength and functionality (Zhang et al. 2018). Chitosan, the most important chitin derivative in terms of uses, is obtained via fractional deacetylation under alkaline conditions. Chitin extraction involves two stages, demineralisation and deproteinisation, which are controlled using chemical or biological methods. Chemical methods require the use of acids and bases, whereas biological methods require the use of microbes. Lactic acid bacteria are widely employed, yet proteolytic bacteria and mixed cultures, combining lactic acid producers and proteolytic microorganisms, have also shown successful applications

Table 1: Summary of source and extraction method of some natural biopolymers.

Natural Biopolymer	Sources	Extraction Method
Alginate	Brown seaweed/algae (Laminaria)	Ion exchange with protons (0.1–0.2 M mineral acid) then neutralisation with alkali and followed by precipitated directly by alcohol, calcium chloride, or mineral acid
Agarose	The cell wall of marine algae; red seaweed (sea kelp)	Hot water extraction (slightly below boiling point for 3–4 hours)
Carrageenan	Red seaweed (Irish moss)	Hot alkaline extraction
Cellulose	Plants (jute, hemp, corn, flasks, rice, wheat straw, and sisal) and bacteria	Plant sources: kraft pulping process Bacterial sources: fermentation
Chitosan	Shrimp shells, crabs, lobster, krill, and mushroom	Demineralisation, deproteination, and deacetylation
Durio zibethinus seed gum	Fruit of Durian (*Durio zibethinus*)	Hot water extraction followed by purification using saturated barium hydroxide
Gelatin	Bovine, porcine and fish collagens	Partial acid or base hydrolysis of collagen
Guar gum	Seeds of *Cyamopsis tetragonoloba*	Boling, cleaning and followed by precipitations with ethanol, copper, barium complex, and finally, re-precipitation with ethanol
Hibiscus sabdariffa fibres	Plant; *Hibiscus sabdariffa* L. (roselle)	Hot water retting
Hyaluronic acid	Animal and fish	Biosynthesised using bacteria Streptococcus zooepidemicus microbial fermentation
Tamarindus indica L. seed gum	Seeds of *Tamarindus indica* L.	Boling, cleaning followed by precipitation using organic solvent
Xanthan gum	Sugar	Aerobic submerged fermentation by *Xanthomonas campestris* bacteria

in various industries. Furthermore, chitin can also be extracted from fungi mycelia and has been gaining significance lately (El Knidri et al. 2019b). Common methods of extraction and processing of chitosan are depicted in (Figure 1).

3.2 Biopolymers Characterisations

A biopolymer is a form of polymer made by living organisms that is made up of repeated monomeric units that are covalently bonded. DNA, RNA, proteins, cellulose, chitin, starch, and other biopolymers are produced by all living creatures (Subramanian et al. 2020). These biopolymers are essential in the body for energy storage, genetic information preservation and transmission, and cellular construction. DNA and RNA are genetic materials that allow all living organisms to store and transmit genetic information. Proteins aid in chemical reactions (enzymes), as well as provide structural support (haemoglobin) (e.g., collagen) (Hennet 2019). Cellulose is

Fermentation

Fermentation in a bioreactor using both batch and fed-batch strategies. Fermentation has been successfully demonstrated for preparation of various biopolymers.

Filtration

cross-flow filtration and pressurised dead-end membrane electro filtration process allows highly concentrated biopolymer dispersion to be filtered

Hydrolysis

Several biopolymers react quickly with high-temperature water, with rapid conversion rates and short residence durations. When biopolymers are treated with sub- or supercritical water, several goals might be achieved.

Esterification

The availability of reactive sites on biopolymers make them suitable substrates for esterification, etherification, oxidation, epoxidation, *N*-acylation, alkylation etc

Poly-condensation

Polycondensation reactions in biopolyamides synthesis can be performed using bio-based dicarboxylic acids and diamines or bifunctional amino acids. Glass transition temperature, crystallinity, melting point, mechanical performance, low-temperature behaviour, and frequently impart specialised functional features, such as adhesive or optical capabilities, are all influenced by the kind and concentration of the unique monomers.

Oxidation

Oxidation process is a potential technique for the recovery of products of industrial value, and successfully used in production of soluble biopolymers.

Dehydration

The steric structure of biopolymers is disrupted after dehydration. The destruction/restoration of the natural structure during dehydration/hydration is usually totally reversible, allowing biopolymers to be stored in a dry form.

Figure 1: Methods for preparation and isolation of biopolymers.

the most important structural component of plant cell walls. Using biopolymers as a key component of biopolymeric products can be a cost-effective strategy in the field of pharmaceuticals as well as food and nutrition. Polysaccharides are a natural choice as they are composed of monosaccharides (sugars) allied together by Oglycosidic links (Sun et al. 2018). Polysaccharides, such as cellulose and chitin, make up the majority of natural polysaccharides. Natural or synthetic polymers are obtained from various sources, and before utilisation in pharmaceuticals or food, industries were extensively characterised based on their qualities. The characterisation techniques involved in the evaluation of natural bio-based polymers in accordance with the American Society for Testing and Materials (ASTM) standards to ensure the product meets the required quality standards are presented in (Figure 2).

3.2.1 Instrumentation and Characterisation for Biopolymers

3.2.1.1 Nuclear Magnetic Resonance (NMR)

NMR is an excellent analytical tool for gaining detailed and especially impartial information on the structural and dynamic properties of various polymer systems. The 2D and 3D NMR spectral analysis reveals information on isotropic chemical shifts (primarily of ^{1}H and ^{13}C). Numerous NMR techniques employ diverse solvent systems for characterising polymer composition, structure, and dynamics, addressing specific material attributes with precision. These methods offer unique insights into polymer behaviour and are vital for comprehensive material analysis and research.

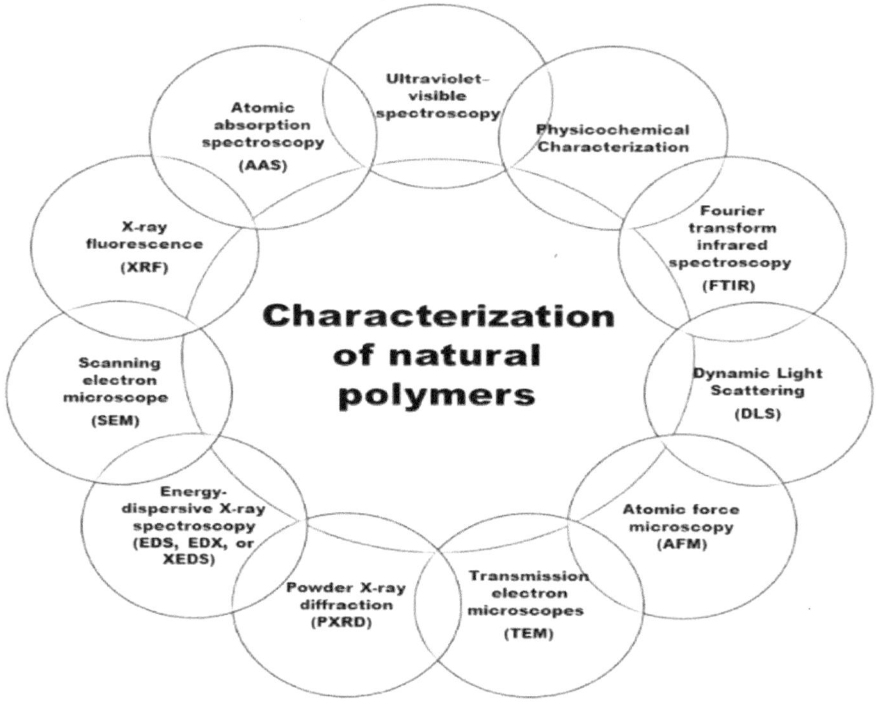

Figure 2: Characterisation techniques involved in the assessment polymers

Characterising polymer systems using NMR is challenging due to the coexistence of processes at various timescales, including the intermediate timescale of the NMR experiment, structural disorder, and dynamic heterogeneity. This complexity makes precise characterisation a formidable task in NMR analysis.

3.2.1.2 Fourier Transform Infrared (FTIR) Spectroscopy

Analytical tools such as FTIR are used to identify the functional and structural properties of biopolymers and evaluate the quality of materials. The FTIR uses interferometry to encode material information and to identify or quantify material, and the Fourier transform decodes the signal into spectra. Moreover, the FTIR is a non-destructive method for determining biomass components in the mid-IR region. FTIR spectroscopy reveals molecular fragments, functional groups, and even the fibre structure (Bhattacharya et al. 2023).

3.2.1.3 X-Ray Diffraction (XRD) Analysis

The size and shape of polymer domains within a sample can be determined using XRD techniques such as low-angle or small-angle scattering. For more advanced structural investigations, the diffractometer is configured to analyse polymer fibres and films (Kotadiya et al. 2016).

3.2.1.4 Thermogravimetric Analysis (TGA)

TGA is a valuable tool for assessing the thermal stability of various materials, including biopolymers and product fabricated using it. This approach measures changes in the weight of a specimen as its incremental temperature rises per unique time (Moin et al. 2002). The moisture and volatile content of biopolymers can be determined using TGA. In a series of research investigations, thermogravimetric (TG) analyses were conducted on natural polymers, including cellulose, chitin, and chitosan, using a PerkinElmer Pyris1TGA instrument equipped with a 20-sample auto-sampler manufactured by PerkinElmer in Waltham, MA. To record the temperature profiles, individuals used a high-resolution mode to record them from 37 to 900°C in a nitrogen atmosphere (20 ml/min) with a constant temperature rise rate of 10°C/min.

3.2.1.5 Scanning Electron Microscopy

Scanning electron microscopy (SEM) is a valuable tool for morphological/topographical characterisation of biopolymers, bridging the gap between high-resolution methods such as AFM, TEM, and light microscopy, which measures large dimensions at low resolution. In addition, SEM used to capture the image of fracture and failure mechanics, the size and shape of particles, the orientation, and dispersion of fillers in polymer matrices, as well as how fillers are oriented and spread out in the matrix.

3.2.1.6 Field Emission Scanning Electron Microscopy (FESEM)

FESEM is an advanced version of SEM that captures images with resolution and a wider energy range than traditional SEM. The latter employs a field emission gun as an electron-generating device, which is one of the key distinctions between SEM and FESEM. This produces highly concentrated electron beams, which increase spatial resolution and allow sample examination at low potentials. Several studies indicate that FESEM has been previously successfully employed to get high-resolution pictures of biopolymer systems instead of SEM.

3.2.1.7 Physicochemical Characterisation

Rheological techniques and procedures have been used in the characterisation of polymers for decades. Rheology estimation is used to evaluate the viscosity and flow behaviour of synthetic polymers (Kumar et al. 2023a). Biopolymer solutions and gels are characterised using a variety of rheological methods. The major classes of biopolymers is proteins and polysaccharides. Depending upon the behaviours of the material under stress conditions, the materials are characterised as a Newtonian or non-Newtonian system. The fundamental sequence structure of biopolymers governs just a small portion of their functional capabilities. The space-filling conformations used are far more significant. The tertiary structure, or the overall form of the macromolecule, follows the secondary structure. Some marine polysaccharides, including carrageenan and agarose, have large double-helical structures (Picout and Ross-Murphy 2003). Table 2 summarises the physical, rheological, and thermal characterisation of natural polymers.

Table 2: Typical morphological and thermal characterisation of natural polymer with its applications.

Source of Polymer	Viscosity (1%) (Ribeiro et al.)	Particle Size (μm)	Thermal Analysis			Application	References
			DSC	Tg	TGA[a]/DTG[b]		
Diospyros melonoxylon	~ 219.1	~ 4.03	Ed_p: 78°C	74°C	70.1°C; 304.7°C; 475.4°C	Mucoadhesive and controlled drug delivery	(Bothara and Singh 2012; Singh and Bothara 2014b)
Buchanania lanzan	~ 112.9	~ 0.95	Ed_p: 89°C	80.93°C–91.96°C	352.6°C; 508.6°C	Mucoadhesive and controlled drug delivery	(Bothara and Singh 2012; Singh S and SB 2014)
Manilkara zapota	~ 219.1	~ 0.5	Ed_p: 138°C	138°C	350.3°C; 614.4°C	Mucoadhesive and controlled drug delivery	(Bothara and Singh 2012; Singh and Bothara 2014a)
Tamarindus indica L.	~ 57	~ 100	(Curve-1: Ed_p: 97.67°C; On_s: 65°C; En_s: 175°C) (Curve-2: Ex_p: 310.86°C; On_s: 245°C; En_s: 400°C)	-	70.1°C; 175.1°C	Pharmaceutical applications include diluents, binders, disintegrators, thickeners, suppository base suspending, and emulsifying agents	(Alpizar-Reyes et al. 2017; Shao et al. 2019)
Chitosan (Mw. 75-160KDa)	20–300	3.2–3.5	Ed_p: 106.25; On_s: 66.81°C; En_s: 130°C	67–118°C	62–80°C; 260–277°C; 500–700°C	Extensively used in various pharmaceutical drug delivery fabrication with wide application on food packaging and preservations	(Pereira et al. 2013; Singh et al. 2021)
κ-carrageenan	5–25	5–10	Ed_p: 52.6; On_s: 40.0°C; En_s: 80.0°C	73.02°C	30–100°C; 150–209°C	Used in the fabrication of biomedical implants, material science, and food industries	(Liew et al. 2017; Shand et al. 1994)

Xanthan gum	1,200–1,600	< 180	Ed_p: 75.0; On_s: 55.0°C; En_s: 79.88°C	60°C	505.15°C–569.15°C; 873.15°C	Pharmaceuticals and additives in food	(Srivastava et al. 2012; Craig et al. 1997)
Guar gum	10,000	46.4–315	(Curve-1:Ed_p: 246.0°C) (Curve-2:Ed_p: 296.0°C) (Curve-3: Ex_p: 310.0°C)	108°C	473.15°C–583°C; 673.15°C	Additive in food, pharmaceuticals, paper, textile, explosive, oil well drilling, and cosmetic industry	(Srivastava et al. 2012; Mudgil et al. 2014; Gliko-Kabir et al. 1999)
Sodium Alginate	< 240 – > 3,500	-	Ex_p: 112.44°C; On_s: 47.90°C	81°C	28.0°C–177.4°C; 177.0 °C–523.5°C; 523.5°C–749.5°C	Food and vegetable coating, gelling, thickening, stabilising, or emulsifying agent in pharmaceuticals and foods	(Singh 2013; Soares et al. 2004)
Cellulose	-	~ 101.6	Ed_p: 96°C; On_s: 60°C; En_s:140°C	220°C and –137 °C	100°C–120°C; 220°C – 300°C; 300 475°C	Versatile use in many industries such as veterinary foods, wood and paper, fibres and clothes, cosmetics, and pharmaceutical industry as excipients	(Szcześniak et al. 2008)
Gelatin	71.8	~ 3.58	(Curve-1:Ed_p: 101.5°C; On_s: 94.9°C; En_s: 125°C) (Curve-2: Ed_p: 207.5°C; On_s: 200.4°C; En_s: 230°C)	~ 76°C	90°C–150°C; 150°C–190°C; 190°C–250°C (213.5°C /211.8°C)	Widely used in food, pharmaceutical, cosmetic, and photographic industries	(Mukherjee and Rosolen 2013)
Agarose	46×10^4	~ 136	Ex_p: 81°C; On_s: 70°C; En_s: 90°C	~ 37°C	30°C–200°C; 265°C–400°C	Frequently used in molecular biology for separation of large molecules and widely used in tissue engineering	(Fernández et al. 2008; Su et al. 2021)

Table 2 contd. ...

...Table 2 contd.

Source of Polymer	Viscosity (1%) (Ribeiro et al.)	Particle Size (μm)	Thermal Analysis			Application	References
			DSC	Tg	TGA[a]/DTG[b]		
Hyaluronic acid	0.73–0.88	60–100	Ed_p: 225°C and 245°C; On_s: 220°C; En_s: 250°C	– 48°C–61°C	30°C–130°C; 200°C–300°C	Widely used in soft tissue augmentation to limit age-related and photo-induced skin ageing, correction of facial lipodystrophy, and prevent the recurrence of hypertrophic scars or keloids	(Lewandowska et al. 2016; Li et al. 2018)
Starch	~ 1,040	10–30	Ed_p: 182°C and 195°C	~ 60°C	Maximum degradation in the range of 227°C–310°C	Used in the food industry, paper and board fabrication, industrial binder, and textile industry and employed as a pharmaceutical tablet binder, diluent, and disintegrants	(Zhong and Sun 2005; Morán et al. 2013)

DSC: differential scanning calorimetry; TGA[a]: mass loss at various temperatures during thermogravimetric analysis; Tg: glass transition temperature, Ed_p: endothermic peak; Ex_p: exothermic peak; On_s: oneset of temperature; En_s: endset temperature; DTG[b]: derivative thermogravimetric peak.

3.3 Biomaterials Fabrication and Application in the Pharmaceutical and Food Industry

Biomaterials have made an impact in the rapidly evolving pharmaceutical and food industry over the last few decades. Biomaterials are materials that are designed to interact with living biological tissue and are utilised for therapeutic and diagnostic purposes as well in diverse fields of the food industry. Biomaterials can be categorised into two origins: naturally occurring and synthetically produced. In a broader classification, they fall into four main groups: metals, ceramics, polymers, and composites, each serving vital roles in various medical and engineering applications (Bagheri et al. 2022). The properties of biomaterials, whether natural or synthetic, are intricately linked to the chemicals used in their preparation and modification. Polymers, with their diverse applications, play a pivotal role in therapeutic methods, from drug delivery to reconstructive techniques, and they even find utility in the food industry (Lastra Ripoll et al. 2021). Biomaterials and their derivatives may aid in developing innovative food packaging materials with an improved barrier, mechanical strength, antibacterial and antioxidant capabilities (Rukmanikrishnan et al. 2020). The present book chapter will review some of the biomaterials, namely chitosan, carrageenan, alginate, cellulose, gelatin, guar gum, xanthan gum, agarose, hyaluronic acid, *Tamarindus indica* L. seed gum, *Hibiscus sabdariffa* fibres, and starch with its fabrication, chemical, and physical properties, and application in the pharmaceutical and food industry. Moreover, the short regulatory status of the US FDA and European Union is also discussed.

3.3.1 Chitosan

The inherent biological compatibility, versatile chemical composition, and tuneable physical characteristics of natural polymers render them ideal for a wide array of applications within the food and pharmaceutical sectors. The outer skeleton of shellfish such as crabs, lobster, and shrimp are the prominent sources of chitosan (Mukherjee Swarupananda et al. 2024; Mohite et al. 2023b; Pawar et al. 2024; Singh et al. 2021). Chitosan (deacetylated chitin) comprises a major portion of the natural polymer used in the food and pharmaceutical industry, and compared to cellulose, chitin as a biopolymer is present in abundance (Aranaz and Acosta 2021). Chitosan, illustrated in Figure 3, is a linear polysaccharide comprising randomly arranged β-(1→4)-linked D-glucosamine (deacetylated) and N-acetyl-D-glucosamine (acetylated) units. It is produced by subjecting chitin to alkaline treatments under rigorous conditions or through enzymatic hydrolysis, often employing chitin deacetylase (Wang et al. 2022). The diverse chemical properties of chitosan, including its aqueous solubility, reactivity, and pH characteristics, are intricately linked to its physical attributes, notably its crystalline structure, molecular weight, degree of deacetylation (Wang et al. 2021), etc.

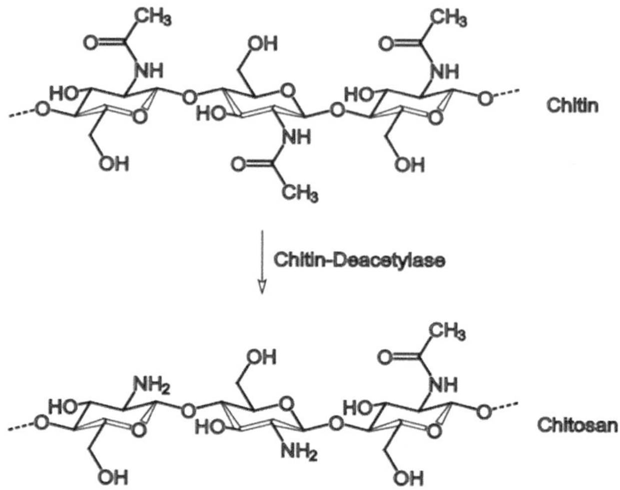

Figure 3: Synthesis of chitosan via deacetylation of chitosan.

3.3.1.1 Crystalline Structure

Partially re-acetylated chitosan exhibits lower crystallinity compared to pure chitosan. In order to achieve less crystalline chitosan, reacetylation of fully deactivated chitosan is favoured over direct solid-state deacetylation of chitin. Moreover, various other process parameters, such as re-precipitation, drying, and lyophilisation, also play a pertinent role in establishing the crystalline state of chitosan (Qiao et al. 2019).

3.3.1.2 Molecular Weight

Chitosan molecular weight is widely varied, and hence they are polydisperse in terms of molecular weight. Depending upon the source of chitin, the molecular weight of chitosan varies considerably. The molecular weight characterisation can be done by a variety of solvents depending upon the concentration of protonation for chitosan. The molecular weight of chitosan can vary significantly due to source and deacetylation conditions, with chitosan from crustacean chitin often exceeding 100,000 daltons. This variability underscores the importance of precise control in chitosan production processes (El Knidri et al. 2019a). Moreover, the aqueous solubility of chitosan is related to the molecular weight of chitosan; the lower the molecular weight, the higher the aqueous solubility (Ru et al. 2019).

3.3.1.3 Degree of Deacetylation

Chitosan, derived from chitin, consists of β-1,4-D-glucosamine units. The degree of deacetylation quantifies the proportion of these units with acetyl groups removed, a critical factor in its properties and applications (Kurita et al. 2000). The degree of deacetylation is a pivotal parameter for chitosan, profoundly influencing its chemical and biological attributes. A higher degree of deacetylation would result in a higher concentration of amino acids, and higher protonation of amino groups would lead to higher aqueous solubility and subsequent higher biological activity (Cheung et al. 2015).

3.3.1.4 Derivatives of Chitosan

Chitosan, a versatile biopolymer, is extensively used in drug delivery and the food industry. In order to tailor its properties for diverse applications, chitosan derivatives are created via modifications with hydrophobic groups, notably acyl, and alkyl groups, a well-researched approach for enhancing its functionality.

3.3.1.4.1 Acylation of Chitosan

The amphiphilic polymer can be made using entirely natural, biodegradable, and renewable ingredients by grafting fatty acid onto chitosan. Acylation reaction can occur in amino groups, hydroxyl groups, or both groups, depending on the reaction circumstances (Zhang et al. 2017). Because of their high hydrophobicity, derivatives with a short chain length and low-to-moderate degree of substitution are soluble in water, whereas derivatives with a long-chain length are practically insoluble in water (Ma et al. 2009).

3.3.1.4.2 Alkylation of Chitosan

Reductive amination of chitosan yields alkyl chitosan derivatives in most cases. The stiffness and interaction properties of modified chitosan are almost unaffected by short alkyl chains (C5), but longer chains promote more efficient hydrophobic contacts and intra-aggregation of the polymer. Furthermore, in forming hydrophobic domains through alkyl-alkyl interactions, the degree of substitution is an important parameter (Sahariah et al. 2015).

3.3.1.4.3 Thiolated Chitosan

The variety of thiolated chitosan with its significant applications has risen dramatically since this chitosan was demonstrated to have considerably improved characteristics than the unmodified polymer (Dalvin 2022). The thiolated chitosan shows appreciable change in various properties of chitosan, such as adhesion to biological surfaces (Wibel et al. 2021), *in situ* gelling properties (Li et al. 2020), swelling behaviours solubility, controlled release of a drug, enhancement of permeation, and cellular uptake (Bastaki et al. 2021). Thiolated chitosan has varied applications, such as therapeutic agents, drug delivery systems, theranostic, tissue engineering, wound material, cosmetics, etc.

3.3.1.5 Regulatory Status

The first topical hemostatic chitosan dressing (HemCon™ Bandage) was cleared by the FDA in 2008 (US FDA 2008). Recently, Axiostat, a chitosan hemostatic dressing, has been found to be substantially equivalent (SESE) by the US FDA and approved for marketing in the US (US FDA 2018). However, the inactive ingredient database does not list or approve any maximum daily exposure for chitosan through any delivery route (US FDA 2022), and hence the use of chitosan for drug delivery must be reviewed on a case-to-case basis. European Food Safety Authority believes that the consumer dietary risk assessment can be waived, and chitosan hydrochloride can be used as a plant elicitor (EFSA 2020).

3.3.2 Carrageenan

Carrageenans are a type of sulfated polysaccharide, also known as carrageenins, derived from red edible seaweeds. Carrageenans are mostly employed in pharmaceuticals, cosmetics, and the food sector because of their unique qualities (thickening and gelling agents) and biological activity (antiviral and anticoagulant). As shown in Figure 4, Carrageenans are sulfated linear polysaccharides made up of repeated disaccharide units with alternating 3-linked -d-galactopyranose and 4-linked -galactopyranose or 3, 6-α anhydrogalactopyranose (Yermak et al. 2020).

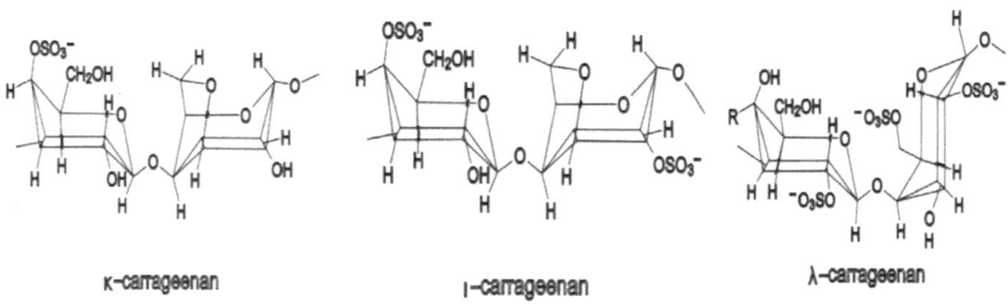

Figure 4: Carrageenan structure.

3.3.2.1 Types and Sources of Carrageenan

There are three kinds of carrageenan depending upon the degree of sulfation: Kappa (κ) carrageenan (Rudhziaha et al. 2015), iota carrageenan (Chitra et al. 2018), and lambda carrageenan (Yermak et al. 2020). Ķ-carrageenan consists of repeating units composed of 3-linked -d-galactose 4-sulfate, and 4-linked 6-anhydro-galactopyranose, with each unit carrying a negative charge. The tropical seaweed *Kappaphycus alvarezii*, sometimes known as *Eucheuma cottonii* in trade, is used to extract κ-carrageenan. The chief species to manufacture iota carrageenan is *Eucheuma spinosum* (or simply spinosum). At the same time, lambda carrageenan is a type of seaweed derived from a variety of species in the *Genera Gigartina* and *Chondrus*.

3.3.2.2 Synthesis of Carrageenan

The first step of synthesis is to harvest the carrageenan-containing seaweeds, which is an expensive process because the seaweed is collected by raking or hand-gathering. Mechanical harvesters have been used, but yields have shown little to no improvement. After that, the seaweed is washed to eliminate any sand, stones, or other contaminants. Then, seaweeds are transported to a production facility, where they are extracted under alkaline conditions to generate high-quality carrageenans at temperatures near the boiling point of the alkaline solution, which lasts for several hours. After that, the extract is filtered and concentrated and then dried and pulverised to the required particle size after being precipitated with a suitable solvent (Yermak et al. 2020).

3.3.2.3 Properties of Carrageenan

The unique property of carrageenan makes them suitable for the pharmaceutical and food industry. All the grades of carrageenan are soluble in hot water. The swelling of carrageenan depends upon the particle density and is varied. Iota carrageenan is sensitive to calcium ions, and it will swell to generate thixotropy, which is caused by creating bivalent calcium bridges between the swelled particles, resulting in a loose 3-D network (Valetti et al. 2013). Protein reactivity is known to exist in all carrageenans. The isoelectric point of a protein, pH of the solution, weight ratio of protein to carrageenan, type of protein, and molecular weight all influence its reactivity (Cui et al. 2018). The interaction of proteins and carrageenan is mediated by a mixture of processes. Divalent bridging and ionic bonding are the mechanisms. Because the carboxylic acid groups of the protein are ionised above the protein's isoelectric point, the carrageenan does not interact with it due to negative-negative repulsions (Cui et al. 2018).

3.3.2.4 Regulatory Status

Carrageenan obtained from *Chondrus crispus, Chondrus ocellatus, Eucheuma cottonii., Eucheuma spinosum, Gigartina pistillata* to *Gigartina stellata*; may be safely used in food as per the given guidelines in the US FDA. Carrageenan may be used orally, nasally, topically, and transdermally, with the maximum daily exposure limits have been mentioned in the database for Inactive Ingredient for Approved Drug Products (US FDA 2022). Moreover, carrageenan is also authorised as a food additive in the European Union with a daily acceptable limit of 75 mg/kg of body weight (Younes et al. 2018).

3.3.3 Xanthan Gum

Polysaccharides are versatile biomolecules sourced from bacteria, plants, and animals, finding applications across diverse industries like food, agriculture, and biomedicine. Their natural abundance, accessibility, biocompatibility, cost-efficiency, and processing ease make them invaluable in fields like tissue engineering and drug delivery, paving the way for innovative solutions (Souza et al. 2022).

Xanthan gum, a polysaccharide, is synthesised through the fermentation process involving gram-negative bacteria Xanthomonas campestris, as reported by Elella et al. This natural production occurs in carbon-rich environments under aerobic conditions, yielding a versatile and widely used thickening agent (Elella et al. 2021).

Due to its exceptional matrix-forming qualities and outstanding biocompatibility, xanthan gum ranks as the second most widely utilised bacterial polysaccharide, following dextran. Its versatile applications span various industries, with a prominent role in the field of biomedicine (Pooja et al. 2014).

3.3.3.1 Chemistry and Physical Properties of Xanthan Gum

Xanthan gum primarily consists of pentasaccharide structures comprising glucose, mannose, and glucuronic acid in a balanced mole ratio of 2:2:1. These core units may also incorporate pyruvyl and o-acetyl residues in varying quantities, providing unique functional properties (Petri 2015). In Figure 5, the backbone consists of a sequence of glucose residues interconnected through (1–4 linkage) while being further connected to a trisaccharide chain through a -mannopyranose unit via (1–3 linkage).

Xanthan gum exhibits distinctive properties, including non-Newtonian fluid behaviour, remarkable thermal resilience at elevated temperatures reaching 90°C, and exceptional stability in acidic (pH 2) and alkaline (pH 12) environments (Lastra Ripoll et al. 2021). An increased presence of pyruvyl groups and reduced acetyl groups promotes the formation of highly viscous solutions, while the opposite composition results in lower viscosity. However, xanthan gum has several drawbacks, such as an uncontrollable degree of thickness, microbial attack, hydration, and delayed dissolution frequency. Various derivatives of xanthan gum have been prepared to address these issues (Riaz et al. 2021). Physical, enzymatic, and chemical approaches are the major broad categories of xanthan gum alterations (Battistel et al. 2011). Chemical modification and degradation have been extensively investigated; however, such modification also influences other uncontrolled aspects of xanthan gum.

The physical modifications are done by sonication, radiation, annealing, using surfactants, etc., and can help to achieve the desired physical and chemical properties of xanthan gum. Xanthan lyase, xanthanase, cellulases, and acetyl esterases can be used for enzymatic modification to achieve the desired properties of xanthan gum. Chemical modification by grafting and crosslinking, etherification and esterification,

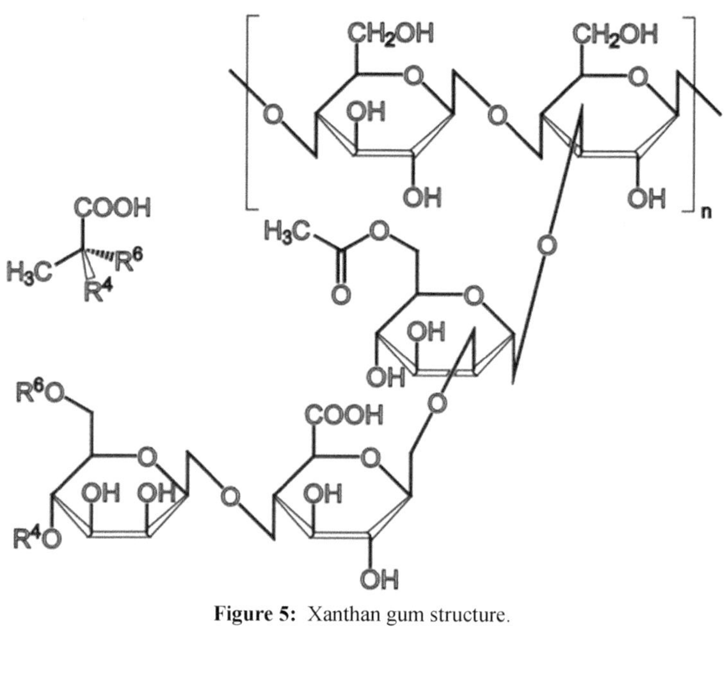

Figure 5: Xanthan gum structure.

acetylation, and pyruvate have been studied extensively to modify xanthan gum chemically (Said et al. 2021) due to its unique physical qualities; xanthan gum has a wide range of applications in the pharmaceutical and food industries. Mixability, pumpability, and pourability of industrially manufactured dressings and sauces are further aided by the shear-thinning flow behaviour. Xanthan gum is frequently used in conjunction with other hydrocolloids to get the required flow behaviour for various semi-prepared food conveniences (Saha and Bhattacharya 2010).

In the cosmetics industry, the most common use of xanthan gum is in the manufacturing of toothpaste. Xanthan gum's shear-thinning flow characteristic enables easy extrusion from the tube or the pump dispenser. It also guarantees that the toothpaste remains stable on the toothbrush (Hublik 2012).

3.3.3.2 Regulatory Status

The US FDA has granted authorisation for xanthan gum due to its non-toxic nature and its ability to enhance the texture and mouthfeel of food products. It must contain no more than 750 parts per million of residual isopropyl alcohol and should yield positive results in both the locust bean gum gel test and pyruvic acid test, as per US FDA regulations. The Inactive Ingredient for Approved Drug Products database suggests that xanthan gum can be administered via various routes, including oral, ophthalmic, rectal, and topical routes, in a variety of dosage forms within the maximum tolerable limit (US FDA). Considering a previous study reported that a 214 mg/kg of xanthan gum repeated oral dose for ten days, the European Food Safety Authority approved it as a food additive (Mortensen et al. 2017).

3.3.4 Alginate

Alginates represent a non-toxic, naturally derived polymer with exceptional versatility. Their unique properties allow for the creation of chelates and pH-responsive gels. With biocompatibility, biodegradability, and a non-antigenic, non-toxic nature, alginates offer a safe and adaptable choice (Singh et al. 2023; Patel et al. 2023). As a result, alginate has gained widespread acceptance as a viable gel carrier choice. Alginate is a structural substance found in the extracellular matrix of brown algae. Laminaria hyperborea, Macrocystis pyrifera, and Ascophyllum nodosum are three of the most common algae sources (Grøndahl et al. 2020).

3.3.4.1 Chemistry and Physical Properties

Alginate polymers consist of alternating blocks of β-D mannuronic acid (M) and α-L-guluronic acid (G), linked by glycosidic bonds. These homopolymeric M and G blocks, along with heteropolymeric MG or GM blocks, define the complex structure of alginates, which is crucial in various applications (Jain and Bar-Shalom 2014). However, high molecular mass polymers are exceedingly viscous, posing difficulty in processing conditions. For instance, high shear rates cause damage when protein and cells are mixed in highly viscous alginate solutions (Kim et al. 2019). Figure 6 represents the structure of alginic acid.

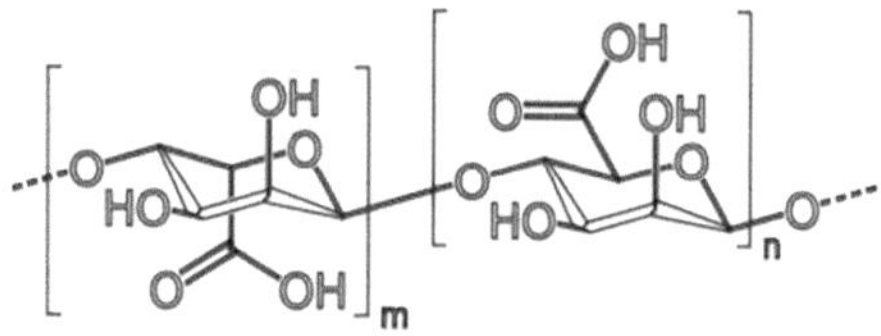

Figure 6: Alginic acid structure.

Guluronic acid containing sodium alginates is water-soluble in a positive manner, whereas mannuronic acid containing sodium alginates is water-soluble in a negative manner. To render alginates water-soluble, three essential conditions must be fulfilled: (a) maintaining an appropriate ionic strength in the alginate medium, (b) introducing gelling ions into the solvent, and (c) ensuring a high pH level in the solvent while deprotonating the carboxyl groups. This combination of factors facilitates the dissolution of alginates in water (Rosiak et al. 2021).

Heavy metals, proteins, and endotoxins, among other contaminants found in alginates, cause an immunological reaction at the implantation or injection site (Lee and Mooney 2012).

3.3.4.2 Chemically Modified Alginate

Chemical modification of alginate biopolymers plays a pivotal role in enhancing their bio-integration and affinity in various applications. This is achieved by introducing highly reactive functional groups, such as aldehyde groups, or incorporating chemical moieties like phosphate or sulfate. These modifications not only improve the biocompatibility of alginate-based products but also expand their versatility in biomedical and industrial contexts (Szabo et al. 2020).

Alginates play a pivotal role in diverse applications, serving as rate-controlling agents in drug delivery systems, providing a versatile matrix for biomolecules and pharmaceutical excipients in localised administration, contributing to advancements in biomedical engineering, and facilitating freeze-drying processes (Gheorghita Puscaselu et al. 2020). Innovative packaging featuring alginate-based films infused with antifungal agents offers a promising solution for safeguarding cheese and fruits from mould contamination. This approach not only enhances food safety and quality but also extends shelf life while preventing product deterioration. Additionally, alginate's versatile applications extend to the biomedical field, encompassing protein and cell delivery, wound care, bone regeneration, and neovascularisation (Sun and Tan 2013).

3.3.4.3 Regulatory Status

Sodium alginate is generally recognised as safe by the US FDA and can be used as a food additive (US FDA). Sodium alginate and other salts of alginate, such as calcium alginate, propylene glycol alginate, ammonium calcium alginate, etc., have been approved in the Inactive Ingredient for Approved Drug Products and can be used within the maximum daily exposure limit (Younes et al. 2017).

3.3.5 Cellulose

The term 'cellulose' was initially coined in 1839, appearing in a report by the French Academy that discussed the research of Payen. This marked the first formal identification of cellulose as a fundamental element within plants (Chidrawar et al. 2023; Singh et al. 2023a; Singh et al. 2023b; Kendre et al. 2023). Cellulose stands as the world's most abundant biomass resource, a timeless and nearly limitless natural polymer deployed for millennia in various forms and applications. This renewable, cost-effective, and biodegradable marvel, produced not only by plants but also by microbes, offers a multitude of advantages. Moreover, the transformation of cellulose into water-soluble derivatives has significantly broadened its scope across diverse fields, making it an invaluable resource for sustainable innovation (Seddiqi et al. 2021).

3.3.5.1 Plant and Bacterial Cellulose

Cellulose can also be derived from species other than plants, such as algae, fungi, and bacteria. This biopolymer is produced by each creature and has specific supramolecular structures, which are extensively researched in studies focusing on the structure, crystallinity, and reactivity of the biomolecule, as well as the potential to build new biomaterials (Freeman et al. 2015). Apart from having a sturdy chain and several distinct functionalities, what distinguishes cellulose from other synthetic polymers is its sensitivity to acetal hydrolysis and oxidation; these features govern its chemistry and manipulation (Reddy et al. 2021).

As shown in Figure 7, cellulose is a linear polysaccharide of long-chain joined by β-1,4 glycosidic bonds. The functional groups in a cellulose monomer are hydroxyls and a methanol group. The absence of side chains or branches indicates that the polymer's structure is organised. The development of crystalline areas results from these interactions between glucose units. It also serves as the foundation for cellulose's enormous hydrogen bond networks, which provide the material with a variety of partly crystalline fibre topologies and morphologies (Reis et al. 2019).

3.3.5.2 Derivatives of Cellulose

Cellulose compounds have been investigated and used extensively in a variety of industries, including wastewater treatment, papermaking, food additives, pharmaceuticals, and textile applications. The functions of many cellulose derivatives are determined by the substituent group. Various derivatives such as hydroxypropyl

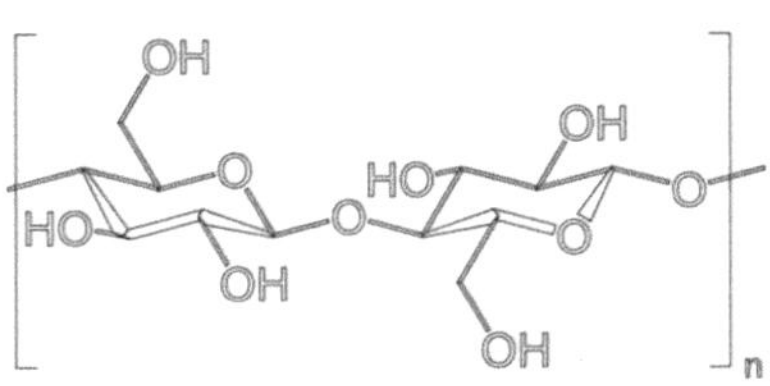

Figure 7: Cellulose structure.

methylcellulose (HPMC) and its various grades, such as cellulose xanthate, cellulose nitrate, cellulose phthalate esters, microcrystalline cellulose, etc., have been prepared to modify the properties of cellulose as per the requirements.

3.3.5.3 Regulatory Status

Methylcellulose and sodium carboxymethylcellulose are generally regarded as safe by the US FDA (US FDA). Ethylcellulose, hydroxypropyl cellulose, and hydroxypropyl methylcellulose have been approved by the US FDA for human consumption (US FDA). Cellulose and its specific derivatives have been approved by the European Food Safety Authority as food additives (Younes et al. 2017).

3.3.6 Gelatin

Gelatin, a natural polymer derived from collagen protein through hydrolysis, boasts unique amino acids that offer numerous health benefits. This water-soluble substance, produced by partially hydrolysing collagen from animal sources, involves the loss of collagen's primary, secondary, and tertiary structures (Singh et al. 2024). Unlike most hydrocolloids, which are polysaccharides, gelatin is a protein containing all essential amino acids except tryptophan, making it a unique and versatile ingredient (Pang et al. 2014). Gelatin, a diverse ingredient, can be derived from collagen sources like cattle bones, hides, pig skins, and fish, offering both agricultural and non-agricultural origins. Notably, gelatin lacks plant sources, and so-called 'vegetable gelatins', like seaweed extracts, are chemically distinct. Various sources, such as a mammalian, fish, and insect origins, yield gelatin with varying properties (Mariod and Adam 2013).

3.3.6.1 Synthesis and Structure of Gelatin

Collagen pre-treatment yields two distinct gelatin types: Type A (pH 6-9) from pigskin with less crosslinking and Type B (pH 5) from bovine hides with complex collagen structures. These gelatins find diverse applications due to their unique properties and origins (Aramwit et al. 2015). Gelatin's primary structure and makeup are like that of parent collagen. This similarity has been established for various tissues and species. The molecular weight of gelatin has been used to explain several characteristics of its behaviour in solutions and gels. Although gelatin is not polydispersed, it does have a distinct molecular weight distribution pattern that correlates to the α-chain and its oligomers. Gelatin is made up of a variety of polypeptide chains, including α-chains, (dimers of α-chain) and γ (trimers of -α chain) (Hanani et al. 2014; Zhang et al. 2020).

3.3.6.2 Functional and Physical Properties of Gelatin

The functional properties of gelatin, such as gel strength and viscosity, are intricately tied to its chemical composition, particularly its molecular weight distribution and amino acid content. Proline and hydroxyproline play pivotal roles in gelatin's renaturation process. The conversion of collagen to gelatin involves the disruption of inter and intramolecular bonds, along with specific peptide bonds. The extent and type of crosslinking found in bones and skins are different. The distribution of molecular weights, shapes, and compositions of gelatine's subunits affect its functional properties (Zhang et al. 2020).

3.3.6.3 Applications of Gelatin

Gelatin's unique gelling properties make it a key ingredient in modern cuisine, and it is used to craft gourmet sweets with texture and stability. In addition to confectionery delights like fruit salads and ice cream, gelatin finds applications in cosmetics and health products, from bath salts to facial creams, offering versatility in various industries (Alipal et al. 2021). In the realm of skincare, collagen and gelatin products play a vital role, serving as key ingredients in functional and nutritional foods. Gelatin finds versatile applications in hydrogels, nanospheres, nanofibers, pharmaceutical additives, and cell transplantation carriers, fostering innovation in skincare formulations (Kamatchi and Leela 2016).

3.3.6.4 Regulatory Status

The Inactive Ingredient for Approved Drug Products lists that gelatin can be used in a variety of dosage forms for drug delivery through different routes within the maximum daily exposure limit (US FDA). However, Subpart B of Part 189 of the Code of Federal Regulations prohibits using cattle materials (gelatin) in manufacturing or processing human food (US FDA).

3.3.7 Guar Gum

Guar gum, derived from the seeds of Cyamopsis tetragonolobus, serves as a white to yellowish-white, high molecular weight seed gum. Its versatile applications span multiple sectors, including food, textiles, paper, cosmetics, and pharmaceuticals (Singh et al. 2022). Its unique qualities, such as biocompatibility and sustainability, make it an excellent candidate in the realm of pharmaceutical research and development (Phadke et al. 2014). The remarkable swelling properties of guar gum in water can be effectively modified through functional group derivatisation, crosslinking, and grafting. These alterations open up diverse possibilities for its utilisation in various biomedical applications, overcoming its inherent limitations.

3.3.7.1 Structure and Properties of Guar Ggum

Guar gums, characterised by their hydrophilic nature, exhibit a distinctive rod-like polymer structure. Within this structure, D-galactose side chains are intricately connected to the D-mannose backbone, maintaining an approximate molecular ratio of 1:2. These polysaccharides feature straight chains of D-mannose units linked together through β (1-4) glycoside linkages, with D-galactose units linked in the opposite orientation (Dionísio and Grenha 2012). As depicted in Figure 8, including hydroxyl groups within the polymer's composition plays a pivotal role in generating diverse derivatives essential for numerous industrial purposes. Guar gum exhibits extraordinary and captivating physical attributes. It remains impervious to hydrocarbons, lipids, alcohols, esters, and ketones, with its exceptional solubility

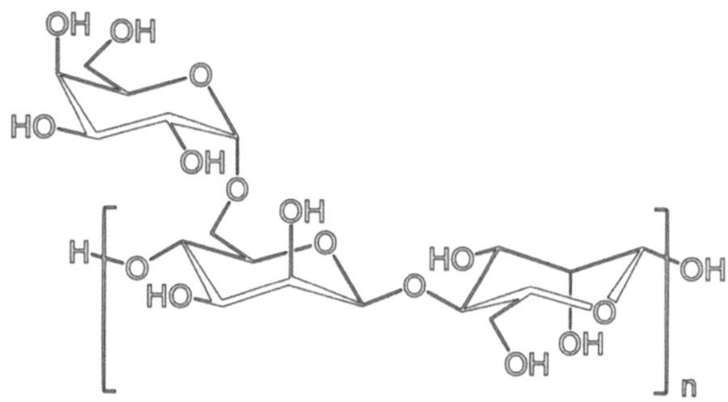

Figure 8: Guar gum structure.

reserved exclusively for water (Kocira et al. 2021). The solution of guar gum exhibits non-Newtonian pseudoplastic flow; increasing the shearing stress decreases the viscosity of the solution. Because guar gum is anionic, it maintains stability and provides consistent viscosity across a wide pH range. The pH range of 6–9 yields the most viscosity, whereas pH 3.5 yields the lowest (Wang et al. 2000).

3.3.7.2 Guar Gum Processing

Commercial extraction of guar gum from seeds employs various mechanical methods such as sieving, polishing, attrition, and roasting. These techniques capitalise on the disparities in hardness among the seed constituents, facilitating the separation of the endosperm from its components. The result is the formation of guar splits once the hull and germ have been successfully isolated. Following heat treatment, the hull can be efficiently separated through attrition milling or impact mills. Sieving facilitates the retrieval of finer germ and hull fractions from the endosperm. Guar gum powder is subsequently acquired via milling. Further refinement involves dissolving it in water, precipitating, and recovery using ethanol or isopropanol (George et al. 2019). Sugars are usually added to clarified guar gum, which is readily available in markets. The viscosity of resultant dispersion is high since a large amount of water is imbibed by guar. The basic characteristics of guar gum are influenced by temperature. The loss of water of hydration around the polymer molecules was caused by a rise in temperature.

3.3.7.3 Chemical Modifications on Guar Gum

Guar gum is frequently combined with other polymers to generate interpenetrating polymer networks (IPN), which provide the otherwise fragile hydrogel with durable mechanical and thermal properties. Guar gum has also been produced in conjunction with pH-sensitive polymers and tested for its efficacy in delivering targeted drugs (Chandrika et al. 2016; Kaur et al. 2013).

3.3.7.4 Regulatory Status

The regulation of guar gum as a food additive is generally documented by the Food and Drug Administration (FDA) (US FDA). It is GRAS-listed and can be used in baked goods and baking mixes, Breakfast cereals, Jams, and jellies at the concentration mentioned in Subpart B of Part 184 of 21CFR184.1339. The Inactive Ingredient for Approved Drug Products states that guar gum can be used in oral, buccal, sublingual, topical, and vaginal dosage forms within the maximum daily exposure limit (US FDA). Guar gum (E 412) is authorised as a food additive in the European Union (EU) (Mortensen et al. 2017).

3.3.7.5 Future Perspectives

Several attempts have been made to introduce functional biomaterials for a variety of applications, spanning from diagnostics to treatment. Recent findings suggest that a thorough understanding of the chemical components and chemistry of biomaterials can be crucial to success in either the food or pharmaceutical industry. Polymers, manufactured specifically to solve the challenges, have been used to circumvent the limitations of traditional dosage forms. Using new polymers has advantages, but they can also be hazardous due to their toxicity and incompatibility. When choosing polymers for a delivery system, extreme caution should be exercised. The most important goal is to develop biocompatible, multifunctional, less toxic polymers that can be used in cost-effective delivery methods.

3.3.8 Agarose

Agarose is a prominent marine polysaccharide representing reversible thermogell behaviour and switchable chemical reactivity for functionalisation. Agarose is isolated and purified from red seaweed, known as agar/agar-bearing algae, or isolated from an agar-peptide with a particular structure and spontaneous gelation. Agarose is an inactive ingredient used in various research fields, including drug delivery, food, DNA electrophoresis, and tissue engineering.

3.3.8.1 Structure and Properties

Structurally, agarose is a natural polymer, consisting of alternative β-D-galactose and 3,6-anhydro-L-galactose units of agarobiose (Figure 9). Agarose, a non-toxic substance, possesses versatile properties ideal for various applications like nucleic acid electrophoresis, chromatography, and cell culture overlays. With a reported molecular weight of approximately 120,000 g/mol and a white appearance, it readily forms a gel in hot water near its upper critical solution temperature (Zarrintaj et al. 2018). Agarose exhibits solubility in various organic solvents, including dimethyl sulfoxide and 1-Butyl-3-methylimidazolium chloride. This diverse solubility profile extends its utility in various scientific and industrial applications, offering versatility in research and biotechnology (Buskila et al. 1997). Agarose gelation involves three distinct stages: induction, gelation, and pseudo-equilibrium, driven by hydrogen bonding and electrostatic interactions that lead to the formation of a helical agarose molecular structure and subsequent gel formation. The melting and gelation temperatures of agarose are influenced by factors like molecular weight, concentration, and lateral groups (Buskila et al. 1997).

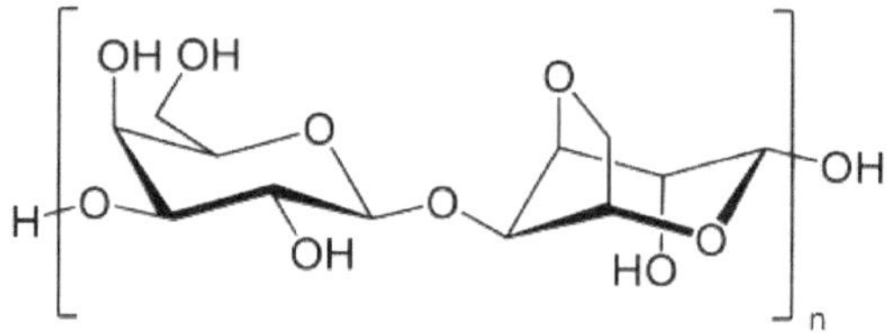

Figure 9: Agarose structure.

Agar is the originating source for agarose, classified by the US FDA as a direct food substance affirmed as GRAS. The FDA has established specific acceptable limits for agar's use in various food categories, such as 0.8% in baked goods, 2.0% in confections and frostings, 1.2% in soft candy, and 0.25% in all other food products. Additionally, agar serves as a stabiliser in animal drugs, feeds, and related products, provided it adheres to good manufacturing and feeding practices, ensuring safety and compliance in these applications (Potter and Hotchkiss 2012). The properties of agarose are tunable by concentration, functionalisation, and blending to mimic the desire tissue performance. Several reports indicated the biomedical application of agarose due to its excellent pharmaceutical additive properties, including controlled self-gelation, water-solubility, adjustable tensile strength, and non-immunogenicity. Moreover, agarose demonstrates skin regeneration efficacy due to the presence of functional groups and water absorption capacity that support cellular adhesion, proliferation, and other cellular activity (Zarrintaj et al. 2017). In addition, the agarose-based hydrogel has been considerably utilised in the investigation of mechanical load reactions for chondrocytes and mesenchymal stem cells (Park et al. 2009). Further, agarose has acquired a wide range of pharmaceutical applications such as drug delivery (Grolman et al. 2019), cancer therapy (Guo et al. 2019), tissue engineering and regenerative medicine (Hafezi et al. 2019), and disease diagnosis, controlling, and treatment (Pearl et al. 1984). Furthermore, agarose has been utilised in various tissue-engineering applications, including neural system (Stokols et al. 2006), bone formation (Sánchez-Salcedo et al. 2008), cardiac regeneration, wound healing, and attachment to tissue such as skin (Miguel et al. 2014), brain, and cornea (Pettikiriarachchi et al. 2010).

3.3.8.2 Regulatory Status and Future Perspectives

Agarose, derived from agar, is a remarkable biopolymer known for its unique attributes. It boasts an exceptional porous structure that facilitates nutrient and oxygen flow, making it ideal for various applications. Notably, agarose is both biodegradable and biocompatible, offering excellent mimicry of human tissue, promoting cell-matrix interaction, exhibiting elasticity, and demonstrating remarkable hydrophilicity. The remarkable biomaterial properties of agarose, including cell migration and versatile mechanical attributes linked to ionic conductivity, position it as a prime candidate for human organ regeneration. Additionally, agarose and its composites play a pivotal role in the regeneration of cartilage tissue, highlighting its significance in biomedical applications. Agarose-based hydrogels, known for mirroring tissue properties and displaying non-immunological, biocompatible, and non-toxic characteristics with strong cell-matrix interaction, show promise in cartilage repair research. Nevertheless, these formulations require further comprehensive evaluation in clinical and real-world settings to realise their full potential.

3.3.9 Hyaluronic Acid

Hyaluronic acid, a naturally-occurring glycosaminoglycan polysaccharide, exhibits a remarkably conserved molecular structure across mammalian species. Hyaluronic acid indicates distinct physiochemical properties, which emphasise its application as a lubricating cushioning, and some researchers reported *in vitro* anti-inflammatory activity.

3.3.9.1 Structure and Properties

Hyaluronic acid, or hyaluronan, is a linear polysaccharide consisting of repeating disaccharide units: β-1,4-D-glucuronic acid and β-1,3-N-acetyl-D-glucosamine. It is a non-sulfated glycosaminoglycan found throughout the body, from the eye's vitreous humour to the extracellular matrix in cartilage tissues (Fraser et al. 1997). Recent research has placed significant emphasis on the modification of three pivotal functional groups. These include glucuronic acid carboxylic acid, primary and secondary hydroxyl groups, and N-acetyl groups. Notably, carboxylates have been subject to amidation, while hydroxyls have undergone modifications such as etherification, divinylsulfone crosslinking, esterification, and bis-epoxide crosslinking, advancing the field of chemical manipulation in a variety of applications (Kuo and Prestwich 2010). These hyaluronic acid derivatives fall into two primary categories: monolithic and living (Prestwich and Kuo 2008). Monolithic hyaluronic acid represents terminally modified HA variants, incapable of forming new bonds amidst cellular or tissue environments. They necessitate processing for versatile applications.

Typical serum hyaluronic acid levels range from 10 to 100 μg/L but significantly rise in patients with osteoarthritis. The elimination half-life has been reported as 2.5 to 5.5 minutes in human plasma (Goa and Benfield 1994). The biomedical application of hyaluronic acid indicates it is an essential part of connective, epithelial, and neural tissues and contributes to cell proliferation and migration.

3.3.10 Tamarindus indica L. *Seed Gum*

Tamarind, a sought-after tropical fruit renowned for its rich pulp, thrives abundantly in the regions of Africa and Asia. *Tamarindus indica*, an evergreen tree within the Fabaceae family, boasts impressive dimensions, reaching heights of up to 24 metres and girths of 7 metres, while its gracefully expansive crown adds to its allure. Tamarind, a nutritional powerhouse, serves as a source of biodegradable polymers from its seeds and kernels. *Tamarindus indica* is abundant in crude protein, carbohydrates, iron, and vitamin A. Virtually, every part of the tree finds diverse applications in food, chemical, pharmaceutical, textile, and other industries, as well as in agriculture and fuel production (Dagar et al. 1995).

3.3.10.1 Structure and Properties

Phytochemical analysis of the *T. indica* tree has revealed a diverse array of bioactive compounds, including phenolics, cardiac glycosides, and organic acids like mallic and tartaric acid. Additionally, the tree's constituents encompass mucilage and pectin, with sugars like arabinose, xylose, galactose, and glucose alongside uronic acid. Furthermore, the ethanolic extract showcases a spectrum of essential elements, albeit in varying quantities, encompassing arsenic, magnesium, potassium, phosphorus, lead, and zinc. Further, the pulp contains organic acids, amino acids, invert sugar (25–30%), pectin, protein, and fat with some pyrazines and thiazoles (2-ethyl thiazole, 2-methyl thiazole). In addition, the leaves are a rich source of triterpenes, lupanone, and lupeol (Imam and Azhar 2007). Additionally, graft structural modification of seed kernel polysaccharides with monomer, acrylamide, and or polyacrylamide followed by ceric ion or free radical ions reported improvement in various pharmaceutical applications (Nandi et al. 2019; Jha et al. 2022).

3.3.10.2 Biomedical and Food Industry Uses

The tamarind fruit pulp is highly prized, finding diverse applications both domestically and industrially. Its tangy pulp is a culinary favourite, enhancing dishes like curries, sauces, and even ice cream, while its seeds serve as valuable by-products stemming from domestic and food industry usage. Tamarind seed polymers have been extensively explored as tablet binder, mucoadhesive drug delivery, sustained or controlled release drug delivery, transdermal film, etc. Moreover, the reported pharmacological use of fruit is laxative, with a spasmolytic effect, and peptic ulcer activity with ameliorative effect against chemical-induced acute. In addition, several pharmacological applications have been reported for various parts of plants, including antimicrobial, antiparasitic, antifungal, antiviral, antinematodal, antidiabetic, antivenom, analgesic, and anti-inflammatory activity. Further, phenolic-rich fruit extract demonstrates antioxidant properties with beneficial effects on cardiovascular health. The chronic long-term different dose toxicity on animals reported no side effect (Silva et al. 2009), with no change in biochemical analysis, liver function, and body weight for isolated seed polysaccharides (Heimbach et al. 2013). In another investigation, the toxic effect was not mentioned; however, a surge in white blood cells with thrombocytes was seen (Bhadoriya et al. 2011).

3.3.11 Starch

Starch is a polysaccharide molecule that is found in nature and is biodegradable, inexpensive, renewable, and abundant. The 'starch' is a Germanic word that means 'Strong, stiff, strengthen, stiffer', while the Greek term for starch is 'amylon', which means 'not milled'. Starch plays a crucial role in various industries where it is enzymatically converted into sugars, such as malted barley in the production of beer and whisky. These sugars are then fermented to yield ethanol, serving a dual purpose in alcoholic beverages and biofuel manufacturing.

3.3.11.1 Structure and Properties

Starch, the second most prevalent organic compound in nature after cellulose, is a polymeric carbohydrate composed of glucose units linked by glycosidic bonds. It serves as the primary energy reserve in the greenest plants. Renowned as the predominant carbohydrate in human diets, starch is abundantly found in staple foods like wheat and potatoes, ensuring a vital role in global nutrition. Plants employ a two-step process to synthesise starch, beginning with the conversion of glucose 1-phosphate to ADP-glucose through the enzyme glucose-1-phosphate adenyltransferase, utilising ATP as an energy source. Subsequently, starch synthase appends ADP-glucose to a growing glucose chain via a 1,4-alpha glycosidic bond, releasing ADP and yielding amylose. Starch branching enzymes then introduce 1,6-alpha glycosidic bonds into amylose, creating the branched structure of amylopectin. Amylose is an α-1,4-dioxane polymer that is approximately linear that can form a strong, isotropic, odourless, tasteless, and colourless film due to its excellent film-forming ability. The amylose has 500 to 6,000 glucose units on average, which are distributed over 1 to 20 chains. Starch is typically composed of around 20–25% amylose and 75–80% amylopectin by weight, with each component contributing to its unique properties. Analysing the molecular structure of starch is crucial to understanding its various applications. The interplay between amylose and amylopectin within starch granules and pastes is a complex relationship, influencing its overall functionality (Zobel 1988).

3.3.11.2 Regulatory Status and Future Perspectives

Meeting current starch demand relies on a limited crop selection, notably potatoes, maize, wheat, and tapioca. Enhancing starch properties for industry involves altering extracted starch chemically and physically, along with influencing starch biosynthesis within the plant. Starch and modified starch are considered safe by the FDA, but their usage must stay within necessary limits and comply with FDA regulations. For safety, food additive containers must be labeled as 'food starch-modified' alongside the required information as per the FDA Act.

Starch's distinctive properties offer numerous possibilities for innovative products. However, a lack of recognition may impede our understanding of granule characteristics. Customising starch for enhanced utility and value is essential, but introducing novel derivatives or substitution levels might face restrictions. Exploring new modified-starch products presents exciting opportunities. While controlling reaction sites within granules and molecules remains challenging, two promising avenues are emerging: leveraging novel commercial starch sources and harnessing biological modifications of existing base starches. These avenues hold significant potential for innovative starch-based products and applications (Bemiller 1997).

4. Conclusions and Future Prospectus

Nature represents a vast reservoir of useful resources that can be employed in various fields. Natural polymers are one such resource and are currently targeted for beneficial applications. The use of polymers sourced from nature opens a new research frontier, with potential applications extending across a variety of sectors. Although, at present, the use of natural polymers in certain fields is still in the developmental stages, there are indications that following the optimisation of relevant factors, natural polymers will soon replace their synthetic counterparts. The use of natural polymers in pharmaceuticals as drug excipients is one area of the application currently being commercialised. However, the fabrication of natural polymer-based packaging materials is still limited by the obvious adverse properties of the polymer. This highlights the need for further studies and investigation into the development and improvement towards enhancing the qualities and characteristics of natural polymers. In addition, the vastness of natural polymers in terms of properties, structure, and sources signifies grey areas that are yet to be explored in the exploration, extraction, purification, modification, and engineering of natural polymers.

References

Maged Younes, Peter Aggett, Fernando Aguilar et al. 2018. Re-evaluation of celluloses E 460(i), E 460(ii), E 461, E 462, E 463, E 464, E 465, E 466, E 468 and E 469 as food additives 16(1): e05047. https://doi.org/https://doi.org/10.2903/j.efsa.2018.5047. https://efsa.onlinelibrary.wiley.com/doi/abs/10.2903/j.efsa.2018.5047.

Maged Younes, Peter Aggett, Fernando Aguilar et al. 2018. Re-evaluation of carrageenan (E 407) and processed Eucheuma seaweed (E 407a) as food additives 16(4): e05238. https://doi.org/https://doi.org/10.2903/j.efsa.2018.5238. https://efsa.onlinelibrary.wiley.com/doi/abs/10.2903/j.efsa.2018.5238.

Mortensen, A., F. Aguilar, R. Crebelli et al. 2017. Re-evaluation of xanthan gum (E 415) as a food additive. EFSA J. 15(7): e04909. https://doi.org/10.2903/j.efsa.2017.4909. https://www.ncbi.nlm.nih.gov/pubmed/32625570.

Mortensen, A., F. Aguilar, R. Crebelli et al. 2017. Re-evaluation of guar gum (E 412) as a food additive. EFSA J 15(2): e04669. https://doi.org/10.2903/j.efsa.2017.4669. https://www.ncbi.nlm.nih.gov/pubmed/32625396.

Additives, Efsa Panel on Food, Food Nutrient Sources added to, M. Younes, P. Aggett, F. Aguilar et al. 2017. Re-evaluation of alginic acid and its sodium, potassium, ammonium and calcium salts (E 400-E 404) as food additives. EFSA J 15(11): e05049. https://doi.org/10.2903/j.efsa.2017.5049. https://www.ncbi.nlm.nih.gov/pubmed/32625343.

Alipal, J., N. A. S. Mohd Pu'ad, T. C. Lee et al. 2021. A review of gelatin: Properties, sources, process, applications, and commercialisation. Materials Today: Proceedings 42: 240–250. https://doi.org/10.1016/j.matpr.2020.12.922.

Alpizar-Reyes, Erik, Hector Carrillo-Navas et al. 2017. Functional properties and physicochemical characteristics of tamarind (*Tamarindus indica* L.) seed mucilage powder as a novel hydrocolloid. Journal of Food Engineering 209: 68–75. https://doi.org/https://doi.org/10.1016/j.jfoodeng.2017.04.021. https://www.sciencedirect.com/science/article/pii/S0260877417301681.

Aramwit, P., N. Jaichawa, J. Ratanavaraporn et al. 2015. A comparative study of type A and type B gelatin nanoparticles as the controlled release carriers for different model compounds, Mater. Express.

Aranaz, I. and N. Acosta. 2021. Chitin- and chitosan-based composite materials. Biomimetics (Basel) 7(1). https://doi.org/10.3390/biomimetics7010001. https://www.ncbi.nlm.nih.gov/pubmed/35076459.

Bagheri, M., M. Validi, A. Gholipour et al. 2022. Chitosan nanofiber biocomposites for potential wound healing applications: Antioxidant activity with synergic antibacterial effect. Bioeng.

Transl. Med. 7(1): e10254. https://doi.org/10.1002/btm2.10254. https://www.ncbi.nlm.nih.gov/pubmed/35111951.

Bastaki, S., S. Aravindhan, N. Ahmadpour Saheb et al. 2021. Codelivery of STAT3 and PD-L1 siRNA by hyaluronate-TAT trimethyl/thiolated chitosan nanoparticles suppresses cancer progression in tumor-bearing mice. Life Sci. 266: 118847. https://doi.org/10.1016/j.lfs.2020.118847. https://www.ncbi.nlm.nih.gov/pubmed/33309720.

Battistel, E., D. Bianchi, M. Fornaroli et al. 2011. Enzymes breakers for viscosity enhancing polymers. J. Pet. Sci. Eng. 77: 10–17.

Bemiller, James N. 1997. Starch modification: challenges and prospects. Starch - Stärke 49(4): 127–131. https://doi.org/https://doi.org/10.1002/star.19970490402. https://onlinelibrary.wiley.com/doi/abs/10.1002/star.19970490402.

Bhadoriya, Santosh Singh, Aditya Ganeshpurkar et al. 2011. *Tamarindus indica*: Extent of explored potential. Pharmacognosy Reviews 5(9): 73.

Bhattacharya, Sankha, Bhupendra G. Prajapati, Nemat Ali et al. 2023. Therapeutic potential of methotrexate-loaded superparamagnetic iron oxide nanoparticles coated with Poly(lactic-co-glycolic acid) and polyethylene glycol against breast cancer: development, characterization, and comprehensive *in vitro* investigation. ACS Omega 8(30): 27634–27649. https://doi.org/10.1021/acsomega.3c03430. https://doi.org/10.1021/acsomega.3c03430.

Bothara, Sunil B. and Sudarshan Singh. 2012. Thermal studies on natural polysaccharide. Asian Pacific Journal of Tropical Biomedicine 2 (2, Supplement): S1031-S1035. https://doi.org/https://doi.org/10.1016/S2221-1691(12)60356-6. https://www.sciencedirect.com/science/article/pii/S2221169112603566.

Buskila, Dan, Lily Neumann, Genady Vaisberg et al. 1997. Increased rates of fibromyalgia following cervical spine injury. A controlled study of 161 cases of traumatic injury. Arthritis & Rheumatism: Official Journal of the American College of Rheumatology 40(3): 446–452.

Chandrika, K. S. V., A. Singh, A. Rathore et al. 2016. Novel cross linked guar gum-gpoly(acrylate) porous superabsorbent hydrogels: characterization and swelling behavior in different environments. Carbohydr. Polym. 149: 175–185.

Cheung, R. C., T. B. Ng, J. H. Wong et al. 2015. Chitosan: An update on potential biomedical and pharmaceutical applications. Mar Drugs 13(8): 5156–86. https://doi.org/10.3390/md13085156. https://www.ncbi.nlm.nih.gov/pubmed/26287217.

Chidrawar, Vijay R., Sudarshan Singh, Titilope John Jayeoye et al. 2023. Porous swellable hypromellose composite fortified with eucalyptus camaldulensis leaf hydrophobic/hydrophilic phenolic-rich extract to mitigate dermal wound infections. Journal of Polymers and the Environment 31(9): 3841–3856. https://doi.org/10.1007/s10924-023-02860-8. https://doi.org/10.1007/s10924-023-02860-8.

Chitra, R., P. Sathya, S. Selvasekarapandian et al. 2018. Synthesis and characterization of iota-carrageenan solid biopolymer electrolytes for electrochemical applications. Ionics. https://doi.org/10.1007/S1158.

Craig, D. Q., A. Kee, S. Tamburic et al. 1997. An investigation into the temperature dependence of the rheological synergy between xanthan gum and locust bean gum mixtures. J. Biomater. Sci. Polym. Ed. 8(5): 377–89. https://doi.org/10.1163/156856297x00164.

Cui, X., F. Yu, Y. Xue et al. 2018. Effects of mixing ratio and pH on the electrostatic interactions of hydrolyzed alaska pollock protein and kappa-Carrageenan. J. Food Sci. 83(8): 2176–2182. https://doi.org/10.1111/1750-3841.14099. https://www.ncbi.nlm.nih.gov/pubmed/30059140.

Dagar, J. C., Gurbachan Singh and N. T. Singh. 1995. Evaluation of crops in agroforestry with teak (*Tectona grandis*), maharukh (*Ailanthus excelsa*) and tamarind (*Tamarindus indica*) on reclaimed salt affected soils. Journal of Tropical Forest Science: 623–634.

Dalvin, Lauren A. 2022. Thiolated chitosan-carboxymethyl dextran nanoparticles: improving intravitreal drug bioavailability for retinoblastoma. Journal of Ophthalmic & Vision Research 17(1): 1–3. https://doi.org/10.18502/jovr.v17i1.10163. https://pubmed.ncbi.nlm.nih.gov/35194489. https://www.ncbi.nlm.nih.gov/pmc/articles/PMC8850858/.

Dionísio, Marita and Ana Grenha. 2012. Locust bean gum: Exploring its potential for biopharmaceutical applications. Journal of Pharmacy & Bioallied Sciences 4(3): 175–185. https://doi.org/10.4103/0975-7406.99013. https://pubmed.ncbi.nlm.nih.gov/22923958. https://www.ncbi.nlm.nih.gov/pmc/articles/PMC3425165/.

EFSA. 2020. Outcome of the consultation with Member States and EFSA on the basic substance application for approval of chitosan hydrochloride for the extension of use in plant protection as an elicitor in horticulture, olive trees, grapes, grass and post-harvest fruit treatment. EFSA Supporting Publications 17(7). https://doi.org/10.2903/sp.efsa.2020.EN-1900.

El Knidri, H., J. Dahmani, A. Addaou et al. 2019a. Rapid and efficient extraction of chitin and chitosan for scale-up production: Effect of process parameters on deacetylation degree and molecular weight. Int. J. Biol. Macromol. 139: 1092–1102. https://doi.org/10.1016/j.ijbiomac.2019.08.079. https://www.ncbi.nlm.nih.gov/pubmed/31404606.

EL Knidri, H., J. Dahmani, A. Addaou et al. 2019b. Rapid and efficient extraction of chitin and chitosan for scale-up production: Effect of process parameters on deacetylation degree and molecular weight. International Journal of Biological Macromolecules 139: 1092–1102. https://doi.org/10.1016/j.ijbiomac.2019.08.079.

Elella, M., H. Abu, E. S. Goda et al. 2021. Xanthan gum-derived materials for applications in environment and eco-friendly materials: a review. Journal of Environmental, Chem. Eng. 9: 104702.

Fernández, Emiliano, Daniel López, Carmen Mijangos et al. 2008. Rheological and thermal properties of agarose aqueous solutions and hydrogels. Journal of Polymer Science Part B: Polymer Physics 46(3): 322–328.

Fraser, J., E. Robert, Torvard C. Laurent et al. 1997. Hyaluronan: its nature, distribution, functions and turnover. Journal of Internal Medicine 242(1): 27–33.

Freeman, Ronit, Job Boekhoven, Matthew B. Dickerson et al. 2015. Biopolymers and supramolecular polymers as biomaterials for biomedical applications. MRS Bulletin 40(12): 1089–1101. https://doi.org/10.1557/mrs.2015.270. https://pubmed.ncbi.nlm.nih.gov/26989295. https://www.ncbi.nlm.nih.gov/pmc/articles/PMC4790466/.

George, Archana, Priyanka A. Shah and Pranav S. Shrivastav. 2019. Guar gum: Versatile natural polymer for drug delivery applications. European Polymer Journal 112: 722–735. https://doi.org/10.1016/j.eurpolymj.2018.10.042.

Gheorghita Puscaselu, R., A. Lobiuc, M. Dimian et al. 2020. Alginate: from food industry to biomedical applications and management of metabolic disorders. Polymers (Basel)12(10). https://doi.org/10.3390/polym12102417. https://www.ncbi.nlm.nih.gov/pubmed/33092194.

Gliko-Kabir, Irit, Adel Penhasi and Abraham Rubinstein. 1999. Characterization of crosslinked guar by thermal analysis. Carbohydrate Research 31(1): 6–13. https://doi.org/https://doi.org/10.1016/S0008-6215(99)00025-7. https://www.sciencedirect.com/science/article/pii/S0008621599000257.

Goa, Karen L. and Paul Benfield. 1994. Hyaluronic acid. Drugs 47(3): 536–566. https://doi.org/10.2165/00003495-199447030-00009. https://doi.org/10.2165/00003495-199447030-00009.

Grolman, Joshua M., Mansher Singh, David J. Mooney et al. 2019. Antibiotic-containing agarose hydrogel for wound and burn care. Journal of Burn Care & Research 40(6): 900–906.

Grøndahl, Lisbeth, Gwendolyn Lawrie, A. Anitha et al. 2020. Applications of alginate biopolymer in drug delivery. In Biointegration of Medical Implant Materials, 375–403.

Guo, Xiaoling, Yong Chen, Weiping Ji et al. 2019. Enrichment of cancer stem cells by agarose multi-well dishes and 3D spheroid culture. Cell and Tissue Research 375(2): 397–408.

Hafezi, Mahshid, Liguo Qin, Pooya Mahmoodi et al. 2019. Osmosis effect on protein sustained release of Agarose hydrogel for anti-friction performance. Tribology International 132: 108–117.

Hanani, Z., A. Nur, Y. H. Roos et al. 2014. Use and application of gelatin as potential biodegradable packaging materials for food products. Int. J. Biol. Macromol. 71: 94–102. https://doi.org/10.1016/j.ijbiomac.2014.04.027.

Heimbach, James T., Hiroshi Egawa, Palma Ann Marone et al. 2013. Tamarind seed polysaccharide: a 28-day dietary study in Sprague-Dawley rats. International Journal of Toxicology 32(3): 198–208.

Hennet, T. 2019. Collagen glycosylation. Curr. Opin. Struct. Biol. 56: 131–138. https://doi.org/10.1016/j.sbi.2019.01.015.

Hublik, G. 2012. Xanthan. pp. 221–229. *In*: Krzysztof Matyjaszewski and Martin Möller (Eds.). Polymer Science: A Comprehensive Reference. Amsterdam: Elsevier.

Imam, Shehla and M. Iqbal Azhar. 2007. TvyO triterpenes lupanone and lupeol, isolated and identified from t4m 4rindus indic4. LINN. Pak. J. Pharm. Sci. 20(2): 125–127.

Jain, D. and D. Bar-Shalom. 2014. Alginate drug delivery systems: application in context of pharmaceutical and biomedical research. Drug Dev. Ind. Pharm. 40(12): 1576–84. https://doi.org/10.3109/03639045.2014.917657. https://www.ncbi.nlm.nih.gov/pubmed/25109399.

Jha, Sheetal, Rishabha Malviya, Shivkanya Fuloria et al. 2022. Characterization of microwave-controlled polyacrylamide graft copolymer of tamarind seed polysaccharide. Polymers 14(5): 1037.

Kamatchi, P. and K. Leela. 2016. Extraction, characterization and application of gelatin from Carcharhinus amblyrhyncho and Sphyraena barracuda. J. Biotechnol. Biochem. 2: 40–49.

Kaur, G., M. Mahajan and P. Bassi. 2013. Derivatized polysaccharides: Preparation, characterization, and application as bioadhesive polymer for drug delivery, Int. J. Polym. Mater. Polym. Biomater. 62: 475–481.

Kendre, Prakash N., Deepak Lokwani, Ajinkya Pote et al. 2023. Potential technologies to develop cellulose beads and microspheres. pp. 145–176. *In*: Regenerated Cellulose and Composites: Morphology-Property Relationship. Springer.

Kim, M. H., Y. W. Lee, W. K. Jung et al. 2019. Enhanced rheological behaviors of alginate hydrogels with carrageenan for extrusion-based bioprinting. J. Mech. Behav. Biomed. Mater. 98: 187–94.

Ko, Hsu-Feng, Charles Sfeir and Prashant N. Kumta. 2010. Novel synthesis strategies for natural polymer and composite biomaterials as potential scaffolds for tissue engineering. Philosophical Transactions of the Royal Society A: Mathematical, Physical and Engineering Sciences 368(1917): 1981–1997.

Kocira, Anna, Katarzyna Kozłowicz, Katarzyna Panasiewicz et al. 2021. Polysaccharides as edible films and coatings: characteristics and influence on fruit and vegetable quality—A review. Agronomy 11(5): 813. https://www.mdpi.com/2073-4395/11/5/813.

Kotadiya, Nirav, Bhupendra G. Prajapati and Sankha Bhattacharya. 2016. Atorvastatin calcium nanoparticles using solvent-anti-solvent precipitation method. e-Journal of Science & Technology 11(4).

Kumar, Akash, Sangeeta Yadav, Jhilam Pramanik et al. 2023a. Chitosan-based composites: development and perspective in food preservation and biomedical applications. Polymers 15(15). https://doi.org/10.3390/polym15153150.

Kumar, Akash, Sangeeta Yadav, Jhilam Pramanik et al. 2023b. Chitosan-based composites: development and perspective in food preservation and biomedical applications. Polymers 15(15): 3150. https://www.mdpi.com/2073-4360/15/15/3150.

Kuo, J. W. and G. D. Prestwich. 2010. Materials of biological origin-Materials analysis and implant uses, Comprehensive biomaterials. Ducheyne, P.

Kurita, K., Y. Kaji, T. Mori et al. 2000. Enzymatic degradation of β-chitin: Susceptibility and the influence of deacetylation. Carbohydrate Polymers 42(1): 19–21. https://doi.org/10.1016/S0144-8617(99)00127-7.

Lastra Ripoll, S. E., S. E. Quintana Martinez and L. A. Garcia Zapateiro. 2021. Rheological and microstructural properties of xanthan gum-based coating solutions enriched with phenolic mango (*Mangifera indica*) Peel Extracts. ACS Omega 6(24): 16119–16128. https://doi.org/10.1021/acsomega.1c02011. https://www.ncbi.nlm.nih.gov/pubmed/34179657.

Lee, K. Y. and D. J. Mooney. 2012. Alginate: properties and biomedical applications. Prog. Polym. Sci. 37: 106–126.

Lewandowska, Katarzyna, Alina Sionkowska, Sylwia Grabska et al. 2016. Surface and thermal properties of collagen/hyaluronic acid blends containing chitosan. International Journal of Biological Macromolecules 92: 371–376.

Li, Haiying, Yuhan Xue, Bei Jia et al. 2018. The preparation of hyaluronic acid grafted pullulan polymers and their use in the formation of novel biocompatible wound healing film. Carbohydrate Polymers 188: 92–100. https://doi.org/https://doi.org/10.1016/j.carbpol.2018.01.102. https://www.sciencedirect.com/science/article/pii/S0144861718301267.

Li, Jingyuan, Hong Xiang, Qian Zhang et al. 2022. Polysaccharide-based transdermal drug delivery. Pharmaceuticals 15(5): 602.

Li, R., Z. Lin, Q. Zhang et al. 2020. Injectable and *in situ*-formable thiolated chitosan-coated liposomal hydrogels as curcumin carriers for prevention of *in vivo* breast cancer recurrence. ACS Appl. Mater. Interfaces 12(15): 17936–17948. https://doi.org/10.1021/acsami.9b21528. https://www.ncbi.nlm.nih.gov/pubmed/32208630.

Liew, Joy, Wei Yi, Kee Shyuan Loh et al. 2017. Synthesis and characterization of modified κ-carrageenan for enhanced proton conductivity as polymer electrolyte membrane. PloS one 12(9): e0185313–e0185313. https://doi.org/10.1371/journal.pone.0185313. https://pubmed.ncbi.nlm.nih.gov/28957374. https://www.ncbi.nlm.nih.gov/pmc/articles/PMC5619736/.

Ma, Guiping, Dongzhi Yang, John F. Kennedy et al. 2009. Synthesize and characterization of organic-soluble acylated chitosan. Carbohydrate Polymers 75(3): 390–394. https://doi.org/https://doi.org/10.1016/j.carbpol.2008.07.035. https://www.sciencedirect.com/science/article/pii/S0144861708003470.

Mariod, A. A. and H. F. Adam. 2013. Review: Gelatin, source, extraction and industrial applications. Acta Sci. Pol. Technol. Aliment. 12: 135–147.

Miguel, Sónia P., Maximiano P. Ribeiro, Hugo Brancal et al. 2014. Thermoresponsive chitosan–agarose hydrogel for skin regeneration. Carbohydrate Polymers 111: 366–373.

Mohite, Popat, Pudji Rahayu, Shubham Munde et al. 2023a. Chitosan-based hydrogel in the management of dermal infections: A review. Gels 9(7). https://doi.org/10.3390/gels9070594.

Mohite, Popat, Pudji Rahayu, Shubham Munde et al. 2023b. Chitosan-based hydrogel in the management of dermal Infections: A Review. Gels 9(7): 594. https://www.mdpi.com/2310-2861/9/7/594.

Mohite, Popat, Sunny R. Shah, Sudarshan Singh et al. 2023a. Chitosan and chito-oligosaccharide: a versatile biopolymer with endless grafting possibilities for multifarious applications. Frontiers in Bioengineering and Biotechnology 11: 1190879.

Mohite, Popat, Sunny R. Shah, Sudarshan Singh et al. 2023b. Chitosan and chito-oligosaccharide: a versatile biopolymer with endless grafting possibilities for multifarious applications. Frontiers in Bioengineering and Biotechnology 11. https://doi.org/10.3389/fbioe.2023.1190879. https://www.frontiersin.org/articles/10.3389/fbioe.2023.1190879.

Moin, K., G. Bhupendra, M. Vishnu et al. 2002. Sodium alginate based insitu gelling system of famotidine: preparation and *in-vivo* characterizations. E-journal of Science & Technology 2(1): 27–42.

Morán, Juan Ignacio, Viviana Paola Cyras et al. 2013. Preparation and characterization of three different derivatized potato starches. Journal of Polymers and the Environment 21(2): 395–404.

Morin-Crini, Nadia, Eric Lichtfouse, Giangiacomo Torri et al. 2019. Applications of chitosan in food, pharmaceuticals, medicine, cosmetics, agriculture, textiles, pulp and paper, biotechnology, and environmental chemistry. Environmental Chemistry Letters 17(4): 1667–1692.

Mudgil, Deepak, Sheweta Barak and Bhupendar Singh Khatkar. 2014. Guar gum: processing, properties and food applications-A Review. Journal of Food Science and Technology 51(3): 409–418. https://doi.org/10.1007/s13197-011-0522-x. https://pubmed.ncbi.nlm.nih.gov/24587515. https://www.ncbi.nlm.nih.gov/pmc/articles/PMC3931889/.

Mukherjee, Indraneil and MaryAnn Rosolen. 2013. Thermal transitions of gelatin evaluated using DSC sample pans of various seal integrities. Journal of Thermal Analysis and Calorimetry 114(3): 1161–1166.

Mukherjee Swarupananda, Karati Dipanjan, Singh Sudarshan et al. 2024. Chitosan-based nanomedicine in the management of age-related macular degeneration: A review. Current Nanomedicine 14(1). https://doi.org/https://dx.doi.org/10.2174/0124681873261772230927074628.

Nagaraja, Kasula, Kummara Madhusudana Rao, KSV Krishna Rao et al. 2022. Dual responsive tamarind gum-co-poly (N-isopropyl acrylamide-co-ethylene glycol vinyl ether) hydrogel: A promising device for colon specific anti-cancer drug delivery. Colloids and Surfaces A: Physicochemical and Engineering Aspects 641: 128456.

Nandi, Gouranga, Abhijit Changder and Lakshmi Kanta Ghosh. 2019. Graft-copolymer of polyacrylamide-tamarind seed gum: Synthesis, characterization and evaluation of flocculating potential in peroral paracetamol suspension. Carbohydrate Polymers 215: 213–225.

Pang, Z., H. Deeth, P. Sopade et al. 2014. Texture and microstructure of gelatin gels with and without milk proteins, Food. Hydrocoll. 35: 484–493. https://doi.org/10.1016/j.foodhyd.2013.07.007.

Park, Kyung Min, Sang Young Lee et al. 2009. Thermosensitive chitosan–Pluronic hydrogel as an injectable cell delivery carrier for cartilage regeneration. Acta Biomaterialia 5(6): 1956–1965.

Patel, Rakesh P., Geeta K. Patel, Nikunjana Patel et al. 2023. Alginate nanoparticles: a potential drug carrier in tuberculosis treatment. pp. 207–234. *In*: Ranjita Shegokar and Yashwant Pathak (Eds.). Tubercular Drug Delivery Systems: Advances in Treatment of Infectious Diseases. Cham: Springer International Publishing.

Patel, Vishnu M., Bhupendra G. Prajapati and Madhabhai M. Patel. 2007. Design and characterization of chitosan-containing mucoadhesive buccal patches of propranolol hydrochloride. Acta pharmaceutica 57(1): 61–72.

Pawar, Anil, Pranjal Lohakane, Ramdas Pandhare et al. 2024. Chitosan fortified repaglinide gastro-retentive mucoadhesive microsphere with improved anti-diabetic attribute. Intelligent Pharmacy.

https://doi.org/https://doi.org/10.1016/j.ipha.2024.01.012. https://www.sciencedirect.com/science/article/pii/S2949866X24000224.

Pearl, Gary S., Irene J. Check and Robert L. Hunter. 1984. Agarose electrophoresis and immunonephelometric quantitation of cerebrospinal fluid immunoglobulins: Criteria for application in the diagnosis of neurologic disease. American Journal of Clinical Pathology 81(5): 575–580.

Pereira, Fernanda Stuani, Deuber Lincon da Silva Agostini, Aldo Eloizo Job et al. 2013. Thermal studies of chitin–chitosan derivatives. Journal of Thermal Analysis and Calorimetry 114(1): 321–327. https://doi.org/10.1007/s10973-012-2835-z. https://doi.org/10.1007/s10973-012-2835-z.

Petri, Denise. 2015. Xanthan gum: A versatile biopolymer for biomedical and technological applications. Journal of Applied Polymer Science 132. https://doi.org/10.1002/app.42035.

Pettikiriarachchi, Jerani T. S., Clare L. Parish, Molly S. Shoichet et al. 2010. Biomaterials for brain tissue engineering. Australian Journal of Chemistry 63(8): 1143–1154.

Phadke, K. V., L. S. Manjeshwar and T. M. Aminabhavi. 2014. Biodegradable polymeric microspheres of gelatin and carboxymethyl guar gum for controlled release of theophylline, Polym. Bull. 71: 1625–1643.

Picout, David R. and Simon B. %J TheScientificWorldJOURNAL Ross-Murphy. 2003. Rheology of biopolymer solutions and gels. 3: 105–121.

Pooja, D., S. Panyaram, H. Kulhari et al. 2014. Xanthan gum stabilized gold nanoparticles: characterization, biocompatibility, stability and cytotoxicity, Carbohydr. Polym 110: 1–9.

Potter, Norman N. and Joseph H. Hotchkiss. 2012. Constituents of foods: properties and significane. In Food science, 30. Ithaca, New York: Springer Science & Business Media.

Prajapati, Bhupendra G. and Krutika K. Sawant. 2009. Poly electrolyte complex of chitosan alginate for local drug delivery. International Journal of ChemTech Research. 1(3): 643–648.

Prestwich, Glenn D. and Jing-wen Kuo. 2008. Chemically-modified HA for therapy and regenerative medicine. Current Pharmaceutical Biotechnology 9(4): 242–245.

Qiao, C., X. Ma, J. Zhang et al. 2019. Effect of hydration on water state, glass transition dynamics and crystalline structure in chitosan films. Carbohydr Polym. 206: 602–608. https://doi.org/10.1016/j.carbpol.2018.11.045. https://www.ncbi.nlm.nih.gov/pubmed/30553363.

Reddy, M. S. B., D. Ponnamma, R. Choudhary et al. 2021. A comparative review of natural and synthetic biopolymer composite scaffolds. Polymers (Basel) 13(7). https://doi.org/10.3390/polym13071105. https://www.ncbi.nlm.nih.gov/pubmed/33808492.

Reis, Davi Teixeira, Anna Karla Dos Santos Pereira, Gessiel Newton Scheidt et al. 2019. Plant and bacterial cellulose: production, chemical structure, derivatives and applications. Orbital: The Electronic Journal of Chemistry 11(5). https://doi.org/10.17807/orbital.v11i5.1349.

Riaz, T., M. W. Iqbal, B. Jiang et al. 2021. A review of the enzymatic, physical, and chemical modification techniques of xanthan gum. Int. J. Biol. Macromol. 186: 472–489. https://doi.org/10.1016/j.ijbiomac.2021.06.196. https://www.ncbi.nlm.nih.gov/pubmed/34217744.

Ribeiro, Sabrina M., Érick D. O. Fratucelli, Paula C. P. Bueno et al. 2019. Antimicrobial and antibiofilm activities of *Casearia sylvestris* extracts from distinct Brazilian biomes against *Streptococcus mutans* and *Candida albicans*. BMC Complementary and Alternative Medicine 19: 1–16.

Rosiak, Piotr, Ilona Latanska, Paulina Paul et al. 2021. Modification of alginates to modulate their physic-chemical properties and obtain biomaterials with different functional properties. Molecules 26(23). https://doi.org/10.3390/molecules26237264.

Ru, G., S. Wu, X. Yan et al. 2019. Inverse solubility of chitin/chitosan in aqueous alkali solvents at low temperature. Carbohydr. Polym. 206: 487–492. https://doi.org/10.1016/j.carbpol.2018.11.016. https://www.ncbi.nlm.nih.gov/pubmed/30553349.

Rudhziaha, S., M. S. A. Rani, A. Ahmad et al. 2015. Potential of blend of kappacarrageenan and cellulose derivatives for green polymer electrolyte application. Ind. Crops Prod. 72: 133–141.

Rukmanikrishnan, B., F. R. M. Ismail, R. K. Manoharan et al. 2020. Blends of gellan gum/xanthan gum/zinc oxide based nanocomposites for packaging application: Rheological and antimicrobial properties. Int. J. Biol. Macromol. 148: 1182–1189. https://doi.org/10.1016/j.ijbiomac.2019.11.155. https://www.ncbi.nlm.nih.gov/pubmed/31760019.

Saha, D. and S. Bhattacharya. 2010. Hydrocolloids as thickening and gelling agents in food: a critical review. J. Food Sci. Technol. 47(6): 587–97. https://doi.org/10.1007/s13197-010-0162-6. https://www.ncbi.nlm.nih.gov/pubmed/23572691.

Sahariah, P., B. E. Benediktssdottir, M. A. Hjalmarsdottir et al. 2015. Impact of chain length on antibacterial activity and hemocompatibility of quaternary N-alkyl and n,n-dialkyl chitosan derivatives. Biomacromolecules 16(5): 1449–60. https://doi.org/10.1021/acs.biomac.5b00163. https://www.ncbi.nlm.nih.gov/pubmed/25830631.

Said, M., B. Haq, D. Al Shehri et al. 2021. Modification of Xanthan gum for a high-temperature and high-salinity reservoir. Polymers (Basel) 13(23). https://doi.org/10.3390/polym13234212. https://www.ncbi.nlm.nih.gov/pubmed/34883714.

Sánchez-Salcedo, S., A. Nieto and M. Vallet-Regí. 2008. Hydroxyapatite/β-tricalcium phosphate/agarose macroporous scaffolds for bone tissue engineering. Chemical Engineering Journal 137(1): 62–71.

Seddiqi, Hadi, Erfan Oliaei, Hengameh Honarkar et al. 2021. Cellulose and its derivatives: towards biomedical applications. Cellulose 28(4): 1893–1931. https://doi.org/10.1007/s10570-020-03674-w. https://doi.org/10.1007/s10570-020-03674-w.

Shand, P. J., J. N. Sofos and G. R. Schmidt. 1994. Differential scanning calorimetry of beef/kappa-carrageenan mixtures. Journal of Food Science 59(4): 711–715. https://doi.org/https://doi.org/10.1111/j.1365-2621.1994.tb08110.x. https://ift.onlinelibrary.wiley.com/doi/abs/10.1111/j.1365-2621.1994.tb08110.x.

Shao, Huimin, Hui Zhang, Yanjun Tian et al. 2019. Composition and rheological properties of polysaccharide extracted from tamarind (*Tamarindus indica* L.) Seed. Molecules (Basel, Switzerland) 24(7): 1218. https://doi.org/10.3390/molecules24071218. https://pubmed.ncbi.nlm.nih.gov/30925745. https://www.ncbi.nlm.nih.gov/pmc/articles/PMC6480175/.

Silva, F. M. V., M. F. Leite, A. C. C. Spadaro et al. 2009. Assessment of the potential genotoxic risk of medicinal *Tamarindus indica* fruit pulp extract using *in vivo* assays. Genetics and Molecular Research: 1085–1092.

Singh, Akhilesh Vikram. 2013. A DSC study of some biomaterials relevant to pharmaceutical industry. Journal of Thermal Analysis and Calorimetry 112(2): 791–793. https://doi.org/10.1007/s10973-012-2638-2. https://doi.org/10.1007/s10973-012-2638-2.

Singh, S. and S. B. Bothara. 2014. Morphological, physico-chemical and structural characterization of mucilage isolated from the seeds of Buchanania lanzan Spreng. International Journal of Health & Allied Sciences 3(1): 33.

Singh, Sudarshan. 2022. Biopolymers towards green and sustainable development. Bentham Science Publishers.

Singh, Sudarshan and Sunil B. Bothara. 2014a. Manilkara zapota (Linn.) Seeds: A Potential Source of Natural Gum. ISRN Pharmaceutics.

Singh, Sudarshan and Sunil B. Bothara. 2014b. Physico-chemical and structural characterization of mucilage isolated from seeds of Diospyros melonoxylon Roxb. Brazilian Journal of Pharmaceutical Sciences 50: 713–725.

Singh, Sudarshan, Vijay R. Chidrawar, Dessy Hermawan et al. 2023. Hypromellose highly swellable composite fortified with psidium guajava leaf phenolic-rich extract for antioxidative, antibacterial, anti-inflammatory, anti-melanogenesis, and hemostasis applications. Journal of Polymers and the Environment. https://doi.org/10.1007/s10924-023-02819-9. https://doi.org/10.1007/s10924-023-02819-9.

Singh, Sudarshan, Warangkana Chunglok, Ozioma F. Nwabor et al. 2023. Porous biodegradable sodium alginate composite fortified with hibiscus *Sabdariffa* L. calyx extract for the multifarious biological applications and extension of climacteric fruit shelf-life. Journal of Polymers and the Environment 31(3): 922–938. https://doi.org/10.1007/s10924-022-02596-x. https://doi.org/10.1007/s10924-022-02596-x.

Singh, Sudarshan, Warangkana Chunglok, Ozioma F. Nwabor et al. 2022. Hydrophilic biopolymer matrix antibacterial peel-off facial mask functionalized with biogenic nanostructured material for cosmeceutical applications. Journal of Polymers and the Environment 30(3): 938–953. https://doi.org/10.1007/s10924-021-02249-5. https://doi.org/10.1007/s10924-021-02249-5.

Singh, Sudarshan, Ozioma F. Nwabor, Dwi M. Sukri et al. 2022. Poly (vinyl alcohol) copolymerized with xanthan gum/hypromellose/sodium carboxymethyl cellulose dermal dressings functionalized with biogenic nanostructured materials for antibacterial and wound healing application. International Journal of Biological Macromolecules 216: 235–250. https://doi.org/https://doi.org/10.1016/j.ijbiomac.2022.06.172. https://www.sciencedirect.com/science/article/pii/S0141813022013861.

Singh, Sudarshan, Ozioma Forstinus Nwabor, Dwi Marlina Syukri et al. 2021. Chitosan-poly(vinyl alcohol) intelligent films fortified with anthocyanins isolated from Clitoria ternatea and Carissa carandas for monitoring beverage freshness. International Journal of Biological Macromolecules 182: 1015–1025. https://doi.org/https://doi.org/10.1016/j.ijbiomac.2021.04.027. https://www.sciencedirect.com/science/article/pii/S0141813021007820.

Singh, Sudarshan, Nassareen Supaweera, Ozioma F. Nwabor et al. 2024. Poly (vinyl alcohol)-gelatin-sericin copolymerized film fortified with vesicle-entrapped demethoxycurcumin/bisdemethoxycurcumin for improved stability, antibacterial, anti-inflammatory, and skin tissue regeneration. International Journal of Biological Macromolecules 258: 129071. https://doi.org/https://doi.org/10.1016/j.ijbiomac.2023.129071. https://www.sciencedirect.com/science/article/pii/S0141813023059706.

Soares, J. P., J. E. Santos, G. O. Chierice et al. 2004. Thermal behavior of alginic acid and its sodium salt. Eclética Química 29(2): 57–64. https://doi.org/https://doi.org/10.1590/S0100-46702004000200009.

Song, Fuyu, Jingwei Gong, Yehan Tao et al. 2021. A robust regenerated cellulose-based dual stimuli-responsive hydrogel as an intelligent switch for controlled drug delivery. International Journal of Biological Macromolecules 176: 448–458.

Souza, Márcio A., Isis T. Vilas-Boas, Jôse M. Leite-da-Silva et al. 2022. Polysaccharides in agro-industrial biomass residues. Polysaccharides 3(1). https://doi.org/10.3390/polysaccharides3010005.

Srivastava, Arti, Vivek Mishra, Pooja Singh et al. 2012. Comparative study of thermal degradation behavior of graft copolymers of polysaccharides and vinyl monomers. Journal of Thermal Analysis and Calorimetry 107(1): 211–223.

Stokols, Shula, Jeff Sakamoto, Chris Breckon et al. 2006. Templated agarose scaffolds support linear axonal regeneration. Tissue Engineering 12(10): 2777–2787.

Su, Ting, Mengying Zhang, Qiankun Zeng et al. 2021. Mussel-inspired agarose hydrogel scaffolds for skin tissue engineering. Bioactive materials 6(3): 579–588.

Subramanian, H., J. Brown and R. Gatenby. 2020. Prebiotic competition and evolution in self-replicating polynucleotides can explain the properties of DNA/RNA in modern living systems. BMC Evol. Biol. 20(1): 75. https://doi.org/10.1186/s12862-020-01641-4.

Sun, B., S. Yu, D. Zhao et al. 2018. Polysaccharides as vaccine adjuvants. Vaccine 36(35): 5226–5234. https://doi.org/10.1016/j.vaccine.2018.07.040.

Sun, Jinchen and Huaping Tan. 2013. Alginate-based biomaterials for regenerative medicine applications. Materials (Basel, Switzerland) 6(4): 1285–1309. https://doi.org/10.3390/ma6041285. https://pubmed.ncbi.nlm.nih.gov/28809210. https://www.ncbi.nlm.nih.gov/pmc/articles/PMC5452316/.

Szabo, L., S. Gerber-Lemaire and C. Wandrey. 2020. Strategies to functionalize the anionic biopolymer Na-alginate without restricting its polyelectrolyte properties. Polymers (Basel) 12(4). https://doi.org/10.3390/polym12040919. https://www.ncbi.nlm.nih.gov/pubmed/32326625.

Szcześniak, Ludwik, Adam Rachocki and Jadwiga Tritt-Goc. 2008. Glass transition temperature and thermal decomposition of cellulose powder. Cellulose 15(3): 445–451. https://doi.org/10.1007/s10570-007-9192-2. https://doi.org/10.1007/s10570-007-9192-2.

US FDA. 21CFR. US FDA. Accessed 04-04-2022. https://www.accessdata.fda.gov/scripts/cdrh/cfdocs/cfcfr/cfrsearch.cfm?fr=172.620.

United State Food and Drug Administration. Inactive Ingredient for Approved Drug Products. US FDA. Accessed 01-04-2022. https://www.accessdata.fda.gov/scripts/cder/iig/index.cfm.

Kevin Hawkins. 2008. HemCon® Bandage. US FDA. Accessed 12-04-2022. https://www.accessdata.fda.gov/cdrh_docs/pdf7/K072486.pdf.

Alan Donald. 2018. AXIOSTAT CHITOSAN HEMOSTATIC DRESSING. Accessed 12-04-2022. https://www.accessdata.fda.gov/cdrh_docs/pdf17/K172324.pdf.

Valetti, N. W., V. Boeris and G. Pico. 2013. Characterization of chymotrypsin-iota-carrageenan complex in aqueous solution: a solubility and thermodynamical stability study. Int. J. Biol. Macromol. 52: 45–51. https://doi.org/10.1016/j.ijbiomac.2012.10.014. https://www.ncbi.nlm.nih.gov/pubmed/23107807.

Wang, J., P. Wang, M. Zhu et al. 2021. Overexpression and biochemical properties of a GH46 chitosanase from marine *Streptomyces hygroscopicus* R1 suitable for chitosan oligosaccharides preparation. Front Microbiol. 12: 816845. https://doi.org/10.3389/fmicb.2021.816845. https://www.ncbi.nlm.nih.gov/pubmed/35173697.

Wang, L., M. Hu and Y. Tao. 2022. Efficient production of D-glucosamine by diacetylchitobiose deacetylase catalyzed deacetylation of N-acetyl-D-glucosamine. Biotechnol. Lett. 44(3): 473–483. https://doi.org/10.1007/s10529-022-03225-2. https://www.ncbi.nlm.nih.gov/pubmed/35072843.

Wang, Qingkui, Peter Ellis and Simon Ross-Murphy. 2000. The stability of guar gum in an aqueous system under acidic conditions. Food Hydrocolloids 14: 129–134. https://doi.org/10.1016/S0268-005X(99)00058-2.

Wibel, R., D. E. Braun, L. Hammerle et al. 2021. *In vitro* investigation of thiolated chitosan derivatives as mucoadhesive coating materials for solid lipid nanoparticles. Biomacromolecules 22 (9): 3980–3991. https://doi.org/10.1021/acs.biomac.1c00776. https://www.ncbi.nlm.nih.gov/pubmed/34459197.

Yermak, I. M., V. N. Davydova, A. O. Kravchenko et al. 2020. Mucoadhesive properties of sulphated polysaccharides carrageenans from red seaweed families Gigartinaceae and Tichocarpaceae. Int. J. Biol. Macromol. 142: 634–642. https://doi.org/10.1016/j.ijbiomac.2019.10.005. https://www.ncbi.nlm.nih.gov/pubmed/31622715.

Zarrintaj, Payam, Behnaz Bakhshandeh, Iraj Rezaeian et al. 2017. A novel electroactive agarose-aniline pentamer platform as a potential candidate for neural tissue engineering. Scientific Reports 7(1): 1–12.

Zarrintaj, Payam, Saeed Manouchehri, Zahed Ahmadi et al. 2018. Agarose-based biomaterials for tissue engineering. Carbohydrate Polymers 187: 66–84.

Zhang, Q. W., L. G. Lin and W. C. Ye. 2018. Techniques for extraction and isolation of natural products: a comprehensive review. Chin. Med. 13: 20. https://doi.org/10.1186/s13020-018-0177-x. https://www.ncbi.nlm.nih.gov/pubmed/29692864.

Zhang, T., R. Sun, M. Ding et al. 2020. Effect of extraction methods on the structural characteristics, functional properties, and emulsion stabilization ability of Tilapia skin gelatins, Food. Chem. 328. https://doi.org/10.1016/j.foodchem.2020.127114.

Zhang, Z., F. Jin, Z. Wu et al. 2017. O-acylation of chitosan nanofibers by short-chain and long-chain fatty acids. Carbohydr Polym. 177: 203–209. https://doi.org/10.1016/j.carbpol.2017.08.132. https://www.ncbi.nlm.nih.gov/pubmed/28962759.

Zhong, Zhikai and X. Susan Sun. 2005. Thermal characterization and phase behavior of cornstarch studied by differential scanning calorimetry. Journal of Food Engineering 69(4): 453–459.

Zobel, H. F. 1988. Molecules to granules: a comprehensive starch review. Starch - Stärke 40(2): 44–50. https://doi.org/https://doi.org/10.1002/star.19880400203. https://onlinelibrary.wiley.com/doi/abs/10.1002/star.19880400203.

5

Polymer Blends and Composites—Biomedical Application

Nikhil Das K.[1] and *Swetha Shaji*[2,*]

1. Introduction

Biomaterials is a fast-growing field in the following past decade (Hench 1998; Langer and Tirrell 2004; Bandyopadhyay 2018). These substances have potential applications in tissue engineering and regenerative medicine for the production of biomedical devices such as steel hip and metal or polymer stents, knee replacements, knee and ligament implants, polymer vascular grafts, ceramic dental implants, polymer sutures, and artificial heart valves, hydrogel contact lenses, etc. (Caló 2015; Lam and Wu 2012; Satarkar and Zach Hilt 2008; Williams 2019; Yang et al. 2018). The developments in the area of regenerative medicine have made biomaterials to be used in the application of healing damaged blood vessels, skin, muscle, bone, and cartilage (Nada 2014). They are divided into metal and metal alloys, organic ceramics, organic and inorganic polymers, and metal, carbon-based materials and composite materials and blends (Sionkowska 2011; Dhandayuthapani et al. 2011; Prasad et al. 2017; Chen and Thouas 2015; Wang and Yeung 2017). Polymers have gained more attention because of their capacity to enhance performance, versatile and adaptable chemical and material designs through unquestionably flexible result methodologies, and potential for customisation based on request requirements.

[1] TGP Bioplastics Pvt. Ltd, Maharashtra, India - 415414, Research & Development division, M.Sc. Biopolymer Science, CIPET - IPT, Kochi, India.
Email: nikhilcheloopadam123@gmail.com

[2] M.Sc. Biopolymer Science, CIPET - IPT, Kochi, India.

* Corresponding author: swethashaji15@gmail.com

The first prerequisite for a polymer to be used in biomedical applications is biocompatibility. The property of being biocompatible is the lack of toxicity or adverse effects on biological systems, or it is a substance's ability to perform in a given application with a sufficient host response (Grainger 1999). Another factor is the biodegradability in various applications. It resolves present difficulties, such as problems with the long-term biocompatibility of revision procedures as well as their practical and moral implications (Nair and Laurencin 2007). Both organic and synthetic polymers have the potential to degrade. Collagen is one example of a natural polymer that degrades and has been utilised for medicinal purposes for a long time (Barbucci 2002). Collagens have been found to be useful as scaffolds in tissue engineering and regenerative medicine applications (Yannas et al. 2010; Kenar et al. 2019). Collagens of various forms are found in a variety of connective tissues, including bone, skin, tendon, cartilage, and other connective tissues (Yue 2014). Collagen is made up of acidic and basic amino acid residues, including glycine, proline, hydroxyproline, and lysine, and it can be positively or negatively charged depending on the environment (pH). Collagen or polyester materials are the basis for the biodegradable polymers utilised in regenerative medicine and tissue engineering applications. Many biocompatible products for biomedical applications can be created by physically or chemically mixing various materials or by employing the right processing techniques.

Polymers utilised in biomedical engineering are divided into two categories: both artificial and natural polymers. Chitosan and other natural polymers exhibit the required biocompatibility, but without the addition of other polymers, their mechanical qualities fall short. The characteristics that repel water of polycaprolactone, a synthetic biopolymer that is commonly employed in tissue engineering, reduces the effectiveness of cell attachment, highlighting the importance of mixing PCL with other polymers to achieve hydrophilicity (Salati et al. 2020).

For instance, PCL is a synthetic polymer that is biocompatible and is frequently used in biomedical engineering. Due to its adequate degradability and biocompatibility, PCL is a promising option for biomedical applications (Sharma et al. 2016; Hejna et al. 2015). PCL has been utilised in the administration of drugs and long-term implants; however, in order to achieve the appropriate mechanical and cell adherence qualities, it needs to be joined with other polymers (Salati et al. 2020). Polyether PEG (polyethylene glycol) is a versatile substance utilised in many industries, including medicine. The attractive process of PEGylation, which involves joining PEG to another molecule, may have therapeutic value. In order to prevent the formation of protein coronas, PEG can prevent protein adsorption, which is significant in medication delivery (Abdolmaleki et al. 2016).

Polymer blending is a common approach for combining the qualities of distinct polymer components or creating new materials that tend to be better. Intermacromolecular interactions may cause such features, affecting macromolecular mobility and microstructure at various scales (Sionkowska 2011; López-Ângulo and Sobral 2016). Because of the biological qualities of their distinct components, natural polymer blends are frequently anticipated for biomedical applications. They also have a number of distinct advantages, including the availability of renewable resources and the ease with which chemical modifications can be made to adapt their features (Guitian Oliveira et al. 2014; Ma et al. 2003). Polymers' variable microstructure allows for more precise creation of biological materials and systems. Several research studies have examined the use of polymers as therapeutic systems, such as drug carriers, diagnostic systems, fluorescent-loaded polymeric nanoparticles, and regenerative scaffolds in tissue engineering (Zarrintaj et al. 2017).

Composite materials are regarded as an indispensable component of regular life because of their exceptional qualities and wide range of applications. A scattered phase and a matrix phase make up a composite material, combining to improve a property (Jose and Joseph 2012). They offer great qualities like high fatigue strength, low weight, and corrosion resistance. There are numerous dental, tissue engineering, wound healing, and regenerative medicine applications of polymer composites (Zafar et al. 2016; Qasim et al. 2018).

Composite materials have a number of advantages compared to ceramics and metal alloys. In radiography, they exhibit unwanted artefacts, and in the case of composite materials, the radio transparency can be modified by mixing contrast medium with the polymer. Diagnostic techniques like CT and MRI are completely compatible with the composite materials. Additionally, they serve as the structural foundation for imaging equipment. In biomedical applications like hard tissue application, composite materials are commonly used, which include bone fracture repair (CF/epoxy composite materials used as external fixators), bone plates (composites of PTFE, PA, and polyester), joint replacement, spine instrumentation, total hip replacement, and intramedullary nails, etc. The soft tissue application includes bulk space fillers, encapsulants and carriers, useful load-bearing and support implants, and other devices like medical equipment. In this chapter, the biomedical application of the polymer blends and composites is given.

2. Natural Polymer Blends

2.1 Blends of Chitosan

Chitosan is a polysaccharide with polyelectrolyte cationic properties. This biopolymer has a high charge density and is biocompatible. Additionally, it is non-toxic, and it has a powerful mucoadhesion. It is possible to create a novel product based on such binary combinations of materials by mixing hyaluronic acid with chitosan polymers. There are numerous biomedical uses for hydrogels, microspheres, sponges, films, and nanoparticles produced containing hyaluronic acid as well as chitosan. Hyaluronic acid and chitosan, as well as their blended films' surface roughness, were changed by blending. The composition of the blend, as well as the solvent used to make the blend, determines the composition of the chitosan-hyaluronic acid mixture (Sionkowska et al. 2020) (Figure 1). PH and ionic strength influence the interactions between chitosan and hyaluronic acid (Ding et al. 2019). Applications for tissue engineering and cartilage regeneration have been investigated using hyaluronic acid and chitosan-based hydrogel mixtures (Sionkowska et al. 2020). *In vivo* cartilage regeneration with chitosan-hyaluronic acid dialdehyde hydrogels has also been investigated (Karabiyik Acar et al. 2018). Coacervate-based scaffolds made of hyaluronic acid and chitosan have been suggested to create cartilage tissue. Additionally, stem cells from bone marrow were enclosed in these coacervates. Zhang et al. reported using

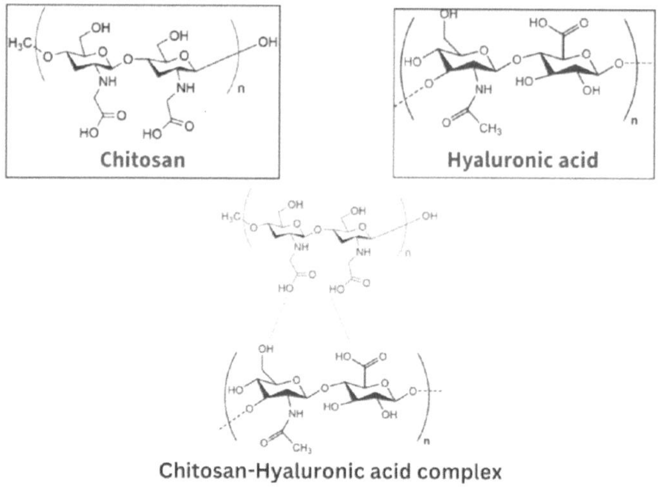

Figure 1: Complexation of chitosan with Hyaluronic acid.

chitosan and hyaluronic acid-based hydrogels for pH-sensitive medication release that are injectable and body temperature sensitive (Zhang et al. 2018). According to the study, the hydrogels' mechanical strength can be increased by the formation of chitosan's protonated amines form hydrogen bonds with one another and the carboxyl group in hyaluronic acid, which also reduces the first rapid drug release from the hydrogel. Additionally, chitosan-based polymers and hyaluronic acid can be employed to create a new generation of wound dressings (Pachuau 2015).

Polymers' variable microstructure allows for more precise creation of biological materials and systems. Several research studies have examined the use of polymers as therapeutic systems, such as drug carriers, diagnostic systems, fluorescent-loaded polymeric nanoparticles, and tissue engineering as regenerative scaffolds (Belluzo et al. 2016). The first need for a polymer to be used in biomedical applications is biocompatibility. There are two types of polymers used in biomedical engineering: synthetic and natural polymers (Belluzo et al. 2016). Chitosan has adequate biocompatibility, but its mechanical qualities are insufficient unless it is blended with polymers. A biocompatible synthetic polymer called polycaprolactone (PCL) is frequently used in tissue engineering; however, its hydrophobic properties reduce the effectiveness of cell adhesion, emphasising the necessity to blend PCL with other polymers to achieve hydrophilicity (Grossen et al. 2017). When it comes to combining polymer pairings, miscibility is crucial. Immiscible polymers are susceptible to phase separation, which modifies blends' surface structures and thus regulates cellular activity (Tipduangta et al. 2021). Rarely are polymer mixes miscible, but by using compatibilisers, they can be somewhat miscible (Tipduangta et al. 2021). Different To promote miscibility, mixtures of polymers have been given compatibilisers (Tipduangta et al. 2021). Maleic anhydride was utilised as a compatibiliser when it was grafted onto polypropylene (PP-g-MAH) (Tipduangta et al. 2021). Electrospinning, gelation, and casting are common ways for fabricating blends; however, the appropriate polymer mixes are specified in the final application and production techniques (Tipduangta et al. 2021).

Due to mechanical defects, natural polymers such as chitosan cannot be used alone in biomedical applications; consequently, blends have been used to compensate for the lack of mechanical qualities (Tipduangta et al. 2021). Natural polymers require a hydrogel by adding a crosslinking agent, making them toxic (Belluzo et al. 2016). In order to overcome these limitations, agarose with self-gelling characteristics was used (Tipduangta et al., 2021). Aldehyde groups were produced on the alginate chain by applying sodium metaperiodate to produce a homogeneous gel. Furthermore, to alter the mechanical properties of the hydrogel, precise agarose concentration selection is necessary (Tipduangta et al. 2021). For biomedical applications, a blend of chitosan and carboxymethyl cellulose. Ultrasonic compatibilisation was used to improve the blend compatibility. The mechanical qualities of the compatibilised blend were found to be improved (Belluzo et al. 2016). For cartilage regeneration, Bhat et al. used a chitosan/agarose/gelatin (CAG) combination. The cryogelation process created porous scaffolds with appropriate mechanical characteristics and enough nutrition penetration. CAG has been shown to produce a three-dimensional environment for cartilage tissue engineering (Bhat et al. 2011). In order to enable the xenotransplantation of a bioartificial pancreas enclosing the islet cell, Yang et al. used an immunoisolative supportive matrix made of chitosan and gelatin (Yang et al. 2008). Miguel et al. created thermosensitive hydrogels for skin regeneration by combining chitosan and agarose. Biocompatibility, porosity, wettability, and cellular proliferation were all good in this product (Miguel et al. 2014). Previously, chitosan/alginate nanoparticles were found to have antibacterial and anti-inflammatory properties and were used for pathogen target therapy (Zarrintaj et al. 2017).

2.2 Blends of Collagen

The use of hyaluronic acid changes the characteristics of collagen (Taravel and Domard 1995). As a structural protein, collagen provides human and animal bodies with basic structural and mechanical support. Natural polymers, such as collagen, are commonly used in biological applications. Examples of collagen application include the creation of synthetic tendons and blood vessels, artificial skin, dental implants, ocular implants, cartilage regeneration, nerve regeneration, and alternatives for bone grafts (collagen composites with inorganic particles) (Taguchi and Tanaka 2002). Collagen and hyaluronic acid may interact in a variety of ways. Taguchi et al. examined hyaluronic acid and collagen, which were blended together, and investigated the electrostatic and hydrophobic interactions between the two polymers. New matrices based on the blends were made (Taguchi and Tanaka 2002). Self-assembly of hyaluronic acid-collagen membranes for regenerative medicine applications is possible (Taguchi et al. 2002). The multipore architectures of the lyophilised matrices allow for the preparation of cartilage regeneration (Taguchi et al. 2002). Crosslinking was carried out using 1-ethyl-3-(3-dimethyl aminopropyl) carbodiimide porous collagen/hyaluronic acid matrices (Park et al. 2003). These scaffolds can be utilized to regenerate cutaneous tissue. Hyaluronic acid and collagen-based porous sponges' mixtures were produced by lyophilising, 1-ethyl-3-(3-dimethylaminopropyl)-carbodiimide (EDC) crosslinking, and freezing together (Segura et al. 2005). The biodegradation kinetics and mechanical strength were influenced by the collagen/hyaluronic acid composition and the appropriate manufacturing method. When this mixture was crosslinked with 1,4-butanedioldiglycidyl ether, substances that could be utilised as soft tissue fillers were produced. The material was shown to have excellent mechanical characteristics, biological stability, and biocompatibility. A three-dimensional network was formed by crosslinking using 1,4-butanediol diglycidyl ether with collagen and HA. This hydrogel is also a promising biomaterial for filling and repairing soft tissues. Hydrogels made of hyaluronic acid and collagen can be employed as a biodegradable medium for effective and regulated gene transfer (Segura et al. 2005). Electrospinning is another method for creating novel materials based on biopolymer mixes. Several research groups have used electrospinning to create a nanofibrous hyaluronic acid/collagen hybrid scaffold (Liu et al. 2014; Qasim et al. 2018)

3. Natural and Synthetic Blends

Synthetic and biocompatible natural polymer mixtures in biomedical applications can have synergistic effects. Synthetic and natural polymers can be combined to mimic tissue behaviour, with the synthetic polymer providing mechanical capabilities and the natural polymer providing ECM-like features (Belluzo et al. 2016). PVA was employed to improve gelatin's mechanical characteristics (Taguchi et al. 2002). Electroactive gelatin was found to be biocompatible and regenerative in bone tissue (Liu et al. 2014). PLA/starch mixes have also received much attention due to their biodegradability, renewability, and biocompatibility. A combination absence of a compatibiliser experiences phase separation due to the PLA's hydrophobic properties and starch's hydrophilic properties, leading to unsuitable mechanical characteristics. Different compatibilisers have been used depending on the application to increase the miscibility of these polymers, including maleic anhydride (MA) and methylene diphenyl diiscyanate (MDI) (Belluzo et al. 2016). Ferrarezi et al. used PEG as a compatibiliser to strengthen the impact and elastic modulus of PLA/thermoplastic starch (TPS) blends. It was discovered that adding a compatibiliser increased impact energy between 55 and 119 J/m (Ferrarezi et al. 2013). In a different study, a PLA/starch mixture had chitosan added to it, which increased elongation at break and antibacterial activity while maintaining a lower tensile strength (Bie et al. 2013). PP has gotten a lot of interest in the biomedical area because it is a common synthetic plastic, yet it has a low level of cellular interaction. Blending PP and chitosan was suggested to improve biocompatibility. Because PP is hydrophobic, it cannot interact with chitosan's polar sites, resulting in an immiscible mixture. In order to enhance blend miscibility and interaction, MA was utilised as a compatibiliser, resulting in a uniform film (Carrasco-Guigón et al. 2017). PE and chitosan were mixed by Quiroz-Castillo et al. to produce a polymer that is biocompatible and has strong mechanical qualities. To improve the compatibility of PE with chitosan, a compatibiliser, polyethylene-g-MA, was utilised. The compatibilised blend was easy to process and had increased elongation at break (Quiroz-Castillo et al. 2014). Agarose/PCL blends are a good choice for medical engineering, but phase separation is a problem. Blends of agarose/PCL and agarose/PEG-PCL were made by Chandel et al. (2016). They discovered that when PEG was added to PCL with agarose, phase separation behaviour was lessened, resulting in improved mechanical characteristics. In addition, the presence of PEG boosted loading capacity while lowering the first burst and controlled drug release rates (López-Ângulo and Sobral 2016). A co-continuous nano-phase morphology with improved mechanical characteristics was produced poly (methyl methacrylate)-b(co)-poly (2-dimethylamino) ethyl methacrylate, a linear triblock, was added by Bera et al. macromolecule to a solution of agarose and PCL (Bera et al. 2015). In a different study, reactive compatibiliser was employed to enhance the miscibility of dextran and agarose with PCL. Reactive compatibilisers such as poly (2-dimethyl amino ethyl methacrylate) and PDMA-b-poly (Nisopropyl acrylamide) copolymers were added to the mixture, which reduced phase separation and sol-fraction while improving mechanical characteristics. Improve miscibility enhanced blood and cytocompatibility compatibility. These scaffolds can be used for tissue engineering and medicine administration (Chandel et al. 2016). Adipic acid, 1,4-Butanediol, and succinic acid combine to form the biocompatible polymer known as poly (PBSA or polybutylene succinate-co-adipate). The cost of PBSA products is quite expensive. It is proposed that the PBSA be blended with other materials to lower the price (Belluzo et al. 2016). To improve the blend's characteristics, we used glycidyl methacrylate as a compatibiliser. Because of polysaccharides and polyphenols in WBTF, adding them to PBSA improved cytocompatibility. Utilising the compatibiliser has slightly decreased cytocompatibility since it improved the miscibility and interface of WBTF and PBSA, which controlled the polysaccharide and polyphenol release. Poly (vinyl chloride) with genistein (G) can both be biomedically compatible with epoxidised soybean oil (ESBO). G/ESBO/PVC is utilised in medical applications that come into touch with blood, such as haemodialysis and blood bag tubing (Belluzo et al. 2016). El-Moghazy et al. developed a chitosanepolyvinyl alcohol (PVA) nanofiber-based electrochemical biosensor. This biosensor recognised Pirimiphos-methyl, which was found in samples of olive oil and recovered to nearly 100% (El-Moghazy et al. 2016). Cadmium enrichment poses a health risk to humans. As an electrochemical biosensor, cadmium was detected using electrospun polyamide 6/chitosan nanofibers with gold nanoparticles (Migliorini et al. 2017).

4. Applications

(I) Polymer Blends for Drug Delivery

Polymer mixing has been touted as a promising method for creating fresh and distinct drug delivery systems with altered physical and chemical properties. Polymer blends mix two or more polymers to produce better, enhanced, or tailored characteristics and functionality, which can lead to considerable benefits in medication delivery. Physical mixtures, core-shell models, and block copolymer models are the three main polymer blending methodologies utilised for drug administration. Changes in the environment's temperature and pH, as well as the physiochemical features of the target organs, can all influence these mixed polymers' ability to deliver drugs. The pattern of drug release from this combined polymeric micro-and nanocarrier can be significantly impacted by the possibility of chemical interactions between medicinal compounds and polymers (Ghasemiyeh and Mohammadi-Samani 2021). When taking into account the effort and expense necessary to obtain regulatory approval for a novel excipient, polymer blends present a compelling alternative for addressing various formulation and drug delivery difficulties. From a functional aspect, polymer blending is done to enhance, customise, or maximise material performance.

While tablets are complicated combinations containing excipients that are not polymeric, like sugars and polyols, there are a few instances where polymer blends give distinctive functionality. Incorporating NaCMC (anionic) and HPMC (non-ionic) has demonstrated synergistic rheological benefits. The increased viscosity is a result of the crosslinking of the polymers' carboxylic acid and hydroxyl groups through the formation of hydrogen bonds. Such synergistic pairings allow for formulations with lower total polymer concentrations, which is helpful when the size of the medicine product or the amount of excipients must be kept to a minimum. Polymer mixes have also been employed to provide particular medication release characteristics in matrix tablets. Researchers blended the appropriate ratios of anionic (metoprolol tartrate and alprenolol hydrochloride) and non-ionic (HPC or methylcellulose) To create zero-order release patterns for two highly soluble medicines using polymers (Ranga Rao et al. 1988). The TIMERx medication delivery system for matrix tablets is based on the advantageous interactions between xanthan gum and locust bean gum. When two polymers with formed secondary structures in solution are combined, the viscosity and gelation are increased, and the resulting gel is significantly more robust than the combined strength of the original polymers (Staniforth and Baichwal 2005).

Tablet, pellet, and capsule film coatings have frequently used polymer combinations. In order to enhance substrate adhesion, modify the coating's mechanical characteristics, enhance moisture resistance, or modulate medicine delivery, a secondary polymer may be used with a main film former. Low-viscosity HPMCs are the most often utilised polymers for non-functional coating. Plasticizers must be added to HPMC when it is used for tablet coating to improve film flexibility. Plasticizers such as polyethylene glycol (PEG) are routinely employed in HPMC films (Harcum et al. 1998). Additionally, polymeric blends are employed in non-functional film coatings to speed up the coating process, provide dry, fully formed coating systems, and boost medicine stability by reducing water vapour permeability. Polymer blends have been widely employed as functional film coatings for applications requiring controlled release. Although organic solvent-based solutions can be used to apply functional film coatings, latex dispersions in aqueous systems are preferred (Cole et al. 1995). These are colloidal dispersions of polymers such as vinyl acetate, ethylcellulose, and polymethacrylates, which are insoluble in water. Polymeric additives can alter mechanical characteristics, and medication release rates can improve the controlled-release film coatings' long-term dissolution profile stability.

(II) Polymer Blends for Tissue Engineering

Electrospinning (Khaled et al. 2014) and 3-D printing (Carson et al. 2016) are two recently developed pharmaceutical processing techniques that utilise polymer blends. Polymer blends for tissue engineering have also been prepared using these techniques. It is critical to consider the fibre's properties, porosity, orientation, surface treatment, and tissue application are all important factors to consider when using fibre meshes and electrospun fibres in tissue engineering applications (Sill and von Recum 2008).

Materials that combine natural and synthetic components in the best possible way to balance between mechanical and biomimetic properties are available (Sill and von Recum 2008). Henry et al. electrospun a slowly disintegrating biodegradable polyesterurethane for developing tissue engineering scaffold, which they produced and described (Henry et al. 2007). They were able to adjust both the mechanical qualities and the pace of deterioration (by hydrolysis) of the scaffolds by altering the proportion of the amorphous and crystalline sections that constitute the polyesterurethane, in addition to the molecular balance of the amorphous division (Henry et al. 2007). Some research has been conducted on applying non-biodegradable materials for tissue engineering (Chuangchote and Supaphol 2006; Kenawy et al. 2003). Kenawy et al. (rubbing alcohol) successfully electrospun poly (ethyleneco-vinyl alcohol) fibrous mat using a 70% isopropanol/water solution (Kenawy et al. 2003). Additionally, the team was able to demonstrate that the aforementioned mats sustained the *in vitro* cultivation of fibroblasts and smooth muscle cells (Sill and von Recum 2008). Natural materials like collagen have also been electrospun in an attempt to replicate better the natural ECM (Zhong et al. 2006; Matthews et al. 2002), chitosan (Matsuda et al. 2007), gelatin (Song et al. 2008; Ayres et al. 2006), fibrinogen (McManus et al. 2007), chitin (Noh et al. 2006), and hyaluronic acid (Ji et al. 2006). Stitzel et al. studied the electrospinning of a mixture that consisted of a blend of PLGA, type I collagen, and elastin to create a vascular substitute (Stitzel et al. 2006). PLGA was added to scaffolds to increase mechanical properties, including burst strength and compliance of the prosthesis when compared with scaffolds primarily of type I collagen and elastin (Stitzel et al. 2006). Park et al. looked into developing novel biodegradable scaffolding materials by electrospinning a hybrid mixture of chitin and PGA (Park et al. 2006). Chitin inclusion in PGA matrix elements speeds up decomposition and is applicable to control how quickly PGA/chitin scaffolds degrade (Park et al. 2006). The ability to construct oriented scaffolds for directional mechanical and biological properties, including the ability to restrict soft muscle cell movement, has led to a passion for tissue-engineered vascular prostheses (Sill and von Recum 2008).

Tissue engineering is a potential technique for healing or recovering damaged soft tissues. Natural polymers such as collagen, fibrin, and elastin were provided with a lot of research attention in the field of soft tissue engineering scaffolds (Sell et al. 2010; Vázquez and Martínez 2019). Because the mechanical qualities are generally significantly lower than those of the targeted tissues, their applicability is limited (Dong and Lv 2016; Al Habis et al. 2020). Biodegradable synthetic polymers offer degradation and durable mechanical qualities as compared to natural polymers (Keun Kwon et al. 2005). Polycaprolactone is a common ingredient in soft tissue engineering, including vascular scaffolds (Mo et al. 2006), urethra (Zhang et al. 2017), and cardiac patches (Asadpour et al. 2018) due to its rubbery qualities (Keun Kwon et al. 2005).

In a recent work, PCL and poly-(lactide-co-caprolactone) were blended using a Thermally Induced Phase Separation (TIPS) method to produce a permeable hollow scaffold at an ideal ratio of 1:3 (respectively) (Sehgal and Ramsden 2008).

Unfortunately, blend scaffolds have low mechanical results compared to raw PCL and PLCL scaffolds. The mechanical and microstructural behaviour of the PCL/PLCL blend scaffolds was investigated by examining the effect of the thermal characteristics of the blend solution. The mechanical properties such as tensile strength, modulus of elasticity, and strain energy density increased together with the rise in temperature of the mixed solution. Mechanical strength improved in direct proportion to strut dimension. The temperature difference in the blend solution impacted the phase separation morphologies. The size of spherulites decreased as the temperature increased. The blend scaffold of PLCL/PCL had improved cell growth surfaces compared to the plain PCL-made scaffold. The highly adaptable technique for modifying the blend solution's temperature prior to the TIPS process can be utilised to raise the mechanical strength of the polymer.

5. Composites

A composite consists of two phases that work together to attain an excellent property. It consists of a matrix part and a scattered part. They serve composite materials with distinctive attributes. They have properties like superior fatigue strength, are lightweight, and have stability against corrosion. The major applications of polymer composites are in regenerative treatment, tissue engineering, and dental care applications. Biocomposite material consists of polymer reinforced with natural fibres. They are compatible with the body. The fibres are protected from mechanical and climatic deterioration by the polymer matrix. The biologically derived fibres, such as those found in wood, regenerated cellulose, and plants, are considered natural fibres. They are ecologically benign, less expensive, lower in weight, entirely recyclable, and renewable sources of composites. Wood and non-wood fibres are the two categories of bio composites.

(1) Classification of Composites

The three composites included in the matrix portion are metal, ceramic, and polymer (Avila et al. 2003; Jose and Joseph 2012). Additionally, they are categorised as particulate, fibrous, and laminate composites depending on the reinforcement involved. Fibrous composites are further subdivided into natural biofibers and synthetic fibre. Biofiber composites can be segmented into degradable and non-biodegradable matrix types (Ilie and Hickel 2009). Green composites are made from biodegradable polymers and natural biofibers. Furthermore, they fall under the categories of textile and hybrid composites. Ones that constitute hybrids are a combination of two or more fibres.

PMCs are classified based on the polymer types, availability, fabrication process, and low cost. For biomedical applications, polymers like copolymers of polylactic acid/ polyglycolide and polyacrylic are used (Ramakrishna et al. 2001). Most commonly used reinforcement structures are the fibres (Mkaddem et al. 2008). The polymers without reinforcement showed insufficient tensile qualities like tensile strength, modulus, and impact resistance.

(2) Processing of Polymer Composites

The manufacturing processes used to create polymer composites, which have several uses, particularly in the biomedical area, include latex technology, electrospinning,

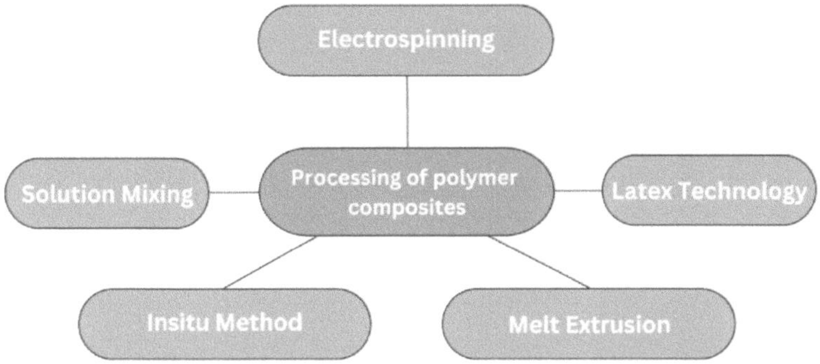

Figure 2: Classifications of polymer processing methods for various polymer blends.

solution mixing, *in situ* techniques, and melt extrusion (Al-Enizi et al. 2018; Kaur et al. 2015).

(A) Electrospinning

By using a strong electric field, fibres with diameters ranging from 2 nanometers to multiple micrometres can be produced during the electrospinning process. This method is applied to prepare nanofibers with greater surface areas. A spinneret, grounded collector, and a high-voltage power supply together form the electrospinning arrangement (Zagho and Elzatahry 2016; Zagho et al. 2018). They are of two types: vertical and horizontal sets (Figure 2). The method involves accelerating the polymeric solution into the collector with the opposed charge. The electrical repulsion forces from the direct current (DC) voltage overcome the interfacial tension of the polymeric compound, resulting in the emission of an electrified jet out of the Taylor pole's tip. Subsequently, the solvent is vaporized to produce the fibres.

(B) Melt Extrusion

It is a relatively widely used method of manufacturing polymeric composites. The twin-screw extruder is responsible for mixing the main resin with the necessary additives under a certain condition to ensure homogeneous melting. To get the final product, the polymer composite, after extrusion, is shaped using compression moulding. By varying the filler content, the thermal and mechanical characteristics of the developed composites can be adjusted. For example, ceramic powder was mixed with polylactic acid using a small compounder and its revolving screw under a nitrogen flow to create ceramic/PLA bioresorbable composites (Mathieu et al. 2006). Multiple combinations of residence time, temperature, and screw rotation speed were tested.

(C) Solution Mixing

It is the process of dissolving polymer with a suitable solvent due to its distinctive thermal character, resulting in the uniform dispersion of additives like montmorillonite clays into the resulting uniform solvent of the polymer. The continual agitation is continued based on calculated time and temperature, as well as mechanical results;

the end product develops within the mould having a particular dimension over a specific thermal profile (Zagho and Khader 2016). For example, by periodic stirring at an appropriate temperature, the PLA material was socked into chloroform to yield PLA reagent. The PLA solution then received the ceramic powder, and it was mixed thoroughly for 30 minutes. To evaporate the solvent, drying at room temperature followed by two days of additional drying over vacuum was applied.

(D) Latex Technology

This technique produces conductive polymer composites by introducing conductive additives into a network of polymers. It provides the benefit of process upscaling, simple processing, and fillers that are evenly distributed throughout the polymer network (Grossiord and Hermant 2016). In the highly viscous polymer, the nanofillers can be incorporated. With this technology, it is possible to create a 3D framework using filler particles (Jurewicz et al. 2011). The method is broken down into three steps: producing a dispersed colloidal form of the nanofiller, combining polymer latex, and curing the resultant colloidal mix. This technology may be utilised to create graphene polymeric composites and carbon nanotubes, which have the potential to be used in biomedical applications. (Anandhan and Bandyopadhyay 2016).

(E) In situ Method

Graphene and modified graphene fillers swell before it is incorporated into the matrix (Kuilla et al. 2010). Initially, in the *in situ* polymerisation process, the liquid monomer is used to swell the filler. In order to begin the polymerisation process, an initiator is added by radiation or heat. This method processes carbon nanotube-polymer composites. It is also used to process polymer composites with high nanotube content (Choi et al. 2022).

6. Application of Composites

(1) Tissue Engineering

The regeneration of damaged tissues with scaffold biomaterials, which support the growth of new cells, is the concept of tissue engineering (O'Brien 2011). Polymer composite has a morphology where cells can migrate, grow, and seed.

(2) Bone

For bone tissue engineering certain degradable polymers which are synthetic or natural, like chitosan (Li et al. 2005; Ho et al. 2005), poly (propylene fumarate) (Vehof et al. 2002), and polyesters (Lu et al., 2003), are used. Scaffolds based on PLAGA/calcium phosphate composite microspheres activate the calcium phosphate reprecipitation process in a bone/implant system (Khan et al. 2007). For repairing bone fractures, a polymer composite containing phosphate glass fibre (PGF) is used, which displays good biological and mechanical characteristics (Ahmed et al. 2008). Bone plates containing unidirectional bio-glass fibres and PLA composites that have been air plasma-treated are used to heal lengthy bone damage (Mehboob et al. 2016). Composite scaffolds made of ethyl cellulose, polylactic acid, and HA are being utilised as load-bearing substitutes (Mao et al. 2018). Polypropylene carbonate, tricalcium phosphate, poly (d)-lactic acid synthetic composites have been used for bone fracture healing (Chang et al. 2017).

(3) Blood

One of the challenging subjects in the biomedical field is the repair of blood vessels. The artificial scaffolds required increase time for patient recovery (Fujihara et al. 2005). The BNC composites blended with PVA are used as artificial blood vessels (Tang et al. 2015). BNC is a bacterial nanocellulose that has properties like excellent crystallinity, chemical quality, water retaining capacity, biocompatibility, and high humid tensile strength (Gatenholm and Klemm 2010; Putra et al. 2008). The BNC tubes are unsuitable for vascular graft models due to water loss and poor wound retention. After blending with PVA, properties like water permeability and tensile characteristics were enhanced, and the mechanical properties were controlled by the BNC tube contents.

For the application of implants, wound dressings, and surgical sutures, the composites of biological biopolymers like hyaluronic acid, chitosan, alginate, collagen, and gelatin were among the materials used (Niklason 2000; Nwe et al. 2009). By fabricating chitosan/gelatin bilayered composites, microporous blood veins were formed (Badhe et al. 2017). With the use of PLA/chitosan nanofibrous scaffolds, the *in vitro* cardiomyocyte cultures for drug screening are used as a platform for creating myocardia (Liu et al. 2017). This composite enhances cardiomyocyte growth and adhesion, as well as the formation of heart tissue, by supporting cardiomyocyte maturation along their transverse direction.

(4) Skin

The skin is the first immune system organ, and it also protects against infections and toxic antigens. It can be vulnerable to diseases, cellular death, and permanent destruction under several stressful situations, such as infection or significant burns (Jeong et al. 2017). Therefore, it is necessary to develop skin-friendly materials that decompose naturally, which is problematic for tissue engineering. The 3D nanofibrous structures developed by moulding polymeric materials have the advantages of mechanical characteristics and an increased surface area-to-volume ratio. It can serve as a carrier for various drug compounds such as antibacterials, antivirals, and anticancer medications (Zagho et al. 2018). Synthetic and natural hydrogels such as alginate, collagen, chitosan and PVA, and polyethylene glycol were used for different kinds of gels to rejuvenate the skin (Chaudhari et al. 2016). Terudermis®, Biobrane®, and Integra® are commercially available fibrous scaffolds that include collagen in combination with medicinal molecules (Chua et al. 2016). For skin regeneration and proliferation, a blend of reconstituted HA and chondroitin sulfate is used (Bhowmick et al. 2017). Alginate-based thin films were used to implant injured and burned skin (Pereira et al. 2013; Thu et al. 2012).

(5) Oral Tissues

The oral tissues are damaged by dental periodontitis and caries (Nanci 2017). These have critical functions in humans. Due to the interaction between the mouth and internal fluids and other cytotoxic material, the material inside the oral cavity gets damaged (Jones 2001). In order to address these challenges, *in vitro* oral tissue engineering has been successfully accomplished (Zafar et al. 2015). Polylactic-co-glycolic acid and polyglycolic acid graft scaffolds are primarily used to create graft scaffolds to maintain the development of full-grown tissues for teeth. Degradable polymer substrates were utilised to create 3D salivary glandular cells. (Duailibi et al. 2004). HA and PLA blends were used to fabricate jaw joints. These blends were infused with specific cells to generate osteochondral tissue with arterial bones (Schek et al. 2005).

(6) Wound Dressing

A biochemical and cellular process called wound healing provides crucial qualities to ward against microbial infections (Archana et al. 2013; Menon and Kligman 2009). Growth factor-secreting cells are important in controlling the wound-healing process (Guo and Dipietro 2010). The key event in an optimal wound healing process is fibroblast movement to the wound site to make collagen fibres for the outermost layer of cells (Bielefeld et al. 2013). The viability and growth of keratinocytes are induced by the fibroblasts (Witte and Kao 2005). The use of wound dressings allows for the accelerated healing of wounds and protection of the wound site (Seaman 2002; Dongargaonkar et al. 2013). The materials for wound dressing include a wide class of materials like foams, hydrofibers, alginate, fibres, colloids, polymeric hydrogels, films, and polymers (Boateng et al. 2008; Mi et al. 2011) (Figure 3).

By incorporating kaolin, polymers' biomechanical and thermodynamic performance can be improved (Chen et al. 2015; Pradhan et al. 2015). In order to improve the properties like drug release and haemostatic characteristics, the polymer dressings can be modified (Li et al. 2016; Loh et al. 2015). Through the integration of kaolin into a mechanically durable PU foam, multifunctional haemostatic composites of polyurethane foam with kaolin were fabricated (Avila et al. 2003). The kaolin content controls both mechanical and pharmacological release qualities. They exhibit

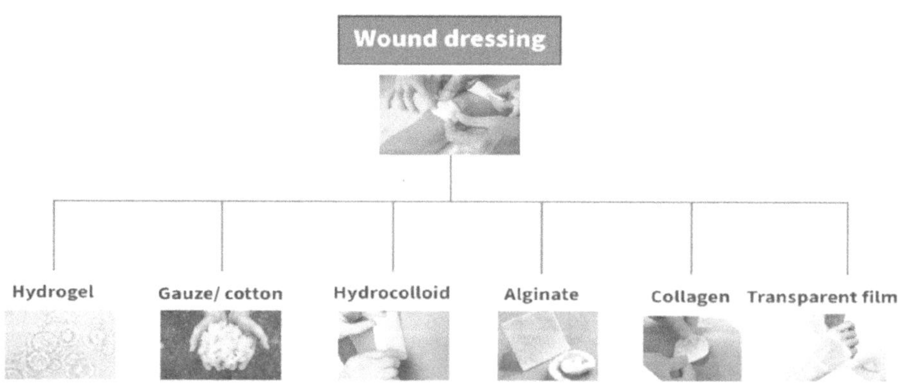

Figure 3: Different wound dressings methods.

properties like exceptional distribution of drugs, cellular compatibility, and absorbent properties. Titanium oxide particles were incorporated as a supporting material into the scaffolds to promote cell growth due to the exceptional antibacterial properties of titanium oxide (Tang et al. 2009). Chitosan/TiO_2 composite membrane is synthesised for wound healing, fibroblast cell's survival, and proliferation (Behera et al. 2017). This composite has excellent antibacterial behaviour against staphylococcus aureus.

Banana peel-based biopolymer possesses antioxidant and antibacterial effects against microorganisms such as bacteria, fungi, and yeast due to its high mineral concentration and phenolic structure (Someya et al. 2002; Mokbel and Hashinaga 2005). Nanocellulose finds applications like dressing treatment for wound care owing to this remarkable material strength, flexibility, and water-holding capabilities (Czaja et al. 2007). Its elevated surface-to-volume ratio promotes and facilitates both cell proliferation and migration since these have limitations that show poor antimicrobial behaviour against pathogenic microorganisms (Lin and Dufresne 2014; Jebali et al. 2014). In the context of *in vitro* wound dressings, a bio-nanocomposite composed of nanocellulose, chitosan, and poly(vinyl pyrrolidone) were employed (Poonguzhali et al. 2017). It exhibits outstanding cytotoxicity, compatibility with blood, and antimicrobial efficacy against both gram-positive and gram-negative bacteria. The electrospun fibres exhibit characteristics like a significant volume ratio to surface area and porosity, along with exceptional compatibility with the extracellular matrix (Lannutti et al. 2007; Meng et al. 2010; Gautam et al. 2013) Additionally, fibre composites demonstrate semipermeability, absorbent properties, and hemostatic qualities. (Zhang et al. 2005). By using this approach, a potent wound dressing was developed by electrospinning thermally responsive poly(N-isopropylacrylamide) incorporated into poly (l-lactic acid-co-ε-caprolactone) along the antibacterial ciprofloxacin (Li et al. 2017). Drug-loaded nanofibers can accelerate the *in vivo* wound healing process. As a wound-healing strategy, electrospinning PCL in conjunction with chitosan and chitosan-caffeic acid is used (Oh et al. 2016). The biocompatible and antibiotic qualities of silk fibroin nanofibers were greatly enhanced by graphene oxide, making graphene oxide/silk fibroin nanofibers having a biologically inspired nano framework an excellent option that promotes wound treatment applications (Wang et al. 2018).

7. Conclusion

Polymer blends have been instrumental in many medical breakthroughs. They can come in a wide variety of forms, and a single synthetic or natural polymer cannot meet every requirement for certain applications. In this regard, research has discovered that the synergic impact of polymers can achieve the goal. Later, blends emerged, and this chapter examined natural and hybrid natural and synthetic blends. Chitosan's multi-property ability makes it suited for a wide range of medical uses. Scientists have discovered that when used in a blend, chitosan works well with both natural and synthetic materials. Researchers are turning to collagen material to achieve mechanical strength, and the results have been promising. When utilised as a blend, synthetic and biodegradable materials such as polyvinyl alcohol, PLA, and PCL are also compatible. A combination of polymers can be utilised for certain medical applications, such as tissue engineering and drug delivery.

The composite materials are promising materials for biomedical applications with superior characteristics and diverse applications. They have the benefit of being affordable as well as available in both synthetic and natural forms. Composites are useful for wound healing, tissue engineering, and dentistry. The natural and synthetic materials that are available for composites are likewise divided into two categories. Depending on the requirements, new composite materials might be added or replaced. Current research studies for new composite advancements are heavily relying on degradable polymers, both natural and synthetic, such as chitosan, PVA, and PLA. Several other applications, including tissue engineering, skin, blood, dental tissue, bone, wound dressing, and others, already incorporate such materials. This chapter discusses several polymer composite processing processes and their important applications.

References

Abdolmaleki, M., T. Tavakoli, O. M. Jazani et al. 2016. Blend membranes based on polyurethane and polyethylene glycol: exploring the impact of molecular weight and concentration of the second phase on gas permeation enhancement. Journal of Polymer Engineering 36: 513–519. https://doi.org/doi: 10.1515/polyeng-2015-0254.

Ahmed, I., A. J. Parsons, G. Palmer et al. 2008. Weight loss, ion release and initial mechanical properties of a binary calcium phosphate glass fibre/PCL composite. Acta Biomaterialia 4: 1307–1314. https://doi.org/10.1016/j.actbio.2008.03.018.

Al Habis, N., A. El Moumen, M. Tarfaoui et al. 2020. Mechanical properties of carbon black/poly (ε-caprolactone)-based tissue scaffolds. Arabian Journal of Chemistry 13: 3210–3217. https://doi.org/10.1016/j.arabjc.2018.10.005.

Al-Enizi, A. M., M. M. Zagho, A. A. Elzatahry. 2018. Polymer-based electrospun nanofibers for biomedical applications. Nanomaterials (Basel) 8: 259. https://doi.org/10.3390/nano8040259.

Anandhan, S. and S. Bandyopadhyay. 2016. Advances in Polymer Materials and Technology.

Archana, D., J. Dutta and P. K. Dutta. 2013. Evaluation of chitosan nano dressing for wound healing: characterization, *in vitro* and *in vivo* studies. Int. J. Biol. Macromol. 57: 193–203. https://doi.org/10.1016/j.ijbiomac.2013.03.002.

Asadpour, S., H. Yeganeh, J. Ai et al. 2018. Polyurethane-polycaprolactone blend patches: scaffold characterization and cardiomyoblast adhesion, proliferation, and function. ACS Biomater. Sci. Eng. 4: 4299–4310. https://doi.org/10.1021/acsbiomaterials.8b00848.

Avila, A. F., P. C. M. Rodrigues, D. B. Santos et al. 2003. A dual analysis for recycled particulate composites: linking micro- and macro-mechanics. Materials Characterization 50: 281–291. https://doi.org/10.1016/S1044-5803(03)00124-4.

Ayres, C., G. L. Bowlin, S. C. Henderson et al. 2006. Modulation of anisotropy in electrospun tissue-engineering scaffolds: Analysis of fiber alignment by the fast Fourier transform. Biomaterials 27: 5524–5534. https://doi.org/10.1016/j.biomaterials.2006.06.014.

Badhe, R. V., D. Bijukumar, D. R. Chejara et al. 2017. A composite chitosan-gelatin bi-layered, biomimetic macroporous scaffold for blood vessel tissue engineering. Carbohydrate Polymers 157: 1215–1225. https://doi.org/10.1016/j.carbpol.2016.09.095.

Bandyopadhyay, A. 2018. Additive manufacturing of biomaterials. Prog. Mater Sci. 93: 45–111. https://doi.org/10.1016/j.pmatsci.2017.08.003.

Barbucci, R. 2002. Integrated biomaterials science. Kluwer Academic/Plenum, New York.

Behera, S., U. Das, A. Kumar et al. 2017. Chitosan/TiO2 composite membrane improves proliferation and survival of L929 fibroblast cells: Application in wound dressing and skin regeneration. International Journal of Biological Macromolecules. https://doi.org/10.1016/j.ijbiomac.2017.02.017.

Belluzo, M. S., L. F. Medina, A. M. Cortizo et al. 2016. Ultrasonic compatibilization of polyelectrolyte complex based on polysaccharides for biomedical applications. Ultrason Sonochem. 30: 1–8. https://doi.org/10.1016/j.ultsonch.2015.11.022.

Bera, A., A. K. Singh Chandel, C. Uday Kumar et al. 2015. Degradable/cytocompatible and pH responsive amphiphilic conetwork gels based on agarose-graft copolymers and polycaprolactone. J. Mater. Chem. B 3: 8548–8557. https://doi.org/10.1039/C5TB01251A.

Bhat, S., A. Tripathi and A. Kumar. 2011. Supermacroprous chitosan-agarose-gelatin cryogels: in vitro characterization and in vivo assessment for cartilage tissue engineering. J. R. Soc. Interface 8: 540–554. https://doi.org/10.1098/rsif.2010.0455.

Bhowmick, S., S. Rother, H. Zimmermann et al. 2017. Biomimetic electrospun scaffolds from main extracellular matrix components for skin tissue engineering application - The role of chondroitin sulfate and sulfated hyaluronan. Mater Sci. Eng. C Mater Biol. Appl. 79: 15–22. https://doi.org/10.1016/j.msec.2017.05.005.

Bie, P., P. Liu, L. Yu et al. 2013. The properties of antimicrobial films derived from poly(lactic acid)/starch/chitosan blended matrix. Carbohydr Polym. 98: 959–966. https://doi.org/10.1016/j.carbpol.2013.07.004.

Bielefeld, K. A., S. Amini-Nik and B.A. Alman. 2013. Cutaneous wound healing: recruiting developmental pathways for regeneration. Cell Mol. Life Sci. 70: 2059–2081. https://doi.org/10.1007/s00018-012-1152-9.

Boateng, J. S., K. H. Matthews, H. N. E. Stevens et al. 2008. Wound healing dressings and drug delivery systems: a review. J. Pharm. Sci. 97: 2892–2923. https://doi.org/10.1002/jps.21210.

Caló, E. 2015. Biomedical applications of hydrogels: A review of patents and commercial products. European Polymer Journal 65: 252–267. https://doi.org/10.1016/j.eurpolymj.2014.11.024.

Carrasco-Guigón, F. J., D. E. Rodríguez-Félix, M. M. Castillo-Ortega et al. 2017. Preparation and characterization of extruded composites based on polypropylene and chitosan compatibilized with polypropylene-graft-maleic anhydride. Materials (Basel) 10: 105. https://doi.org/10.3390/ma10020105.

Carson, D., Y. Jiang and K. A. Woodrow. 2016. Tunable release of multiclass anti-HIV drugs that are water-soluble and loaded at high drug content in polyester blended electrospun fibers. Pharm. Res. 33: 125–136. https://doi.org/10.1007/s11095-015-1769-0.

Chandel, A., A. Bera, B. Nutan et al. 2016. Reactive compatibilizer mediated precise synthesis and application of stimuli responsive polysaccharides-polycaprolactone amphiphilic co-network gels. Polymer 99. https://doi.org/10.1016/j.polymer.2016.07.033.

Chang, G. W., C.-L. Tseng, Y.-S. Tzeng et al. 2017. An *in vivo* evaluation of a novel malleable composite scaffold (polypropylene carbonate/poly (D-lactic acid)/tricalcium phosphate elastic composites) for bone defect repair. Journal of the Taiwan Institute of Chemical Engineers 80. https://doi.org/10.1016/j.jtice.2017.06.022.

Chaudhari, A. A., K. Vig, D. R. Baganizi et al. 2016. Future Prospects for Scaffolding Methods and Biomaterials in Skin Tissue Engineering: A Review. Int. J. Mol. Sci. 17: 1974. https://doi.org/10.3390/ijms17121974.

Chen, Q. and G. A. Thouas. 2015. Metallic implant biomaterials. Materials Science and Engineering: R: Reports 87: 1–57. https://doi.org/10.1016/j.mser.2014.10.001.

Chen, X., X. Sang and Q. Zhang. 2015. Preparation and characterization of polyurethane-imide/kaolinite nanocomposite foams. RSC Adv. 5: 53211–53219. https://doi.org/10.1039/C5RA06438D.

Choi, S.-M., E.-J. Shin, S.-M. Zo et al. 2022. Revised manuscript with corrections: polyurethane-based conductive composites: from synthesis to applications. Int. J. Mol. Sci. 23: 1938. https://doi.org/10.3390/ijms23041938.

Chua, A. W. C., Y. C. Khoo, B. K. Tan et al. 2016. Skin tissue engineering advances in severe burns: review and therapeutic applications. Burns Trauma 4, 3. https://doi.org/10.1186/s41038-016-0027-y.

Chuangchote, S. and P. Supaphol. 2006. Fabrication of Aligned Poly(vinyl alcohol) Nanofibers by Electrospinning. Journal of Nanoscience and Nanotechnology 6: 125–9. https://doi.org/10.1166/jnn.2006.17916.

Cole, G., J. Hogan and M. Aulton. 1995. Pharmaceutical Coating Technology. https://doi.org/10.3109/9780203014356.

Czaja, W. K., D. J. Young, M. Kawecki et al. 2007. The future prospects of microbial cellulose in biomedical applications. Biomacromolecules 8: 1–12. https://doi.org/10.1021/bm060620d.

Dhandayuthapani, B., Y. Yoshida, T. Maekawa et al. 2011. Polymeric scaffolds in tissue engineering application: A review. International Journal of Polymer Science 2011, 290602. https://doi.org/10.1155/2011/290602.

Ding, L., Y. Huang, X. Cai et al. 2019. Impact of pH, ionic strength and chitosan charge density on chitosan/casein complexation and phase behavior. Carbohydrate Polymers 208: 133–141. https://doi.org/10.1016/j.carbpol.2018.12.015.

Dong, C. and Y. Lv. 2016. Application of collagen scaffold in tissue engineering: recent advances and new perspectives. Polymers (Basel) 8: 42. https://doi.org/10.3390/polym8020042.

Dongargaonkar, A. A., G. L. Bowlin and H. Yang. 2013. Electrospun blends of gelatin and gelatin-dendrimer conjugates as a wound-dressing and drug-delivery platform. Biomacromolecules 14: 4038–4045. https://doi.org/10.1021/bm401143p.

Duailibi, M. T., S. E. Duailibi, C. S. Young et al. 2004. Bioengineered teeth from cultured rat tooth bud cells. J. Dent. Res. 83: 523–528. https://doi.org/10.1177/154405910408300703.

El-Moghazy, A. Y., E. A. Soliman, H. Z. Ibrahim et al. 2016. Biosensor based on electrospun blended chitosan-poly (vinyl alcohol) nanofibrous enzymatically sensitized membranes for pirimiphos-methyl detection in olive oil. Talanta 155: 258–264. https://doi.org/10.1016/j.talanta.2016.04.018.

Ferrarezi, M. M. F., M. De Oliveira Taipina, L. C. Escobar Da Silva et al. 2013. Poly(Ethylene Glycol) as a Compatibilizer for Poly(Lactic Acid)/Thermoplastic Starch Blends. J. Polym. Environ. 21: 151–159. https://doi.org/10.1007/s10924-012-0480-z.

Fujihara, K., W.-E. Teo, T.-C. Lim et al. 2005. An Introduction to Electrospinning and Nanofibers, World Sci. https://doi.org/10.1142/5894.

Gatenholm, P. and D. Klemm. 2010. Bacterial nanocellulose as a renewable material for biomedical applications. MRS Bulletin. 35: 208–213. https://doi.org/10.1557/mrs2010.653.

Gautam, S., A. K. Dinda and N. C. Mishra. 2013. Fabrication and characterization of PCL/gelatin composite nanofibrous scaffold for tissue engineering applications by electrospinning method. Mater Sci. Eng. C Mater. Biol. Appl. 33: 1228–1235. https://doi.org/10.1016/j.msec.2012.12.015.

Ghasemiyeh, P. and S. Mohammadi-Samani. 2021. Polymers blending as release modulating tool in drug delivery. Frontiers in Materials 8. https://doi.org/10.3389/fmats.2021.752813.

Grainger, D. W. 1999. The Williams dictionary of biomaterials. Materials Today 2: 29. https://doi.org/10.1016/S1369-7021(99)80066-2.

Grossen, P., D. Witzigmann, S. Sieber et al. 2017. PEG-PCL-based nanomedicines: A biodegradable drug delivery system and its application. J. Control Release 260: 46–60. https://doi.org/10.1016/j.jconrel.2017.05.028.

Grossiord, N. and M.-C. Hermant. 2016. Latex Technology for Conductive Polymer Nanocomposites. https://doi.org/10.1002/0471440264.pst506.pub2.

Guitian Oliveira, N., T. Sirgado, L. Reis et al. 2014. *In vitro* assessment of three dimensional dense chitosan-based structures to be used as bioabsorbable implants. J. Mech. Behav. Biomed. Mater 40: 413–425. https://doi.org/10.1016/j.jmbbm.2014.09.014.

Guo, S., Dipietro, L.A., 2010. Factors affecting wound healing. J. Dent. Res. 89: 219–229. https://doi.org/10.1177/0022034509359125.

Harcum, W. W., P. E. Barnum, G. W. Skinner et al. 1998. Pharmaceutical applications of naturally occurring water-soluble polymers. Pharmaceutical Science & Technology Today. https://doi.org/10.1016/S1461-5347(98)00072-8.

Hejna, A., K. Formela and M. Saeb. 2015. Processing, mechanical and thermal behavior assessments of polycaprolactone/agricultural wastes biocomposites. Industrial Crops and Products 76: 725–733. https://doi.org/10.1016/j.indcrop.2015.07.049.

Hench, L. L. 1998. Biomaterials: a forecast for the future. Biomaterials 19: 1419–1423. https://doi.org/10.1016/s0142-9612(98)00133-1.

Henry, J. A., M. Simonet, A. Pandit et al. 2007. Characterization of a slowly degrading biodegradable polyester-urethane for tissue engineering scaffolds. J. Biomed. Mater. Res. A 82: 669–679. https://doi.org/10.1002/jbm.a.31094.

Ho, M.-H., D.-M. Wang, H.-J. Hsieh et al. 2005. Preparation and characterization of RGD-immobilized chitosan scaffolds. Biomaterials 26: 3197–3206. https://doi.org/10.1016/j.biomaterials.2004.08.032.

Ilie, N. and R. Hickel. 2009. Macro-, micro- and nano-mechanical investigations on silorane and methacrylate-based composites. Dent. Mater 25: 810–819. https://doi.org/10.1016/j.dental.2009.02.005.

Jebali, A., S. Hekmatimoghaddam, A. Behzadi et al. 2014. Antimicrobial activity of nanocellulose conjugated with allicin and lysozyme. Cellulose. https://doi.org/10.1007/s10570-013-0084-3.

Jeong, K.-H., D. Park and Y.-C. Lee. 2017. Polymer-based hydrogel scaffolds for skin tissue engineering applications: a mini-review. J. Polym. Res. 24:112. https://doi.org/10.1007/s10965-017-1278-4.

Ji, Y., K. Ghosh, X. Z. Shu et al. 2006. Electrospun three-dimensional hyaluronic acid nanofibrous scaffolds. Biomaterials 27: 3782–3792. https://doi.org/10.1016/j.biomaterials.2006.02.037.

Jones, F. H. 2001. Teeth and bones: applications of surface science to dental materials and related biomaterials. Surface Science Reports 42: 75–205. https://doi.org/10.1016/S0167-5729(00)00011-X.

Jose, J. and K. Joseph. 2012. Advances in Polymer Composites: Macro- and Microcomposites - State of the Art, New Challenges, and Opportunities. pp. 1–16. https://doi.org/10.1002/9783527645213.ch1.

Jurewicz, I., P. Worajittiphon, A. A. K. King et al. 2011. Locking carbon nanotubes in confined lattice geometries--a route to low percolation in conducting composites. J. Phys. Chem. B 115: 6395–6400. https://doi.org/10.1021/jp111998p.

Karabiyik Acar, O., A. B. Kayitmazer and G. Torun Kose. 2018. Hyaluronic Acid/Chitosan Coacervate-Based Scaffolds. Biomacromolecules 19: 1198–1211. https://doi.org/10.1021/acs.biomac.8b00047.

Kaur, G., R. Adhikari, P. Cass et al. 2015. Electrically conductive polymers and composites for biomedical applications. RSC Adv. 5: 37553–37567. https://doi.org/10.1039/C5RA01851J.

Kenar, H., C. Y. Ozdogan, C. Dumlu et al. 2019. Microfibrous scaffolds from poly(l-lactide-co-ε-caprolactone) blended with xeno-free collagen/hyaluronic acid for improvement of vascularization in tissue engineering applications. Mater Sci. Eng. C Mater. Biol. Appl. 97: 31–44. https://doi.org/10.1016/j.msec.2018.12.011.

Kenawy, E. R., J. M. Layman, J. R. Watkins et al. 2003. Electrospinning of poly(ethylene-co-vinyl alcohol) fibers. Biomaterials 24: 907–913. https://doi.org/10.1016/s0142-9612(02)00422-2.

Keun Kwon, I., S. Kidoaki and T. Matsuda. 2005. Electrospun nano- to microfiber fabrics made of biodegradable copolyesters: structural characteristics, mechanical properties and cell adhesion potential. Biomaterials 26: 3929–3939. https://doi.org/10.1016/j.biomaterials.2004.10.007.

Khaled, S. A., J. C. Burley, M. R. Alexander et al. 2014. Desktop 3D printing of controlled release pharmaceutical bilayer tablets. Int. J. Pharm. 461: 105–111. https://doi.org/10.1016/j.ijpharm.2013.11.021.

Khan, Y., E. Cushnie, J. Kelleher et al. 2007. *In situ* synthesized ceramic-polymer composites for bone tissue engineering: bioactivity and degradation studies. Journal of Materials Science 42: 4183–4190. https://doi.org/10.1007/s10853-006-0636-0.

Kuilla, T., S. Bhadra, D. Yao et al. 2010. Recent advances in graphene based polymer composites. Progress in Polymer Science 35: 1350–1375. https://doi.org/10.1016/j.progpolymsci.2010.07.005.

Lam, M. T. and J. C. Wu. 2012. Biomaterial applications in cardiovascular tissue repair and regeneration. Expert Rev. Cardiovasc. Ther. 10: 1039–1049. https://doi.org/10.1586/erc.12.99.

Langer, R. and D. A. Tirrell. 2004. Designing materials for biology and medicine. Nature 428: 487–492. https://doi.org/10.1038/nature02388.

Lannutti, J., D. Reneker, T. Ma et al. 2007. Electrospinning for tissue engineering scaffolds. Materials Science and Engineering: C 27: 504–509. https://doi.org/10.1016/j.msec.2006.05.019.

Li, H., G. R. Williams, J. Wu et al. 2017. Poly(N-isopropylacrylamide)/poly(l-lactic acid-co-ε-caprolactone) fibers loaded with ciprofloxacin as wound dressing materials. Mater. Sci. Eng. C Mater. Biol. Appl. 79: 245–254. https://doi.org/10.1016/j.msec.2017.04.058

Li, Z., H. R. Ramay, K. D. Hauch et al. 2005. Chitosan-alginate hybrid scaffolds for bone tissue engineering. Biomaterials 26: 3919–3928. https://doi.org/10.1016/j.biomaterials.2004.09.062.

Li, Z., E. Ye, David, null, Lakshminarayanan et al. 2016. Recent advances of using hybrid nanocarriers in remotely controlled therapeutic delivery. Small 12: 4782–4806. https://doi.org/10.1002/smll.201601129.

Lin, N. and A. Dufresne. 2014. Nanocellulose in biomedicine: Current status and future prospect. European Polymer Journal 59: 302–325. https://doi.org/10.1016/j.eurpolymj.2014.07.025.

Liu, Y., H. Cui, X. Zhuang et al. 2014. Electrospinning of aniline pentamer-graft-gelatin/PLLA nanofibers for bone tissue engineering. Acta Biomater. 10: 5074–5080. https://doi.org/10.1016/j.actbio.2014.08.036.

Liu, Y., S. Wang and R. Zhang. 2017. Composite poly(lactic acid)/chitosan nanofibrous scaffolds for cardiac tissue engineering. Int. J. Biol. Macromol. 103: 1130–1137. https://doi.org/10.1016/j.ijbiomac.2017.05.101.

Loh, X. J., T.-C. Lee, Q. Dou et al. 2015. Utilising inorganic nanocarriers for gene delivery. Biomater. Sci. 4: 70–86. https://doi.org/10.1039/C5BM00277J.

López-Ângulo, D. and P. Sobral. 2016. The effect of processing parameters and solid concentration on the microstructure and pore architecture of gelatin-chitosan scaffolds produced by freeze-drying. Materials Research 19. https://doi.org/10.1590/1980-5373-MR-2015-0793.

Lu, H. H., M. D. Kofron, S. F. El-Amin et al. 2003. *In vitro* bone formation using muscle-derived cells: a new paradigm for bone tissue engineering using polymer-bone morphogenetic protein matrices. Biochem. Biophys. Res. Commun. 305: 882–889. https://doi.org/10.1016/s0006-291x(03)00858-1.

Ma, L., C. Gao, Z. Mao et al. 2003. Collagen/chitosan porous scaffolds with improved biostability for skin tissue engineering. Biomaterials 24: 4833–4841. https://doi.org/10.1016/s0142-9612(03)00374-0.

Mao, D., Q. Li, N. Bai et al. 2018. Porous stable poly (lactic acid)/ethyl cellulose/hydroxyapatite composite scaffolds prepared by a combined method for bone regeneration. Carbohydr. Polym. 180: 104–111. https://doi.org/10.1016/j.carbpol.2017.10.031.

Mathieu, L. M., P.-E. Bourban and J.-A. E. Månson. 2006. Processing of homogeneous ceramic/polymer blends for bioresorbable composites. Composites Science and Technology 66: 1606–1614. https://doi.org/10.1016/j.compscitech.2005.11.012.

Matsuda, A., G. Kagata, R. Kino et al. 2007. Preparation of chitosan nanofiber tube by electrospinning. J. Nanosci. Nanotechnol. 7: 852–855. https://doi.org/10.1166/jnn.2007.520.

Matthews, J., G. Wnek, D. Simpson et al. 2002. Electrospinning of collagen nanofibers. Biomacromolecules 3: 232–8. https://doi.org/10.1021/bm015533u.

McManus, M. C., E. D. Boland, D. G. Simpson et al. 2007. Electrospun fibrinogen: Feasibility as a tissue engineering scaffold in a rat cell culture model. Journal of Biomedical Materials Research Part A 81A: 299–309. https://doi.org/10.1002/jbm.a.30989.

Mehboob, H., J.-H. Bae, M.-G. Han et al. 2016. Effect of air plasma treatment on mechanical properties of bioactive composites for medical application: Composite preparation and characterization. Composite Structures 143: 23–32. https://doi.org/10.1016/j.compstruct.2016.02.012.

Meng, Z. X., Y. S. Wang, C. Ma et al. 2010. Electrospinning of PLGA/gelatin randomly-oriented and aligned nanofibers as potential scaffold in tissue engineering. Materials Science and Engineering: C 30: 1204–1210. https://doi.org/10.1016/j.msec.2010.06.018.

Menon, G. K. and A. M. Kligman. 2009. Barrier functions of human skin: a holistic view. Skin Pharmacol Physiol 22: 178–189. https://doi.org/10.1159/000231523.

Mi, L., H. Xue, Y. Li et al. 2011. A thermoresponsive antimicrobial wound dressing hydrogel based on a cationic betaine ester. Advanced Functional Materials 21. https://doi.org/10.1002/adfm.201100871.

Migliorini, F. L., R. C. Sanfelice, A. Pavinatto et al. 2017. Voltammetric cadmium(II) sensor based on a fluorine doped tin oxide electrode modified with polyamide 6/chitosan electrospun nanofibers and gold nanoparticles. Microchim. Acta 184: 1077–1084. https://doi.org/10.1007/s00604-017-2082-x.

Miguel, S. P., M. P. Ribeiro, H. Brancal et al. 2014. Thermoresponsive chitosan-agarose hydrogel for skin regeneration. Carbohydr. Polym. 111: 366–373. https://doi.org/10.1016/j.carbpol.2014.04.093.

Mkaddem, A., I. Demirci and M. E. Mansori. 2008. A micro–macro combined approach using FEM for modelling of machining of FRP composites: Cutting forces analysis. Composites Science and Technology 68: 3123.

Mo, X., H.-J. Weber and S. Ramakrishna. 2006. PCL-PGLA Composite Tubular Scaffold Preparation and Biocompatibility Investigation. Int. J. Artif. Organs 29: 790–799. https://doi.org/10.1177/039139880602900809.

Mokbel, M. S. and F. Hashinaga. 2005. Antibacterial and antioxidant activities of banana (Musa, AAA cv. Cavendish) Fruits Peel. American Journal of Biochemistry and Biotechnology 1: 125–131. https://doi.org/10.3844/ajbbsp.2005.125.131.

Nada, A. 2014. Natural and Synthetic Biomedical Polymers. https://doi.org/10.1016/C2011-0-07330-1.

Nair, L. S. and C. T. Laurencin. 2007. Biodegradable polymers as biomaterials. Progress in Polymer Science 32: 762–798. https://doi.org/10.1016/j.progpolymsci.2007.05.017.

Nanci, A. 2017. Ten Cate's Oral Histology-e-book: development, structure, and function. Elsevier Health Sciences.

Niklason, L. E. 2000. Engineering of bone grafts. Nat. Biotechnol. 18: 929–930. https://doi.org/10.1038/79394.

Noh, H. K., S. W. Lee, J.-M. Kim et al. 2006. Electrospinning of chitin nanofibers: degradation behavior and cellular response to normal human keratinocytes and fibroblasts. Biomaterials 27: 3934–3944. https://doi.org/10.1016/j.biomaterials.2006.03.016.

Nwe, N., T. Furuike and H. Tamura. 2009. The mechanical and biological properties of chitosan scaffolds for tissue regeneration templates are significantly enhanced by chitosan from *Gongronella butleri*. Materials (Basel) 2: 374–398. https://doi.org/10.3390/ma2020374.

O'Brien, F. 2011. Biomaterials & scaffolds for tissue engineering. Materials Today 14: 88–95. https://doi.org/10.1016/S1369-7021(11)70058-X.

Oh, G.-W., S.-C. Ko, J.-Y. Je et al. 2016. Fabrication, characterization and determination of biological activities of poly(ε-caprolactone)/chitosan-caffeic acid composite fibrous mat for wound dressing application. Int. J. Biol. Macromol. 93: 1549–1558. https://doi.org/10.1016/j.ijbiomac.2016.06.065.

Pachuau, L. 2015. Recent developments in novel drug delivery systems for wound healing. Expert Opinion on Drug Delivery 12: 1895–1909. https://doi.org/10.1517/17425247.2015.1070143.

Park, K. E., H. K. Kang, S. J. Lee et al. 2006. Biomimetic nanofibrous scaffolds: preparation and characterization of PGA/chitin blend nanofibers. Biomacromolecules 7: 635–643. https://doi.org/10.1021/bm0509265.

Park, S.-N., H.-J. Lee, K.-H. Lee et al. 2003. Biological characterization of EDC-crosslinked collagen-hyaluronic acid matrix in dermal tissue restoration. Biomaterials 24: 1631–1641. https://doi.org/10.1016/s0142-9612(02)00550-1.

Pereira, R., A. Carvalho, D. C. Vaz et al. 2013. Development of novel alginate based hydrogel films for wound healing applications. Int. J. Biol. Macromol. 52: 221–230. https://doi.org/10.1016/j.ijbiomac.2012.09.031.

Poonguzhali, R., S. K. Basha and V. S. Kumari. 2017. Synthesis and characterization of chitosan-PVP-nanocellulose composites for *in-vitro* wound dressing application. Int. J. Biol. Macromol. 105: 111–120. https://doi.org/10.1016/j.ijbiomac.2017.07.006.

Pradhan, A. K., P. K. Rana and P. K. Sahoo. 2015. Biodegradability and Swelling capacity of Kaolin based Chitosan-g-PHEMA Nanocomposite hydrogel. International Journal of Biological Macromolecules 74: 620.

Prasad, K., O. Bazaka, M. Chua et al. 2017. Metallic Biomaterials: Current Challenges and Opportunities. Materials (Basel) 10. https://doi.org/10.3390/ma10080884.

Putra, A., A. Kakugo, H. Furukawa et al. 2008. Tubular bacterial cellulose gel with oriented fibrils on the curved surface. Polymer. 49: 1885.

Qasim, S. B., M. S. Zafar, S. Najeeb et al. 2018. Electrospinning of Chitosan-Based Solutions for Tissue Engineering and Regenerative Medicine. Int. J. Mol. Sci. 19. https://doi.org/10.3390/ijms19020407.

Quiroz-Castillo, J. M., D. E. Rodríguez-Félix, H. Grijalva-Monteverde et al. 2014. Preparation of extruded polyethylene/chitosan blends compatibilized with polyethylene-graft-maleic anhydride. Carbohydr. Polym. 101: 1094–1100. https://doi.org/10.1016/j.carbpol.2013.10.052.

Ramakrishna, S., J. Mayer, E. Wintermantel et al. 2001. Biomedical applications of polymer-composite materials: a review. Composites Science and Technology 61: 1189–1224. https://doi.org/10.1016/S0266-3538(00)00241-4.

Ranga Rao, K. V., K. P. Devi and P. Buri. 1988. Cellulose Matrices for Zero-Order Release of Soluble Drugs. Drug Development and Industrial Pharmacy 14: 2299–2320. https://doi.org/10.3109/03639048809152017.

Salati, M. A., J. Khazai, A. M. Tahmuri et al. 2020. Agarose-based biomaterials: opportunities and challenges in cartilage tissue engineering. Polymers (Basel) 12. https://doi.org/10.3390/polym12051150.

Salati, M. A., J. Khazai, A. M. Tahmuri et al. 2020. Agarose-Based Biomaterials: Opportunities and Challenges in Cartilage Tissue Engineering. Polymers (Basel) 12. https://doi.org/10.3390/polym12051150.

Satarkar, N. S. and J. Zach Hilt. 2008. Hydrogel nanocomposites as remote-controlled biomaterials. Acta Biomater. 4: 11–16. https://doi.org/10.1016/j.actbio.2007.07.009.

Schek, R., J. Taboas, S. Hollister et al. 2005. Tissue engineering osteochondral implants for temporomandibular joint repair. Orthodontics & Craniofacial Research 8: 313–319. https://doi.org/10.1111/j.1601-6343.2005.00354.x.

Seaman, S. 2002. Dressing selection in chronic wound management. J. Am. Podiatr. Med Assoc. 92: 24–33. https://doi.org/10.7547/87507315-92-1-24.

Segura, T., P. H. Chung and L. D. Shea. 2005. DNA delivery from hyaluronic acid-collagen hydrogels via a substrate-mediated approach. Biomaterials 26: 1575–1584. https://doi.org/10.1016/j.biomaterials.2004.05.007.

Sehgal, A. and A. Ramsden. 2008. Treating hypotension in the preterm infant: when and with what: a critical and systematic review. J. Perinatol. 28: 167; author reply 168. https://doi.org/10.1038/sj.jp.7211883.

Sell, S. A., P. S. Wolfe, K. Garg et al. 2010. The use of natural polymers in tissue engineering: a focus on electrospun extracellular matrix analogues. Polymers 2: 522–553. https://doi.org/10.3390/polym2040522.

Sharma, C., A. K. Dinda, P. D. Potdar et al. 2016. Fabrication and characterization of novel nano-biocomposite scaffold of chitosan-gelatin-alginate-hydroxyapatite for bone tissue engineering. Mater Sci. Eng. C Mater. Biol. Appl. 64: 416–427. https://doi.org/10.1016/j.msec.2016.03.060.

Sill, T. J. and H. A. von Recum. 2008. Electrospinning: applications in drug delivery and tissue engineering. Biomaterials 29: 1989–2006. https://doi.org/10.1016/j.biomaterials.2008.01.011.

Sionkowska, A. 2011. Current research on the blends of natural and synthetic polymers as new biomaterials: Review. Progress in Polymer Science 36: 1254–1276. https://doi.org/10.1016/j.progpolymsci.2011.05.003.

Sionkowska, A., M. Gadomska, K. Musiał et al. 2020. Hyaluronic acid as a component of natural polymer blends for biomedical applications: a review. Molecules 25(18). https://doi.org/10.3390/molecules25184035.

Someya, S., Y. Yoshiki and K. Okubo. 2002. Antioxidant compounds from bananas (Musa Cavendish). Food Chemistry 79: 351–354. https://doi.org/10.1016/S0308-8146(02)00186-3.

Song, J.-H., H.-E. Kim and H.-W. Kim. 2008. Production of electrospun gelatin nanofiber by water-based co-solvent approach. J. Mater Sci. Mater. Med. 19: 95–102. https://doi.org/10.1007/s10856-007-3169-4.

Staniforth, J. N. and A. R. Baichwal. 2005. TIMERx®: novel polysaccharide composites for controlled/programmed release of drugs in the gastrointestinal tract. Expert Opinion on Drug Delivery 2: 587–595. https://doi.org/10.1517/17425247.2.3.587.

Stitzel, J., J. Liu, S. J. Lee et al. 2006. Controlled fabrication of a biological vascular substitute. Biomaterials 27: 1088–1094. https://doi.org/10.1016/j.biomaterials.2005.07.048.

Taguchi, T., T. Ikoma and J. Tanaka. 2002. An improved method to prepare hyaluronic acid and type II collagen composite matrices. Journal of Biomedical Materials Research 61: 330–336. https://doi.org/10.1002/jbm.10147.

Taguchi, T. and J. Tanaka. 2002. Swelling behavior of hyaluronic acid and type II collagen hydrogels prepared by using conventional crosslinking and subsequent additional polymer interactions. J. Biomater. Sci. Polym. Ed. 13: 43–52. https://doi.org/10.1163/156856202753525927.

Tang, C., N. Chen, Q. Zhang et al. 2009. Preparation and properties of chitosan nanocomposites with nanofillers of different dimensions. Polymer Degradation and Stability 94: 124–131. https://doi.org/10.1016/j.polymdegradstab.2008.09.008.

Tang, J., L. Bao, X. Li et al. 2015. Potential of PVA-doped bacterial nano-cellulose tubular composites for artificial blood vessels. J. Mater. Chem. B 3: 8537–8547. https://doi.org/10.1039/C5TB01144B.

Taravel, M. N. and A. Domard. 1995. Collagen and its interaction with chitosan: II. Influence of the physicochemical characteristics of collagen. Biomaterials 16: 865–871. https://doi.org/10.1016/0142-9612(95)94149-F.

Thomas, S., K. Joseph, S. K. Malhotra et al. 2012. Polymer Composites, Macro- and Microcomposites, Polymer Composites. Wiley.

Tipduangta, P., P. Belton, W. J. McAuley et al. 2021. The use of polymer blends to improve stability and performance of electrospun solid dispersions: The role of miscibility and phase separation. Int. J. Pharm. 602:120637. https://doi.org/10.1016/j.ijpharm.2021.120637.

Vázquez, J. J. and E. S. M. Martínez. 2019. Collagen and elastin scaffold by electrospinning for skin tissue engineering applications. Journal of Materials Research 34: 2819–2827. https://doi.org/10.1557/jmr.2019.233.

Vehof, J. W. M., J. P. Fisher, D. Dean et al. 2002. Bone formation in transforming growth factor beta-1-coated porous poly(propylene fumarate) scaffolds. J. Biomed. Mater. Res. 60: 241–251. https://doi.org/10.1002/jbm.10073.

Wang, S.-D., Q. Ma, K. Wang et al. 2018. Improving antibacterial activity and biocompatibility of bioinspired electrospinning silk fibroin nanofibers modified by graphene oxide. ACS Omega 3: 406–413. https://doi.org/10.1021/acsomega.7b01210.

Wang, W. and K. W. K. Yeung. 2017. Bone grafts and biomaterials substitutes for bone defect repair: A review. Bioactive Materials 2: 224–247. https://doi.org/10.1016/j.bioactmat.2017.05.007.

Williams, D. F. 2019. Challenges with the development of biomaterials for sustainable tissue engineering. Front Bioeng. Biotechnol. 7: 127. https://doi.org/10.3389/fbioe.2019.00127.

Witte, R. P. and W. J. Kao. 2005. Keratinocyte-fibroblast paracrine interaction: the effects of substrate and culture condition. Biomaterials 26: 3673–3682. https://doi.org/10.1016/j.biomaterials.2004.09.054.

Yang, C., R. Xue, Q. Zhang et al. 2017. Nanoclay cross-linked semi-IPN silk sericin/poly(NIPAm/LMSH) nanocomposite hydrogel: An outstanding antibacterial wound dressing. Mater. Sci. Eng. C Mater. Biol. Appl. 81: 303–313. https://doi.org/10.1016/j.msec.2017.08.008.

Yang, J., G. L. Koons, G. Cheng et al. 2018. A review on the exploitation of biodegradable magnesium-based composites for medical applications. Biomed. Mater 13: 022001. https://doi.org/10.1088/1748-605X/aa8fa0.

Yang, K.-C., C.-C. Wu, F.-H. Lin et al. 2008. Chitosan/gelatin hydrogel as immunoisolative matrix for injectable bioartificial pancreas. Xenotransplantation 15: 407–416. https://doi.org/10.1111/j.1399-3089.2008.00503.x.

Yang, C., R. Xue, Q. Zhang et al. 2017. Nanoclay cross-linked semi-IPN silk sericin/poly(NIPAm/LMSH) nanocomposite hydrogel: An outstanding antibacterial wound dressing. Mater Sci. Eng. C Mater. Biol. Appl. 81: 303–313. https://doi.org/10.1016/j.msec.2017.08.008.

Yannas, I. V., D. S. Tzeranis, B. A. Harley et al. 2010. Biologically active collagen-based scaffolds: advances in processing and characterization. Philos. Trans. A Math Phys. Eng. Sci. 368: 2123–2139. https://doi.org/10.1098/rsta.2010.0015.

Yue, B. 2014. Biology of the extracellular matrix: an overview. J. Glaucoma 23: S20-23. https://doi.org/10.1097/IJG.0000000000000108.

Zafar, M., S. Najeeb, Z. Khurshid et al. 2016. Potential of electrospun nanofibers for biomedical and dental applications. Materials (Basel) 9. https://doi.org/10.3390/ma9020073.

Zafar, M. S., Z. Khurshid and K. Almas. 2015. Oral tissue engineering progress and challenges. Tissue. Eng. Regen. Med. 12:387–397. https://doi.org/10.1007/s13770-015-0030-6.

Zagho, M. and A. Elzatahry. 2016. Recent Trends in Electrospinning of Polymer Nanofibers and their Applications as Templates for Metal Oxide Nanofibers Preparation. https://doi.org/10.5772/65900.

Zagho, M. M., E. A. Hussein and A. A. Elzatahry. 2018. Recent overviews in functional polymer composites for biomedical applications. Polymers (Basel) 10: 739. https://doi.org/10.3390/polym10070739.

Zagho, M. M. and M. M. Khader. 2016. The impact of clay loading on the relative intercalation of Poly(Vinyl Alcohol)-clay composites. Journal of Materials Science and Chemical Engineering 4: 20–31. https://doi.org/10.4236/msce.2016.410003.

Zarrintaj, P., A. S. Moghaddam, S. Manouchehri et al. 2017. Can regenerative medicine and nanotechnology combine to heal wounds? The search for the ideal wound dressing. Nanomedicine (Lond) 12: 2403–2422. https://doi.org/10.2217/nnm-2017-0173.

Zhang, K., Q. Fu, J. Yoo et al. 2017. 3D bioprinting of urethra with PCL/PLCL blend and dual autologous cells in fibrin hydrogel: An *in vitro* evaluation of biomimetic mechanical property and cell growth environment. Acta Biomater 50: 154–164. https://doi.org/10.1016/j.actbio.2016.12.008.

Zhang, W., X. Jin, H. Li et al. 2018. Injectable and body temperature sensitive hydrogels based on chitosan and hyaluronic acid for pH sensitive drug release. Carbohydr. Polym. 186: 82–90. https://doi.org/10.1016/j.carbpol.2018.01.008.

Zhang, Y., C. T. Lim, S. Ramakrishna et al. 2005. Recent development of polymer nanofibers for biomedical and biotechnological applications. J. Mater. Sci. Mater Med. 16: 933–946. https://doi.org/10.1007/s10856-005-4428-x.

Zhang, Y.-Q. 2002. Applications of natural silk protein sericin in biomaterials. Biotechnol. Adv. 20: 91–100. https://doi.org/10.1016/s0734-9750(02)00003-4.

Zhong, S., W. E. Teo, X. Zhu et al. 2006. An aligned nanofibrous collagen scaffold by electrospinning and its effects on *in vitro* fibroblast culture. J. Biomed. Mater. Res. A 79: 456–463. https://doi.org/10.1002/jbm.a.30870.

6

Biopolymer Complexation for the Development of Nano-Carriers and Nano-Drug Delivery Systems in the Food Industry

Titilope J. Jayeoye,[1,*] *Oladipupo O. Olatunde*,[2] *Ozioma F. Nwabor*[3] and *Sudarshan Singh*[4]

1. Introduction

Nanotechnology is an interdisciplinary area of research and development that involves manufacturing, processing, and applying substances or devices with one or more dimensions in the nanometer range (Mozafari 2007). The world's nanotechnology value as of 2020 is worth 54.2 billion US dollars, projected to reach about 127 billion US dollars in 2027. With the popularity of nanotechnology soaring consistently, the adoption and integration of nanomaterials in the food industry are gaining traction. Nanotechnology in food has been applied to improve the taste, flavor, color, texture,

[1] Department of Chemistry, Faculty of Physical Sciences, Alex-Ekwueme Federal University, Ndufu-Alike, P.M.B.1010, Abakaliki, Ebonyi State, Nigeria.

[2] Department of Food and Human Nutritional Sciences, Faculty of Agricultural and Food Sciences, University of Manitoba, Winnipeg, Manitoba, R3T 2N2, Canada.

[3] Division of Infectious Diseases, Department of Internal Medicine, Faculty of Medicine, Prince of Songkla University, Hat Yai, Songkhla 90112, Thailand.

[4] School of Allied Health Sciences, Walailak University, Nakhon Si Thammarat, 80160, Thailand.

* Corresponding author: titilope12@gmail.com

and consistency of foodstuffs, increase bioavailability, absorption, and stability of nutraceuticals compounds, novel food packaging materials development, as well as to develop nano-sensors for traceability and monitoring the status of food during transport and storage (Sekhon 2010). Food nanotechnology has received increasing interest, which opens a new frontier of possibilities for the food industry.

Biopolymers, which are large molecular structures synthesized by a living organism (Schmitt et al. 2009), have a more complex chemical structure when compared to synthetic polymers (Hosseini et al. 2017a). The low level toxicity of compatibility and biodegradability of biopolymers over synthetic polymers have gained the attention of researchers (Esfanjani et al. 2016). Biopolymers have also been extended to include bioplastics, such as polyglycolic acid, polylactic acid, and their block copolymer (Enderle 2012). These polymers, as an essential functional ingredient, cut across a diverse range of industrial applications not limited to pharmaceuticals and foods (Hosseini et al. 2017a; Patel et al. 2010; Kumar et al. 2018). In foods, biopolymers have been used for shelf-life extension, stability, enhanced textural properties, as well as improved sensory properties (Olatunde et al. 2020a; Olatunde et al. 2019; Olatunde et al. 2020b; Corredig et al. 2011; Hosseini et al. 2017a). For food packaging, biopolymers have been employed in the development of edible biopolymer-based coatings and films, which provide an additional protective layer or coating for raw materials, thus extending their shelf life (Park et al. 2014). The complexation of natural polymers has attracted much attention in the last few decades due to its potential in the nanotechnology area, particularly in the development of nano-drugs and nano-carriers. The complexation of several polymers and surfactants, including polylactic-co-glycolic acid, chitosan, and alginic acid, has been investigated for the targeted delivery and controlled release of bioactive compounds (Hosseini et al. 2017b). However, the biopolymer complex nano-drugs in foods are still in its infancy. Thus, this chapter gathers recent information on the complexation of biopolymers for the production or development of nano-drugs or nano-carriers for food applications

2. Biopolymers Complexation in Nano-Drugs Formulation

Biopolymers' complexation in nano-drugs formulation relies on the use of intramolecular and intermolecular forces of attraction between polymer layers to create new materials with improved physicochemical properties. In this regard, molecular interactions such as covalent bonding, Coulombic forces, hydrogen bonding, hydrophobic interaction, and Van der Waal forces are exploited for the creation of supramolecular networks (Hosseini et al. 2017a), with improved and enhanced capacity for nano-drugs immobilization and delivery. Covalent bonds are strong intermolecular forces in the range of 25–250 kcal/mol, often realized through some ordered chemical reactions between biopolymers and materials to be infused or conjugated. Common examples are using free radical graft polymerization of biopolymers such as chitosan with phenolic compounds. For instance, Zhu et al. (2014) successfully grafted a biologically efficient flavonoid compound onto chitosan to improve the anti-microbial and anti-diabetic properties of the resulting material. Similar to the free radical polymerization method for polyphenols grafting to biopolymers is the use of the carbodiimide chemistry between EDC and NHS, with concomitant generation of the highly effective amide bond, for instance, between the amino groups of chitosan and the carboxyl groups of polyphenols, have been widely reported (Xu et al. 2021; Liu et al. 2016; İlyasoğlu et al. 2019).

Furthermore, the use of Genipen (a bi-functional, water-soluble cross-linker) can provide strong covalent bonding between biopolymers, and this is thought to be more environmentally sustainable than the use of glutaldehyde (Muzzarelli 2009). Genipen can also react with amino group-containing compounds, such as chitosan and proteins, to produce a bluish color fluorescent adduct (Zhou et al. 2016). Moreover, Coulombic interactions, which can be repulsive or attractive, are equally strong interactions between 25–250 kcal/mol, like a covalent bond. They are often generated through the alteration of pH and ionic strength.

Hydrogen bonding and hydrophobic interactions are other prominent forces between biopolymers of about 2–10 kcal/mol and 1–2 kcal/mol, respectively. A hydrogen bond is generated between polar/hydrophilic moieties, such as the interaction between an amino group and a hydroxyl group. Hydrophobic interaction, on the other hand, is interaction generated between hydrophobes or non-polar groups. For instance, structural modification of protein via physical, chemical, and enzymatic modification methods exposes the hydrophobic regions in the polypeptide backbone for interaction. Van der Waal forces are weak electrostatic forces of attraction between neutral molecules. Chemical complexation between biopolymers is designed to act as carriers or platforms for the immobilization of nano-drugs for onward delivery to the target medium. As such, a supramolecular assembly carrier system is developed comprising of the nano-drugs of interest interspersed within the biopolymers network structure. These systems generally take the following forms: carbohydrate-carbohydrate-based complexation, protein-protein-based complexation, carbohydrate-protein-based complexation, and lipid-based complexation.

2.1 Carbohydrate-Carbohydrate-Based Complexation

Here, the interaction between two or more biopolymers, specifically carbohydrates, in addition to the nano-drug, is combined to generate a supramolecular assembly network. In this category, single polysaccharide biopolymers (e.g., sodium alginate and carboxymethyl cellulose) or synthetic biopolymers (PVA, PEG, PVP, etc.) are applied as the platform for the immobilization of nano-drugs. They are used alone or blended/combined depending on the nanocarrier platform being developed. For instance, using the ionic gelation method for chitosan nanoparticle synthesis, El-Alfy et al. (2020) investigated a novel delivery system of antibiotics loaded on biocompatible chitosan nanoparticles to inhibit Gram-positive and Gram negative bacteria. Due to the sheer interest in chitosan nanoparticles, particularly for food and pharmaceutical applications, the following reviews are worth mentioning such as Yanat et al. 2021; Singh et al. 2021a. Moreover, Ribeiro et al. (2020) investigated chitosan-chitosan nanoparticles by applying already formulated tripolyphosphate-mediated chitosan nanoparticles to prepare a Pickering emulsion, which was further characterized as a potential food-grade product. To offset some notable limitations of biopolymers, especially in film fabrication, the blending of biopolymers has emerged as a promising strategy. For instance, starch, a common biopolymer, is water-sensitive; thus, its film-forming capacity is heavily dependent on its moisture content, contributing to the low mechanical resistance of starch-based products. Furthermore, starch is limited by its poor thermal processability, arising from its high melting point and low thermal decomposition temperature (Wang et al. 2003).

To maximize the full usage of starch in food product-related applications, blending starch with biopolymers is now popular. Wang et al. (2021) investigated the rheological and gelation effect on the phase structure of the hydrocolloid mixed system of hydroxypropyl methylcellulose and hydroxypropyl starch. The study was aimed at understanding parameters that would determine and affect the mixing capacity of the biopolymers. In a bid to gain insight into the physico-mechanical properties of edible film fabricated from the blending of starch, agar and maltodextrin, Wongphan et al. (2020) reported the roles played by the contributing polymers in the overall film quality of the blends.

2.2 Protein-Protein-Based Complexation

Proteins are another class of biopolymers that have been well utilized as bio-based platforms for nano-drugs delivery for food applications. Common protein biopolymers include collagen, gelatin, silk fibroin, whey, zein, soy and gluten; milk proteins such as whey and casein are equally popular. For instance, zein, the prominent protein present in corn, has been well established for its ability to encapsulate bioactive compounds for food applications, such as resveratrol (Davidov-Pardo et al. 2015), lutein (Hu et al. 2012), and curcumin (Gómez-Estaca et al. 2012). Owing to its poor solubility in most food solvents, zein is processed into nanoparticle forms using the anti-solvent procedure with water as the anti-solvent agent (Li et al. 2013). Zein nanoparticles have found wide applications in the food area, ranging from the encapsulation and delivery of epigallocatechin gallate (EGCG) (Liang et al. 2017) to the fabrication of edible film loaded with bioactive compounds (Jiang et al. 2021). For a further understanding of protein biopolymer chemistry, see the review by Nasrollahzadeha et al. (2021).

2.3 Carbohydrate-Protein-Based Complexation

Under this complexation strategy, carbohydrates in the form of polysaccharides and protein interaction are adopted in a bid to obtain a delivery platform for nano-drugs while also playing important roles as functional components of food colloids. Polysaccharide-protein complex formation capitalizes on the key inherent characteristics of the different biopolymers, for instance, with the knowledge that proteins are surface-active molecules and thus can interact with polysaccharides (hydrophilic molecules) through electrostatic and hydrophobic interactions. The major behavior of polysaccharide-protein complex in solution, such as solubility and stability in terms of formation and deformation properties of the complex, is dependent on several factors, including temperature, pH, and ionic strength of the medium, charge density and distribution, and biopolymers mixing ratios (Warnakulasuriya

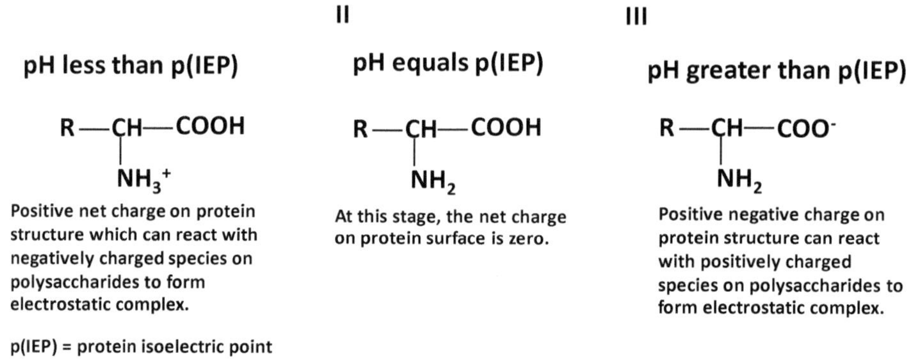

Figure 1: Interplay between solution pH and protein isoelectric point.

et al. 2018). The isoelectric point (IEP) of proteins determines protein behavior in solutions, thus it is a critical factor in protein complexation with polysaccharides.

Three prominent scenarios could play out between protein IEP and solution pH. As depicted in Figure 1, at stage I, where the solution pH is less than the p(IEP), positive charges dominate the net charge of proteins and are thus readily available for interaction with the negatively charged functional groups of polysaccharide surfaces, forming a strong electrostatic complex. At stage II, where the solution pH is almost equal to the p(IEP), the net charge on the protein surface is zero. At this stage, the protein is least soluble, with poor stability, arising from weaker complex formation between the polysaccharide and protein. However, at stage III, when the solution pH is greater than the p(IEP), there are more negative charges on protein surfaces, which are thus readily complex with positively charged groups of polysaccharides. From the foregoing, it can be observed that tuning the pH and ionic strength of solutions presents a powerful strategy for modulating polysaccharide-protein complexation for the design and formation of novel food emulsions (Ghosh et al. 2012).

2.4 Lipid-Based Complexation

Nano-liposomes, or nanometric versions of liposomes, are colloidal structures formed by the input of energy to phospholipids and cholesterol (stabilizer) in aqueous environments (Mozafari et al. 2008). It is postulated that the specific interactions between these compounds in biological membranes are crucial in forming complexes. The compatibility and safety of these lipid vesicles have attracted particular attention in pharmaceutical, cosmetic, and food industries as nanocarrier systems for the stability and delivery of several bioactive agents. The main component of the naturally occurring bilayers is the phospholipids, which are amphiphilic (Bruni et al. 2017). The amphiphilic nature of lipid vesicles encourages their utilization in the entrapment, delivery, and release of lipid-soluble, water-soluble, and amphiphilic materials. Generally, in food, lipid vesicles have shown promise for use in the stabilization of compounds and protection of sensitive compounds against environmental stress, in augmenting antimicrobial and antioxidative efficacy, and in increasing targetability. Additionally, certain mixtures of lipid molecules can

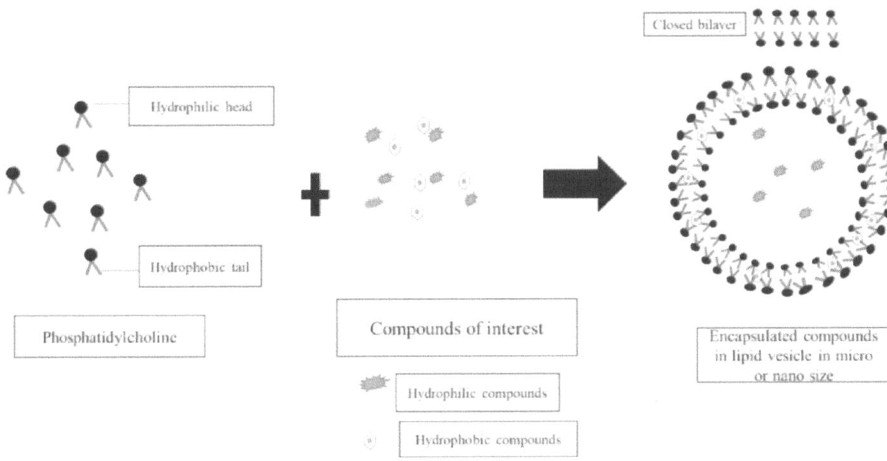

Figure 2: Mechanism of encapsulation of bioactive compounds based on lipid complexation.

result in non-liposomal structures, such as hexagonal, lamellar, cubic, or micellar phases (Siegel et al. 2008). The application of liposomes in foods without any safety concerns can be attributed to the source of the phospholipids, which are usually from natural sources, such as soy, egg, or milk (Mozafari et al. 2008). The application of lipid vesicles for encapsulation enables stability and timely and targeted release of bioactive compounds, which broadens its application. Targeting lipid vesicles can be achieved mainly via active (e.g., by incorporation of active compounds) or passive (e.g., targeting based on particle size) mechanisms (Mozafari et al. 2005).

The complexation of lipids is based on a simple phenomenon involving the aggregation of amphiphilic molecules in an aqueous environment via the shielding of their hydrophobic regions (Figure 2). Nevertheless, the aggregated phospholipids can arrange themselves to form an organized and closed bilayer vesicle (i.e., liposomes or nano-liposomes) when sufficient energy is provided (Mozafari et al. 2008). The entrapment of compounds occurs during this process. The preparation method of liposomes determines their structural characteristics based on size, type, and number of bilayers (Table 1) (Mozafari 2007; Zagana et al. 2008). For a detailed illustration of the methods of liposome complexation and their impact on its structural properties, the author recommends the review article by Mozafari et al. (2008).

3. Nano-Drugs and Nano-Carriers in Food Applications

Nanotechnology innovations have transformed several scientific and industrial areas, particularly the food industry. The application of nanotechnology spans different areas of food science, including functional food development, food packaging, food processing, detection of foodborne pathogens, food safety, and shelf-life extension. The following sections discuss the complexation of biopolymers for developing nano-carriers and nano-drugs in food. The mechanisms of the preparation of different nano-drugs or carriers commonly used in food are also discussed briefly, along with some insights on safety issues and regulatory concerns on nano-processed food

Table 1: The main types of lipid vesicles and their characteristics.

Types of Vesicles	Sizes	Number of Lipid Bilayers
Small unilamellar vesicles (SUV)	20 – 100 nm	1
Large unilamellar vesicles (LUV)	> 100 nm	1
Double-bilayer vesicles (DBV)	> 100 nm	2
Multilamellar vesicles (MLV)	> 0.5 μm	Approximately 5–20
Oligolamellar vesicles (OLV)	0.1 – 1 μm	Approximately 5
Multivesicular vesicles (MVV)	>1 μm	Multicompartmental structure

Source: Mozafari et al. (2008)

products. For this section, the nano-drugs or carriers of interest in the food industry are limited to:

(a) Nanoparticles from bioactive compounds of plant origin, such as curcumin nanoparticles;
(b) Nanoparticles obtained from biopolymers, such as zein nanoparticles, curcumin-zein nanoparticles, whey protein nanoparticles/isolates and casein nanoparticles;
(c) Nanoemulsions/Pickering emulsions;
(d) Metal nanoparticles including Ag, ZnO, TiO_2 and SiO_2 nanoparticles.

3.1 Biopolymers Nanoparticles as Nano-Drugs in Food

3.1.1 Curcumin Nanoparticles

Curcumin is the active phenolic compound extracted from the rhizome of turmeric (*Curcuma longa*). Curcumin has been found to possess several functional and biological properties ranging from anti-diabetic, anti-amyloid, anti-inflammatory, antioxidant, antimicrobial, anticancer, and anti-fibrosis properties (Karthikeyan et al. 2020). Limiting the tremendous potential of curcumin in biological and food-related applications is its poor solubility in water. Hence, efforts are geared toward preparing water-processable curcumin, of which curcumin nanoparticles or nano-curcumin have yielded satisfactory outcomes. Basniwal et al. (2011) prepared water-soluble and stable nano-curcumin by first dissolving curcumin in an organic solvent (dichloromethane). Then, the organic fraction curcumin solution was injected into boiling water at a controlled rate under ultra-sonication. Afterwards, the organic solvent was expelled using a rotary evaporator, and the solution was freeze-dried. The orange pellet obtained is the curcumin nanoparticles, which can be readily re-solubilized in water to yield an orange suspension. This method of nano-curcumin preparation is known as "solvent anti-solvent precipitation" and has been further tuned using other organic solvents, such as ethanol (Aditya et al. 2017; Devara et al. 2015; Duong et al. 2020; Pandit et al. 2015), in a bid to obtain easily controllable curcumin nanoparticles for diverse applications. Another common method of nano-curcumin preparation is the ionic gelation technique (Suryani et al. 2017). In this approach, calcium chloride ($CaCl_2$) and sodium alginate are used to reinforce the ionic interaction between chitosan and tripolyphosphate (TPP) by modulating the various mixing ratios of the reacting mixtures. While investigating the encapsulation of nano-curcumin in notable biopolymers prepared by the pH-shift method, Peng et al. (2020) tested caseinate, whey protein, soy protein, and gum arabic as common stabilizers of nanocurcumin. The authors reported that the biopolymer-loaded curcumin nanoparticles displayed high stability based on their dissolution in salts such as NaCl and $CaCl_2$.

3.1.2 Curcumin-Zein Nanoparticles

Curcumin-zein nanoparticles are another important nano-drugs in the food industry. It is a blend of two poorly soluble compounds, curcumin (a polyphenolic compound) and zein (the major protein of maize). Patel et al. (2010) reported the preparation of zein-curcumin nanoparticles by the solubilization of the blend in ethanol/water. The authors equally investigated the roles of stabilizers such as polyvinyl pyrrolidone (PVP) and protein (sodium caseinate) on the stability of the formulated nanoparticles. Furthermore, Fereydouni et al. (2021) investigated an oil-in-water (O/W) nano-emulsion of nano-curcumin immobilized in a zein nanofibrous membrane. Other biological applications of the prepared nanofiber were investigated, such as cytotoxicity, intracellular reactive oxygen species (ROS), and antibacterial activity. Feng et al. (2020a) investigated the role of biopolymer (chondroitin sulfate, CS) stabilization on curcumin-zein nanoparticles through pH adjustment. Their results showed that adding CS before the anti-solvent effect of zein at pH 7.0 produces colloidal dispersion with the best solubility. Similarly, Wei et al. (2020) investigated the effect of micro-fluidization on the formation of curcumin-loaded zein-propylene glycol alginate complex nanoparticles. It was concluded that high-pressure micro-fluidization (HPM) facilitates the *in vitro* release of curcumin during simulated gastrointestinal digestion and thus presents a novel food-grade nanoparticle with desirable stability and digestive properties.

3.1.3 Whey Protein Isolates (WPI) Nanoparticles

Milk is normally sourced from animals such as cows, goats, sheep, and buffalo and contains two major classes of proteins. Casein accounts for about 80% (w/w) of milk and is often extracted by isoelectric precipitation or through enzymatic processing. Whey protein, on the other hand, accounts for about 20% (w/w) of milk and is the byproduct of casein extraction (Madureira et al. 2007). Casein contains αs1-casein, αs2-casein, β-casein and κ-casein (Bhat et al. 2016), while whey protein contains fractions such as α-lactalbumin, β-lactoglobulin, protease peptone, glycolmacropeptide, immunoglobulins, and serum albumin. Owing to its rich nutrient density, whey protein has been popularized as a food supplement. Generally, whey protein isolates (WPI) are obtained from the extraction of milk whey fraction using purification methods such as micro-filtration and ion-exchange. WPI has found significant usability in developing packaging materials through film casting, either alone or in combination with polysaccharides (Janjarasskul et al. 2020; Çakmak et al. 2020; Tavares et al. 2021). Aside from its use as a biopolymer for the formulation of edible films, WPI can be processed into nanoparticles owing to its good functionality through the pH-cycling process (Giroux et al. 2010; Abbasi et al. 2017), organic solvent de-solvation process (Gülseren et al. 2012). WPI nanoparticles have been investigated and validated as effective nano-carriers for bioactive compounds such as curcumin (Teng et al. 2014) and other polyphenols (Bagheri et al. 2013). It is worth mentioning that peptides with multifunctional properties can also be produced from WPI (Zhao et al. 2020).

3.1.4 Soy Protein (SP) or Soy Protein Isolates (SPI) Nanoparticles

Soy protein is a globular protein with two prominent subunits, conglycinin and glycinin, which are rich in amino acids such as leucine, aspartate and glutamate (Peles et al. 2013). Soy protein isolate nanoparticles have been prepared by Teng and colleagues through the desolvation method, using ethanol as the desolvating agent (Teng et al. 2012). The method involves solubilizing SPI in water followed by the addition of ethanol. Glutaraldehyde was then added as a cross-linking agent, followed by the addition of ethanol, evaporation, dilution, and centrifugation to form the SPI nanoparticles. SPI nanoparticles in combination with polysaccharide biopolymers (carboxymethyl chitosan, CMC) have been investigated for the delivery of vitamin D3, a typical hydrophobic micronutrient (Teng et al. 2013), through a simple ionic gelation method. It was found that in comparison with the single polymer, CMC and CMC/SPI complex required a lower concentration of Ca^{2+} for nanoparticle formation, which resulted in better particle formation and stability and enhanced loading efficiency of vitamin D3. Similarly, Yuan et al. (2021) synthesized PSPI nanoparticles through controlled partial enzymatic hydrolysis for the encapsulation and delivery of curcumin. Here, pH-controlled enzymatic hydrolysis of SPI was initiated using pepsin and pancreatic enzymes.

3.1.5 Casein Nanoparticles

Casein nanoparticles are often synthesized using calcium chloride or phosphate, resulting in the formation of colloidal nanoparticles, also known as casein micelles or nanoparticles. Barick et al. (2021) investigated the capacity of casein nanoparticles as a nanocarrier for the delivery of curcumin with satisfactory loading efficacy. Moreover, natamycin (a food preservative), encapsulated on casein-zein nanoparticles and further immobilized on gelatin-based nanocomposite film, was investigated by Mo et al. (2021). It was noted that natamycin-loaded casein-zein nanoparticles increased the mechanical properties of gelatin film with sustained release of the anti-fungal drug. In a similar account, casein nanoparticles were investigated for the oral delivery of vitamin B9 or folic acid (Penalva et al. 2015). In summary, the casein nanoparticles were prepared using the coacervation method, further stabilized by amino acids lysine or arginine and then dried by spray drying. *In vitro* release studies demonstrated that the synthesized casein nanoparticles functioned as gastro-resistant material capable of reliably releasing folic acid under a simulated intestinal condition.

3.1.6 Gelatin Nanoparticles (Gel NPs)

The most important uses of gelatin are related to its ability to form heat-reversible sol-gels. As a result, gelatin is applied as a thickener, stabilizer, emulsifier, foaming agent, and crystal modifier. Gelatin has been well established for its film-forming capacity, either alone or in combination with polysaccharides. Syahida et al. (2021) investigated a gelatin/palm wax/lemongrass essential oil-coated kraft paper as packaging material for the shelf-life extension of ground beef. The authors concluded that the active compounds in the composite film delayed microbial growth and lipid oxidation, thus highlighting the capacity of the film to function in food packaging. Ahmad et al. (2012) equally investigated packaging film for sea bass slices by incorporating lemongrass essential oil in the prefabricated gelatin film. The authors reported that the incorporation of lemongrass essential oil increased the antimicrobial and antioxidative capacity of the film and thus ensured the shelf-life extension of sea bass slices by over 12 days compared to the controls during refrigerated storage. A comprehensive review of the application of gelatin as a biodegradable packaging material in food products has been published (Hanani et al. 2014).

An additional application of gelatin-based materials is their use in the fabrication of gelatin nanoparticles. Gelatin nanoparticles are synthesized through the desolvation method using ethanol or acetone and coacervation. Geh et al. (2016) reported the optimization steps for the synthesis of Gel NPs through the one-step desolvation method. The protocol comprises the dissolution of Gel in water at 50°C with stirring, followed by pH adjustment and the addition of acetone to the solution to initiate desolvation and nanoparticle formation. Glutaraldehyde was finally added to the solution to cross-link the formed nanoparticles to prevent particle agglomeration. Recently, a two-step synthesis approach was proposed as an improvement over the single-step protocol. In this light, Carvalho et al. (2018) reported the synthesis of Gel NPs through a two-step desolvation approach. In a bid to improve the purity of Gel NPs, the two-step approach included additional use of acetone, with further partitioning and repeated centrifugation. Furthermore, Gel NPs synthesis based on coacervation or phase separation method has been reported. In this approach, homogenous solutions of charged macromolecules are subjected to liquid-liquid phase separation, resulting in the formation of a dense, polymer-rich part at the bottom of the reaction tube and a transparent solution at the top. Notably, these two liquid phases are immiscible, non-compatible and are in equilibrium. Adding alcohol and salts often promotes coacervation, thus assisting in nanoparticle formation.

Furthermore, Gel NPs synthesis via simple coacervation has been reported (Mohanty et al. 2005). Another Gel NPs synthesis method via nanoprecipitation has also been reported (Naskar et al. 2019). Gel NPs are excellent nanomaterials for a wide range of applications, including the delivery of plant secondary metabolites (Karimi et al. 2020) and as addictive for the enhancement of Pickering emulsion (Tan et al. 2021; Feng et al. 2020b). Other notable applications of Gel NPs can be found in the detailed review by Elzoghby (2013). Several other protein-based nanoparticles can be fabricated under appropriate synthesis conditions. These examples have been reviewed elsewhere (DeFrates et al. 2018).

3.1.7 Liposome Nanoparticles

Several methods have been used for the encapsulation of molecules in lipid vesicles. The major and first step involves the solubilization of lipids in an appropriate organic solvent (chloroform or methanol). The most common method used for liposome preparation is the thin film method, which involves the removal of the organic solvent used in dissolving the lipid ingredients (phosphatidylcholine and cholesterol) to produce thin films (Bruni et al. 2017). The aqueous phase containing the active compound is added and used to rehydrate the films, followed by a sufficient amount of mechanical or thermal energy, which causes bilayer sheets of the hydrophobic components to separate from the bulk to form liposomes (Mozafari et al. 2008). The major limitations of this method are the use of organic solvent, lack of a continuous production process, and low encapsulation efficiencies (Lasic et al. 2001). To address the limitation of scaling up liposome preparation for food application, several methods, including inclusion coacervation, complexation, spray drying, spray cooling/chilling, extrusion, centrifugal suspension separation, co-crystallization lyophilization, emulsion, fluidized bed coating, and liposome entrapment, have been employed (Fang et al. 2010; Mozafari 2006). Liposomes have been used to encapsulate several bioactive compounds (Chorachoo et al. 2013; Mazandrani et al. 2016). The encapsulation of polyphenols from *Thymus* spp. in liposome was reported to improve antioxidant and antimicrobial properties, as well as stability, compared to its unencapsulated counterpart (Gortzi et al. 2006). Colletier et al. (2002) documented the encapsulation of Pronase in liposomes using the thin film method. Other bioactive compounds encapsulated in liposomes using different methods are summarized in Table 2.

3.1.7 Nanoemulsions/Pickering Emulsion as Food Nano-Drug

The emulsion is a mixture of two immiscible liquids, such as oil and water. Emulsions with droplet size in the nano range (< 100 nm) are known as nanoemulsions. In simple terms, the addition of oil to water would generate a highly immiscible mixture that is thermodynamically stable. However, with the addition of an emulsifier, a nano-emulsion is formed where small-sized liquid droplets (dispersed phase) are dispersed in a liquid medium (continuous phase). Common types available are oil-in-water (O/W), water-in-oil (W/O), and oil-in-oil emulsions. From the foregoing, emulsifiers are the indispensable constituents central to the formation of an emulsion. Emulsifiers are materials such as surfactants, proteins, lipids, or polysaccharides, possessing the inherent property of weakening the interfacial tensions (surface energy per unit area), keeping the oil and the water phases of an emulsion (Gupta et al. 2016). Polysaccharide-based Pickering emulsion has been covered in detail elsewhere (Cui et al. 2021). Modified polysaccharides are also synthesized to address some limitations associated with using native polysaccharides and thus extend the basic applications of the available polysaccharides for Pickering emulsion formulation in modern food applications. Emulsifiers also act as stabilizing agents for nanoemulsions through steric hindrance and electrostatic repulsion. A special class of emulsifiers with huge biological and food-related applications is the "Pickering emulsion". They are solid particles with the ability to accumulate at the oil-water interphase and stabilize the

Table 2: Different methods of nanoliposome-encapsulated compound preparation.

Bioactive Compounds	Lipid Ingredients	Method of Preparation	Size (nm)	Findings	References
Quercetin	Phosphatidylcholine, cholesterol	Thin film hydration	146	Improved stability and bioavailability	Jangde et al. (2016)
Fish gelatin hydrolysate	Cholesterol and 1,2-dipalmitoyl-sn-glycero-3-phosphocholine (DPPC)	Thin film hydration	134–621	Improved stability and bioactivities	Hosseini et al. (2017b)
Bovine serum albumin	soybean phospholipid, cholesterol, Tween-80 and vitamin E	Thin film hydration	1,187	Improved stability	Liu et al. (2015)
Quercetin	Lipoid E80, cholesterol, cyclodextrins, Phospholipon 90H, Lipoid S100	Ethanol injection	123–296	Regardless of lipid ingredients, all prepared liposome was physically stable after 1 year of storage at 4°C	Azzi et al. (2018)
Ethanolic coconut husk extract	Phosphatidylcholine, cholesterol	Ethanol injection	210–459	Enhanced stability and antibacterial properties	Olatunde et al. (2019)
Ethanolic coconut husk extract	Phosphatidylcholine, cholesterol, glycerol and Tween 80	Ethanol injection	261–676	Liposomes with glycerol or cholesterol could serve as a delivery-based system for improving the antibacterial properties with high stability	Olatunde et al. (2020c)
Curcumin	Phosphatidylcholine, cholesterol	Detergent removal	36–70	Improved stability and solubility in aqueous medium	De Leo et al. (2018)
Epigallocatechin gallate (EGCG)	Cholesterol, deoxycholic acid	Extrusion	131–378	Prevention of degradation, improved activity of EGCG against basal cell carcinomas	Fang et al. (2006)
Algae methanolic extract	Phosphatidylcholine, glycerol	Heating	86–118	Improved antioxidant stability and controlled release	Savaghebi et al. (2020)
Cyclosporin A	Phosphatidylcholine, cholesterol	Supercritical anti-solvent	166–244	High stability and reproductivity	Karn et al. (2013)
Essential oils, lipids and dietary supplements	L-α-Phosphatidylcholine	Supercritical reverse-phase evaporation	196–234	High encapsulation efficiency and stability regardless of the nature of the compounds	Trucillo et al. (2019)
Polyphenol-rich green tea extract	Soy phosphatidylcholine	Micro-fluidization	35–38	Enhanced bioactivity and bioavailability	Dag et al. (2019)
Shrimp	Soy phosphatidylcholine and glycerol	Ultrasonication	40–284	High oxidative stability and better retention of EPA and DHA	Gulzar et al. (2020)
Curcumin	Phosphatidylcholine, cholesterol	Thin film hydration	339.3	Improved permeability and bioavailability	(Patel et al. 2009)

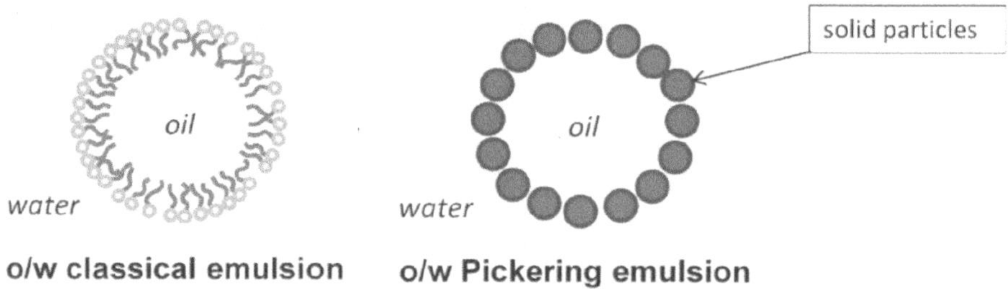

Figure 3: Comparison between classical surfactant-based emulsion and Pickering emulsion. solid particles adsorbed at the oil–water interface stabilize the droplets in place of the surfactant molecules. Adapted from (Chevalier et al. 2013).

fabricated emulsion. They still retain the basic properties of the classic surfactant-based emulsion and thus can be used as substitutes in most applications. The solid particle stabilization offered by Pickering emulsion imbues them with additional properties such as resistance to coalescence, tunable permeability, and improved elastic responses (Tang et al. 2019), manifesting in higher stability. The relationship between surfactant-based emulsion and Pickering emulsion is shown in Figure 3.

Recently, Alehosseini et al. (2021) investigated a D-limonene-loaded Pickering emulsion stabilized with chitosan nanoparticles. D-limonene was protected against harsh conditions through encapsulation with the Pickering emulsion. It was also found that introducing D-limonene led to a change in the microstructural property of chitosan nanoparticles, resulting in the formulation of a product with an improved thermal property. Similarly, Xiao et al. (2021) reported the emulsifying capacity of xanthan gum on the ovalbumin-stabilized oil-in-water emulsion. The emulsifying ability of ovalbumin as the major component of egg white protein has been investigated. The incorporation of xanthan gum improved the physicochemical properties of the emulsion arising from the interaction between ovalbumin and xanthan gum. This ultimately led to a significant increase in the formulated emulsion stability, thus highlighting the potential of xanthan gum as an emulsion stabilizer in egg white-derived food products. Similarly, Li et al. (2021) investigated the role of arabinoxylan from different sources (wheat bran, corn bran, rice bran, and rye bran) on oil-in-water whey protein isolate emulsion. The authors reported that arabinoxylan from various sources decreased the interfacial charge arising from the adsorption of the polysaccharide onto the oil droplets. In contrast, natural biopolymers are well investigated for Pickering to produce a modified polysaccharide-based emulsion formulation. Table 3 provides examples of nanoemulsion formulations and their applications in food-related products.

3.1.8 Biopolymers Mediated Nano-Synthesis of Metal Nanoparticles

Metal nanoparticles (MNPs), particularly of Ag, Au, Cu, and related metal oxide nanoparticles (MONPs) (ZnO, CuO, TiO2, etc.), owing to their unique properties, have found strong applications in the food industry, particularly as antimicrobial agents for packaging materials. MNPs and MONPs are produced by reducing their

Table 3: Compositions, properties and applications of common nanoemulsions.

S/N	Composition	Droplet Sizes (nm)	Remarks	Applications	References
1	3% Tween 20 solution as the aqueous continuous phase, plus EOs from clove, cinnamon and lavender	82–3,472	EOs composition played active role in determining the droplet size and the stability of the resulting formulations	Anti *E. coli* activity in foods and beverage products over 20 days of storage	Falleh et al. (2021)
2	Phase inversion temperature (PIT) based synthesis of cinnamon oil O/W nano-emulsion	100–815	10 wt %, oil phase, 10 wt% surfactant (Tween 80), 80 wt% water combinations under PIT treatment. The oil phase property determined the property and stability of the formulation	Used as an antimicrobial agent against common food-grade bacteria	Chuesiang et al. (2019)
3	Nano-emulsion from EOs of cinnamon, rosemary and oregano. Continuous aqueous phase comprised (5 g Inulin/100 g and 3 g Tween 80/100 g solution)	226–546	Nano-emulsion was prepared using the ultrasonic homogenizer	Formulated nano-emulsion was used as an antibacterial agent (*Escherichia coli* and *Listeria monocytogenes*) toward the preservation of celery vegetables	Dávila-Rodríguez et al. (2019)
4	Cellulose nanofibril and chitin nanofibril aqueous phase combined with corn oil	3,000–7,000	Nano-emulsion of single polysaccharides and blends were obtained by blending the aqueous and oil phase together	Formulated nano-emulsion showed good prospects for edible particle-based emulsifiers for food products	Lv et al. (2021)
	Jujube gum and nettle oil (NE) EOs nano-emulsion	63–240	Nano-emulsion prepared through deep homogenization of the components	Formulated nano-emulsion was applied with antibacterial and antioxidant agents to extend the shelf life of fish (Beluga sturgeon fillet) over 15 days	Gharibzahedi et al. (2017)
5	Mustard oil EOs nano-emulsion using Span 80 and Tween 80 as ionic surfactant	87–213	Ultrasound-assisted preparation of the various blends results in the formation of nano-emulsions with varying properties	The investigation of nano-emulsion property demonstrated its potential for food and cosmetic-based applications	Carpenter et al. (2017)
6	Emulsion produced from using different emulsifiers (Tween 80 or whey protein isolate) in combination with EOs (cinnamon leaf, lemon, and bergamot)	79–200	High-pressure homogenization of the various combinations led to the formation of different nano-emulsions for trials	The various formulations were tested as anti-fungal agents against *Aspergillus niger*	Ribes et al. (2017)
7	Cinnamon and black pepper EOs are prepared using different methods	< 100	High-pressure homogenization and ultrasound method	Ultrasound-obtained cinnamon oil nano-emulsion demonstrated high antimicrobial properties against *E. coli* and *L. monocytogenes* for food safety and preservation	Jiménez et al. (2018)

EOs = Essential oil

respective precursor salts using strong reducing agents. This process is achieved mostly in an aqueous system, and successful synthesis is dependent on several factors, such as pH, temperature, the type of reductant used and synthesis time. MNPs and MONPs produced using chemical reductants (sodium borohydride, hydrazine hydrate and organic solvents like dimethylformamide DMF) are not environmentally friendly due to their toxicity. In a bid to circumvent this issue, environmentally benign materials have been explored for MNPs and MONPs synthesis, a process known as green synthesis, sustainable synthesis, or environmentally friendly synthesis. These synthesis approaches are convenient, simple, economical and easily scalable (Bhattarai et al. 2018; Chandra et al. 2020) for industrial production, as green starting materials are abundant and readily available.

The properties of nanoparticles are quite different from those of the parent materials. As a result, nanoparticles show better properties, attributed to their high surface-area-to-volume ratio, high loading capacity, and tunable surfaces, compared to their starting materials. MNPs and MONPs synthesis realized through green materials have been documented as a prominent way to reduce the inherent toxicity load of nanomaterials, thus the increasing popularity of green synthesized materials used for medical and food-related products. Under the green synthesis of MNPs and MONPs, the use of extracts of plant parts as environmentally friendly reductants has been widely investigated. Plant seeds, seed pods, aerial parts, stems, flowers, and leaves have all been reported to contain essential bioactive metabolites capable of supplying electrons to reduce metal salts to produce metabolite-stabilized nanoparticles (Kumar et al. 2021). The mechanism of such reduction has been reported to follow the general trend involving small nuclei formation (nucleation) arising from the injection of the biomass-based reductant into the MNP and MONP salts. The formed nuclei are unstable and vulnerable to crystal growth, where the small nuclei formed are fused to form bigger particles in the cluster formation stage. They are further molded to maturation in the reaction environment comprising the biomass reductant and finally the bioactive components of the reductant are wrapped around the matured particles to prevent facile particle aggregation (Eze et al. 2019; Jayeoye et al. 2021a; Eze et al. 2021).

Besides using plant parts to synthesise MNPs and MONPs, biodegradable polymers for the reduction and stabilization of nanoparticles also fall within the purview of green synthesis. Jayeoye et al. (2021a) reported synthesizing conducting polymer-doped sodium alginate-stabilized gold nanoparticles for biological applications. Similarly, several other studies have reported the use of biopolymers and synthetic polymers (polyvinyl alcohol (PVA), polyethylene glycol (PEG), polyvinyl pyrrolidone (PVP), etc.) for the preparation of MNPs and MONPs (Ortega et al. 2021; Jayeoye et al. 2020a; Jayeoye et al. 2020b; Jayeoye and Rujiralai 2020; Chawla et al. 2020).

Generally, MNPs and MONPs are applied in food-related materials for their high antimicrobial efficacy. Owing to their ability to release metal ions (M^{+}) and high surface area-to-volume ratio, the nanoparticles are highly effective at inducing deleterious effects in bacteria, fungi, and viruses. MNPs and MONPs mediated antimicrobial action has been reported to involve, first, the release of (M^+), e.g., Ag+ from the various metal nanoparticles, which can bind electron donors in the microbial cell, such as sulfhydryl, amino, imidazole, and carbonyl groups. These functional groups are components of membrane proteins; as a result, the membrane becomes permeable to more M^+ penetration, resulting in cell death. Moreover, the entry of M^+ into the microbial cells induces damage to cellular components (RNA, DNA, and metabolic enzymes), which alters microbial processes, such as transcription and translation (Shaikh et al. 2019; Slavin et al. 2017). Secondly, the accumulation of MNPs and MONPs in microbial cells, resulting from cell permeability, causes microbial membrane rupturing. The third mechanism involves the generation of ROS and free radicals (e.g., $OH^{\cdot}$, O_2^-, H_2O_2, HClO), which are strong oxidizing agents with the capacity to initiate damage to microbial genetic materials, enzymes and other cellular components (Zou et al. 2018). From the foregoing, MNPs and MONPs can be used as active antimicrobial components of food materials, especially in food packaging, as discussed later.

4. Nano-Carriers and Nano-Drugs for Food Application

4.1 Bioactive Compounds Stability and Improved Bioactivities

Biopolymer complexation for developing nano-drugs has shown potential for use in food products. Several micronutrients (essential trace elements and vitamins) have been added to food after their encapsulation in polymer complexes. The instability of bioactive compounds from food origin is a major drawback affecting their food industry application. In addition, the inability of these bioactive molecules to properly reach the target site or cells in the body also remains a challenge (Mazandrani et al. 2016). Polymer complexation via nanoencapsulation of these active drugs has been considered an effective technique for improving the bioavailability and biostability of the bioactive compounds (Shoji et al. 2004). The encapsulation could help reduce the interaction between bioactive compounds and the environment (light, oxygen, and water), reduce the transfer rate, and improve the half-life, bioavailability and handling ability of bioactive compounds (Fang et al. 2010). Gulzar et al. (2020) documented improved eicosapentaenoic acid (EPA) and docosahexaenoic acid (DHA) stability in shrimp oil after encapsulation in liposomes. Lipid oxidation and microbial proliferation were retarded in mayonnaise incorporated with nanoliposome-encapsulated algae extract (Savaghebi et al. 2021). Reduction in EPA and DHA in fish treated with high-voltage cold plasma was circumvented with the combined treatment of liposomal encapsulated plant extract (Olatunde et al. 2020b). Likewise, Liu et al. (2015) reported higher stability for BSA encapsulated in nano-liposomes during *in vitro* digestion than the unencapsulated counterparts. Barick et al. (2021) reported an enhanced bioavailability and anticancer efficacy for curcumin-encapsulated casein nanoparticles when compared to the non-encapsulated compound, while Yuan et al. (2021) reported that SPI nanoparticles were able to protect the loaded curcumin against degradation under simulated gastric-intestinal condition.

4.2 Food Packaging

The development of food packaging materials through fortification with nanoparticles has garnered traction in the recent decade. The environmental contamination/ pollution associated with using non-biodegradable plastics has been on the rise, and the need for sustainable alternatives cannot be overemphasized. It has been estimated that 4.8–12.7 million tons of landed plastics end up in the ocean yearly (Jambeck et al. 2015). Plastics constitute huge problems to the environment by damaging its aesthetics, which have also persisted for years. In the aquatic environment, plastics can be fragmented into microplastics, thus fueling the general apprehension about risks to organisms and humans (Thompson et al. 2009) arising from the indiscriminate usage of plastics. To reduce the negative effects associated with the use of plastics, particularly in the food packaging industry, the use of biopolymers-based products has caught the attention of researchers.

Biodegradable food packaging materials are produced using a simple blending method, with the addition of appropriate plasticizers, to impart flexibility, reduce film brittleness and improve film toughness (Jantrawut et al. 2017). Common examples include glycerol, polyethylene glycol (PEG), sorbitol, and polypropylene glycol (PG). The solution casting technique is facile, easily scalable for large-scale production, and ensures sufficient interaction between constituent blends. Kumar et al. (2018) investigated the fabrication of a biodegradable nanocomposite film of chitosan/gelatin and AgNPs as an active packaging material. AgNPs used for the film fabrication were synthesized using the fruit extract of the *Mimuosops elengi* plant. The presence of sharp absorbance at 427 nm in the colloidal AgNPs showed the successful synthesis of highly stable and dispersed AgNPs. It was further noted that adding AgNPs to chitosan/gelatin solution to realize the nanocomposite film through the solution casting technique improved the mechanical property and decreased light transmittance in the visible region. The fabricated film was tested as a packaging material for red grapes, where it greatly enhanced the shelf life of the fruit in comparison with the control groups. Table 4 shows some more reported works between biopolymers-based food packaging materials fortified with various MNPs and MONPs through nanocomposite formation. The possibility of generating synergistic actions between the biopolymer-based film layers and the doped MNPs/ MONPs is often proposed to help increase the activity of the materials against food spoilage pathogens.

Another popular and reliable method of fabricating biodegradable food packaging materials is electrospinning. Electrospinning uses a high-voltage electric field to treat a solution of polymeric material, resulting in the formation of thin, fine nanofibers on a collector. The polymer solution is oftentimes impregnated with a bioactive material, nanoparticles, and other antibacterial/antimicrobial agents to imbue the resulting nanofiber with improved properties for biological, pharmaceutical, and food applications. Electrospun nanofibers possess high porosity and surface area-to-volume ratio and thus can serve as effective nanocarriers and delivery vehicles for active biofunctional materials. The recent popularity of the technique is attributed to the fact that the process can be realized at room temperature; hence, the protection

Table 4: Biodegradable polymer films as packaging materials fortified with various metal and metal oxide nanoparticles.

Polymer Combinations for Film	Filler Nanomaterial/Synthesis	Properties	Applications	References
PVA/Nanocellulose	AgNPs (synthesized using starch as reductant)	Tensile strength of PVA was increased by 8 wt% from NC addition	Investigation of film properties for packaging, with high antimicrobial properties with no cytotoxic effect on HepG2 cell	Sarwar et al. (2018)
Starch-PVA	AgNPs synthesis was achieved via a UV-light mediated reduction inside starch-PVA polymer matrix		The fabricated nanocomposite film exhibited potent antimicrobial activity against two bacteria (*E. coli* and *L. innocua*) and tow fungi (*Apergillus niger* & *Penicillium expansum*)	Cano et al. (2016)
Sandwich Chitosan/rGO	AgNPs (synthesized through corn stalk reduction of GO-$AgNO_3$ to yield rGO-AgNPs)	The infusion of rGO and AgNPs increased the antimicrobial properties, mechanical, oxygen barrier and water properties of film	Film demonstrated good antibacterial activity against *E. coli* and *S. aureus*	Gu et al. (2021)
Agar	CuSNPs (synthesized through copper acetate and thiourea mixture in the presence of corn starch and ammonia as stabilizing and hydrolyzing agents, respectively)	Spherical CuSNPs with size 2–10 nm	Nanocomposite showed antibacterial activity against *E. coli* and *L. monocytogenes*, without cytotoxicity on skin fibroblast L929 cell line	Roy et al. (2019)
Different biopolymers: agar, SA, k-carrageenan, chitosan and carboxymethyl cellulose	CuONPs (synthesized through mixing of copper chloride and NaOH, under heating at 70ºC)	Inclusion of CuONPs improved the water and UV-light barrier, with enhanced mechanical properties	CuONPs blending ensured potent antibacterial properties against *E. coli* and *L. monocytogenes*	Shankar et al. (2017)

Table 4 contd. ...

...*Table 4 contd.*

Polymer Combinations for Film	Filler Nanomaterial/Synthesis	Properties	Applications	References
Chitosan/gelatin	ZnONPs (synthesized using *Casia fistula* fruit extract as the reductant ensured the formation of nanoparticles with a size of about 29 nm)	ZnONPs inclusion improved elasticity, thermal stability and enhanced structural properties of nanocomposite film	Potent antibacterial activity against *E. coli*	Kumar et al. (2020)
Starch/Kefiran	ZnONPs (purchased from a company, with size 20–30 nm and area 35–50 m^2/g)	ZnONPs inclusion improved the mechanical and UV protection properties of nanocomposite film,	The physical and mechanical property enhancement of film validated its potentiality as packaging material	Babaei-Ghazvini et al. (2018)
Crosslinked PVA-GO Film	AuNPs (synthesized through the normal sodium citrate reduction of gold salt)	The dispersion of GO and AuNPs in crosslinked PVA enhanced the physical and mechanical properties of film	Film demonstrated good antimicrobial capacity and was used for the shelf-life extension of banana	Chowdhury et al. (2020)
Various biopolymers: agar, carrageenan and carboxymethyl cellulose	ZnONPs (synthesized by NaOH-mediated reduction of zinc acetate salt under heating)	ZnONPs ensured increased color, moisture content, hydrophobicity, UV barrier, elongation and thermal strength	Improved antimicrobial activity against *E. coli* and L. monocytogenes. Improved food packaging properties	Kanmani et al. (2014)

GO = graphene oxide; rGO = reduced graphene oxide; CuSNPs = copper sulfide nanoparticles, SA = odium alginate, CuONPs = copper oxide nanoparticles; ZnONPs = zinc oxide nanoparticles; AuNPs = gold nanoparticles

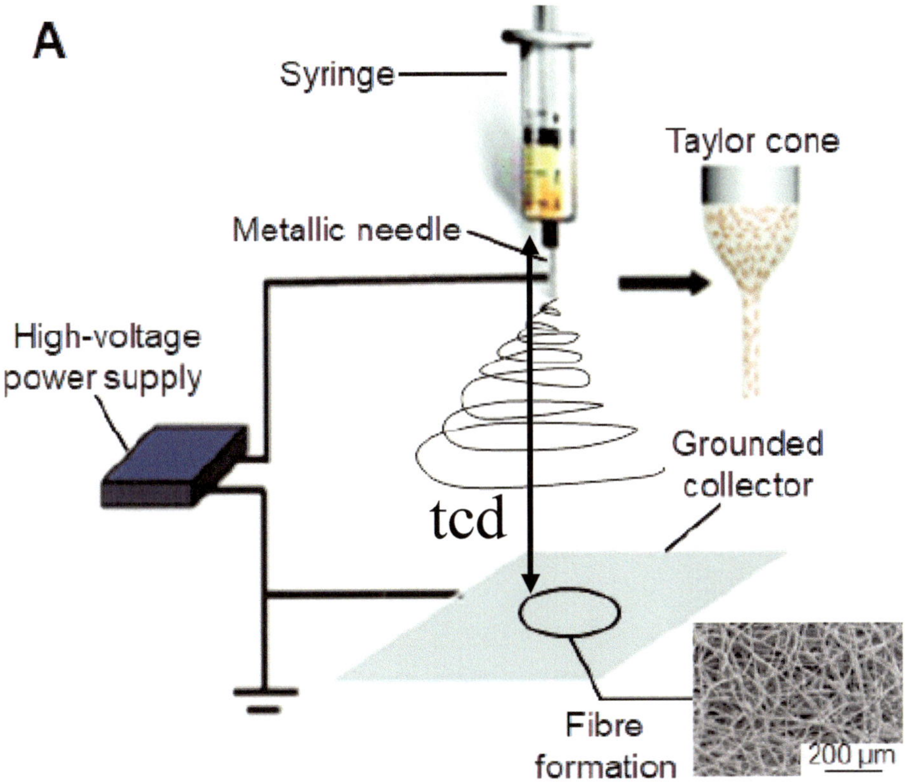

Figure 4: Typical electrospinning set up for the fabrication of biodegradable nanofibers. Adapted from (Zhang et al. 2014), with modification, tcd is the (tip-to-collector distance).

of heat-sensitive bioactive material is guaranteed, with high encapsulation efficiency and potency (Yuan et al. 2017). Typical electrospinning set-up is shown in Figure 4.

As depicted in Figure 4, the tip-to-collector distance (tcd) is an important parameter affecting the quality of fabricated fiber. It plays a significant role in determining the morphology, fiber diameter, and length of the generated nanofiber. Other parameters like solution properties (viscosity, conductivity, temperature, humidity, polymer concentration, etc.) and process parameters (voltage, needle diameter) (Hemmati et al. 2021) have to be optimized to achieve optimal spinning conditions. With the increasing application of biopolymers as a base material for nanofibers fabrication, diverse biodegradable electrospun nanofibers have been reported, including use as food packaging material, as summarized in Table 5.

5. Future Perspectives and Conclusions

Biopolymers complexation presents a versatile avenue for the maximization of diverse intermolecular and intramolecular interactions within a single biopolymer or with others. Their distinct structural chemistry can be exploited for the immobilization of nano-drugs; as a result, they can be of use for the delivery of important compounds

Table 5: Polymer-based fabrication of biodegradable nanofibrous materials for antimicrobial activity for food packaging.

Polymeric Material	Immobilized Active Component	Remarks	Application	References
CMC/PEO	AgNPs (synthesized using CMC as reductant under heating at 60°C)	Electrospinning solution comprised of CMC/PEO/CMC stabilized AgNPs	Fabricated nanofiber was applied as antimicrobial agent against *E. Coli, S. aureus, P. aeruginosa*	Fouda et al. (2013)
Feather keratin/PVA/PEO	AgNPs (synthesized through PVP stabilization using Tannic acid as reductant at RT)	Water-soluble AgNPs was mixed with feather keratin, PVA and PEO solution and then spun	AgNPs loaded nanofiber showed improved mechanical properties such as tensile strength and elongation at break	He et al. (2020)
Pullulan/PVA blend	AgNPs (standard NPs purchased form company)	AgNPs was blended with pullulan/ PVA and then spun	Enhanced antimicrobial activity with demonstrable property for preservation	Islam et al. (2013)
PEO/HIO/SA	Stable and dispersed HIO was purchased from company	HIO imparted improved tensile strength on nanofiber	Antimicrobial activity against bacterial strain *Vibro vulnificus*	Moon et al. (2013)
Chitosan/Polycaprolactone/ Chlo-HNT	HNT was purchased from the company, and Chlo was loaded under ultrasonication in ethanol	The solution of chitosan/PCL and chlo-HNT was prepared and spun	Prepared nanofiber demonstrated reduced mechanical properties with increased water vapor barrier properties; sustained release of chlorogenic acid was obtained, which will increase the shelf life of food products	Zou et al. (2020)
HPMC/PEO	Gallic acid (GA)	Gallic acid was loaded into PEO/ HPMC solution and spun	The fabricated nanofiber showed good antioxidant activity arising from GA loading; the formed material was tested as a packaging material for enhancing the shelf life of packaged walnut	Aydogdu et al. (2019)

Chitosan/PEO	Oregano essential oil (OEO)	OEO was added into the prepared solution of Chitosan/PEO blends and spun	Spun fiber exhibited good mechanical properties and improved water vapor barrier performance; the fabricated electro-spun nanofibrous material showed distinct antimicrobial properties, suitable for food packaging	Ardekani-Zadeh et al. (2019)
Chitosan-poly (vinyl alcohol)	Natural anthocyanin-rich extracts	Significant increase in surface roughness of film formed by casting with the addition of hydrophobic anthocyanin-rich extract	Excellent colorimetric response of food beverage due to microbial spoilage	Singh et al. (2021b)

CMC = carboxymethyl cellulose; PEO = poly(ethylene oxide); PVA = poly(vinyl alcohol), RT = room temperature; HIO = hydrated iron oxide; SA = sodium alginate; Chlo-HNT = chlorogenic acid loaded halloysite nanotube; HPMC = hydroxypropyl methyl cellulose in food formulations. In this regard, carbohydrate-carbohydrate, carbohydrate-protein, protein-based and lipid-based interactions have been discussed. These interactions between biopolymers confer additional properties between biopolymer layers and the infused nano-drugs possessing different physico-chemical properties. Research in this area has not reached a saturation point, thus the need to expand the frontier of scientific hypothesis in this field cannot be over-emphasized. Thus, nanomaterials infused with biopolymers-complexed materials can be adopted to realize supramolecular materials with distinct properties.

Although the adoption of metal nanoparticles (MNPs) fortification into food-based materials has assisted in no small measure in increasing the antimicrobial properties of the resulting materials, the issue of MNPs leaching into foods and their attendant toxicity is a controversial area subject to more investigation. As it is generally accepted, the immobilization of MNPs in biopolymer matrixes and, more importantly, the use of biopolymers in MNP nano-synthesis has been accepted as a way of reducing nanoparticle toxicity load. The highly ordered structure of biopolymers presents a compact avenue for MNP immobilization, which can further improve its diverse biological and food-related applications. MNP properties are significantly dependent on the chemical properties of the capping or stabilizing materials, such as the use of biopolymers, with attractive biocompatibility and low cytotoxicity, which have received widespread popularity in the synthesis of nanoparticles with potential applications in food-related products. MNPs and other nanoparticles are special compared to other materials owing to their high surface-area-to-volume ratio. This confers the ability of the materials to be transported into the internal crevices of animals and thus may elicit deleterious health complications. We support the general reservations on the suitability of nanoparticles and MNPs for food-related applications and thus recommend further evaluation of the safety of the nanoparticles in animal models. This will serve a holistic purpose for informed decision-making and guide researchers in choosing materials for nano-fabrications, especially those designed specifically for food applications.

Abbreviations

EDC = Ethyl-3-(3-dimethylaminopropyl) carbodiimide hydrochloride
NHS = N-hydroxy succinimide
PVA = Polyvinyl alcohol
PEG = Polyethylene glycol
EGCG = Epigallocatechin gallate
CMC = Carboxymethyl chitosan
AgNPs = Silver nanoparticles

Acknowledgments

Jayeoye T. J. and Olatunde O. O. would like to acknowledge the support of the Center of Excellence for Innovation in Chemistry and Division of Physical Science, Prince of Songkla University, Thailand, and the Richardson Centre for Functional Foods and Nutraceuticals, University of Manitoba, Canada, respectively, for the award of post-doctoral fellowship, while this work lasted. Sudarshan Singh would like to thank the Food Technology and Innovation Research Center of Excellence, Institute of Research and Innovation, Walailak University, Thailand, for extending the library facility and Scopus access for the award of a research fellowship.

References

Abbasi, A., Z. Emam-Djomeh and M. Ebrahimzadeh Mousavi. 2017. Stability of whey protein nanoparticles at various protein concentrations. Int. J. Nutr. Sci. 2(3): 165–169.

Aditya, N., I. E. Hamilton and I. T. Norton. 2017. Amorphous Nano-curcumin stabilized oil in water emulsion: physico chemical characterization. Food Chem. 224: 191–200.

Ahmad, M., S. Benjakul, P. Sumpavapol et al. 2012. Quality changes of sea bass slices wrapped with gelatin film incorporated with lemongrass essential oil. Int. J. Food Microbiol. 155(3): 171–178.

Alehosseini, E., S. M. Jafari and H. S. Tabarestani. 2021. Production of d-limonene-loaded pickering emulsions stabilized by chitosan nanoparticles. Food Chem. 354: 129591.

Ardekani-Zadeh, A. H. and S. F. Hosseini. 2019. Electrospun essential oil-doped chitosan/poly (ε-caprolactone) hybrid nanofibrous mats for antimicrobial food biopackaging exploits. Carbohydr. Polym. 223: 115108.

Aydogdu, A., G. Sumnu and S. Sahin. 2019. Fabrication of gallic acid loaded Hydroxypropyl methylcellulose nanofibers by electrospinning technique as active packaging material. Carbohydr. Polym. 208: 241–250.

Azzi, J., A. Jraij, L. Auezova et al. 2018. Novel findings for quercetin encapsulation and preservation with cyclodextrins, liposomes, and drug-in-cyclodextrin-in-liposomes. Food Hydrocoll 81: 328–340.

Babaei-Ghazvini, A., I. Shahabi-Ghahfarrokhi and V. Goudarzi. 2018. Preparation of UV-protective starch/kefiran/ZnO nanocomposite as a packaging film: Characterization. Food Package Shelf Life 16: 103–111.

Bagheri, L., A. Madadlou, M. Yarmand et al. 2013. Nanoencapsulation of date palm pit extract in whey protein particles generated via desolvation method. Food Res. Int. 51(2): 866–871.

Barick, K., A. Tripathi, B. Dutta et al. 2021. Curcumin encapsulated casein nanoparticles: Enhanced bioavailability and anticancer efficacy. J. Pharm. Sci. 110(5): 2114–2120.

Basniwal, R. K., H. S. Buttar, V. Jain et al. 2011. Curcumin nanoparticles: preparation, characterization, and antimicrobial study. J. Agric. Food Chem. 59(5): 2056–2061.

Bhat, M. Y., T. A. Dar and L. R. Singh. 2016. Casein proteins: structural and functional aspects. Milk proteins–from structure to biological properties health aspects. InTech, Rijeka 1–17.

Bhattarai, B., Y. Zaker and T. P. Bigioni. 2018. Green synthesis of gold and silver nanoparticles: Challenges and opportunities. Curr. Opin. Green Sustain. Chem. 12: 91–100.

Bruni, N., B. Stella, L. Giraudo et al. 2017. Nanostructured delivery systems with improved leishmanicidal activity: a critical review. Int. J. Nanomedicine. 12: 5289.

Çakmak, H., Y. Özselek, O. Y. Turan et al. 2020. Whey protein isolate edible films incorporated with essential oils: Antimicrobial activity and barrier properties. Polym. Degrad. 179: 109285.

Cano, A., M. Cháfer, A. Chiralt et al. 2016. Development and characterization of active films based on starch-PVA, containing silver nanoparticles. Food Packag Shelf Life 10: 16–24.

Carpenter, J. and V. K. Saharan. 2017. Ultrasonic assisted formation and stability of mustard oil in water nanoemulsion: Effect of process parameters and their optimization. Ultrason. Sonochem. 35: 422–430.

Carvalho, J. A., A. S. Abreu, V. T. P. Ferreira et al. 2018. Preparation of gelatin nanoparticles by two step desolvation method for application in photodynamic therapy. J. Biomater. Sci. Polym. Ed. 29(11): 1287–1301.

Chandra, H., P. Kumari, E. Bontempi et al. 2020. Medicinal plants: Treasure trove for green synthesis of metallic nanoparticles and their biomedical applications. Biocatal Agric. Biotechnol. 24: 101518.

Chawla, P., N. Kumar, A. Bains et al. 2020. Gum arabic capped copper nanoparticles: Synthesis, characterization, and applications. Int. J. Biol. Macromol. 146: 232–242.

Chevalier, Y. and M.-A. Bolzinger. 2013. Emulsions stabilized with solid nanoparticles: Pickering emulsions. Colloids Surf A: Physicochem. Eng. Asp. 439: 23–34.

Chorachoo, J., T. Amnuaikit and S. P. Voravuthikunchai. 2013. Liposomal encapsulated rhodomyrtone: a novel antiacne drug. Evid Based Complement Alternat. Med. 2013.

Chowdhury, S., Y. L. Teoh, K. M. Ong et al. 2020. Poly (Vinyl) alcohol crosslinked composite packaging film containing gold nanoparticles on shelf life extension of banana. Food Packag Shelf Life 24: 100463.

Chuesiang, P., U. Siripatrawan, R. Sanguandeekul et al. 2019. Antimicrobial activity and chemical stability of cinnamon oil in oil-in-water nanoemulsions fabricated using the phase inversion temperature method. LWT 110: 190–196.

Colletier, J.-P., B. Chaize, M. Winterhalter et al. 2002. Protein encapsulation in liposomes: Eficiency depends on interactions between protein and phospholipid bilayer. BMC Biotech. 2(1): 1–8.

Corredig, M., N. Sharafbafi and E. Kristo. 2011. Polysaccharide–protein interactions in dairy matrices, control and design of structures. Food Hydrocol. 25(8): 1833–1841.

Cui, F., S. Zhao, X. Guan et al. 2021. Polysaccharide-based Pickering Emulsions: Formation, Stabilization and Applications. Food Hydrocol 106812.

Dag, D., S. Guner and M. H. Oztop. 2019. Physicochemical mechanisms of different biopolymers' (lysozyme, gum arabic, whey protein, chitosan) adsorption on green tea extract loaded liposomes. Int. J. Biol. Macromol. 138: 473–482.

Davidov-Pardo, G., I. J. Joye and D. J. McClements. 2015. Encapsulation of resveratrol in biopolymer particles produced using liquid antisolvent precipitation. Part 1: Preparation and characterization. Food Hydrocol. 45: 309–316.

Dávila-Rodríguez, M., A. Lopez-Malo, E. Palou et al. 2019. Antimicrobial activity of nanoemulsions of cinnamon, rosemary, and oregano essential oils on fresh celery. LWT 112: 108247.

De Leo, V., F. Milano, E. Mancini et al. 2018. Encapsulation of curcumin-loaded liposomes for colonic drug delivery in a pH-responsive polymer cluster using a pH-driven and organic solvent-free process. Molecules 23(4): 739.

DeFrates, K., T. Markiewicz, P. Gallo et al. 2018. Protein polymer-based nanoparticles: fabrication and medical applications. Int. J. Mol. Sci. 19(6): 1717.

Devara, R. K., H. U. R. Mohammad, B. Rambabu et al. 2015. Preparation, optimization and evaluation of intravenous curcumin nanosuspensions intended to treat liver fibrosis. Turk. J. Pharm. Sci. 12: 8–24.

Duong, B. H., H. N. Truong, Q. A. Phan Nguyen et al. 2020. Preparation of curcumin nanosuspension with gum arabic as a natural stabilizer: Process optimization and product characterization. Processes 8(8): 970.

El-Alfy, E. A., M. K. El-Bisi, G. M. Taha et al. 2020. Preparation of biocompatible chitosan nanoparticles loaded by tetracycline, gentamycin and ciprofloxacin as novel drug delivery system for improvement the antibacterial properties of cellulose based fabrics. Int. J. Biol. Macromol. 161: 1247–1260.

Elzoghby, A. O. 2013. Gelatin-based nanoparticles as drug and gene delivery systems: reviewing three decades of research. J. Control Release 172(3): 1075–1091.

Enderle, J. 2012. Introduction to biomedical engineering: Academic press.

Esfanjani, A. F. and S. M. Jafari. 2016. Biopolymer nano-particles and natural nano-carriers for nano-encapsulation of phenolic compounds. Colloids Surf B Biointerfaces 146: 532–543.

Eze, F. N., T. J. Jayeoye and A. J. Tola. 2021. Fabrication of label-free and eco-friendly ROS optical sensor with potent antioxidant properties for sensitive hydrogen peroxide detection in human plasma. Colloids Surf B Biointerfaces 111798.

Eze, F. N., A. J. Tola, O. F. Nwabor et al. 2019. Centella asiatica phenolic extract-mediated bio-fabrication of silver nanoparticles: characterization, reduction of industrially relevant dyes in water and antimicrobial activities against foodborne pathogens. RSC Adv. 9(65): 37957–37970.

Falleh, H., M. B. Jemaa, M. A. Neves et al. 2021. Formulation, physicochemical characterization, and anti-*E. coli* activity of food-grade nanoemulsions incorporating clove, cinnamon, and lavender essential oils. Food Chem 129963.

Fang, J.-Y., W.-R. Lee, S.-C. Shen et al. 2006. Effect of liposome encapsulation of tea catechins on their accumulation in basal cell carcinomas. J. Dermatol. Sci. 42(2): 101–109.

Fang, Z. and B. Bhandari. 2010. Encapsulation of polyphenols – a review. Trends Food Sci. Technol. 21(10): 510–523.

Feng, S., Y. Sun, D. Wang et al. 2020a. Effect of adjusting pH and chondroitin sulfate on the formation of curcumin-zein nanoparticles: Synthesis, characterization and morphology. Carbohydr. Polym. 250: 116970.

Feng, X., H. Dai, L. Ma et al. 2020b. Properties of Pickering emulsion stabilized by food-grade gelatin nanoparticles: influence of the nanoparticles concentration. Colloids Surf B Biointerfaces 196: 111294.

Fereydouni, N., J. Movaffagh, N. Amiri et al. 2021. Synthesis of nano-fibers containing nano-curcumin in zein corn protein and its physicochemical and biological characteristics. Sci. Rep. 11(1): 1–15.

Fouda, M. M., M. El-Aassar and S. S. Al-Deyab. 2013. Antimicrobial activity of carboxymethyl chitosan/polyethylene oxide nanofibers embedded silver nanoparticles. Carbohydr. polym. 92(2): 1012–1017.

Geh, K. J., M. Hubert and G. Winter. 2016. Optimisation of one-step desolvation and scale-up of gelatine nanoparticle production. J. Microencapsul. 33(7): 595–604.

Gharibzahedi, S. M. T. and S. Mohammadnabi. 2017. Effect of novel bioactive edible coatings based on jujube gum and nettle oil-loaded nanoemulsions on the shelf-life of Beluga sturgeon fillets. Int. J. Biol. Macromol. 95: 769–777.

Ghosh, A. K. and P. Bandyopadhyay. 2012. Polysaccharide-protein interactions and their relevance in food colloids. In The complex world of polysaccharides: InTech Croatia.

Giroux, H. J., J. Houde and M. Britten. 2010. Preparation of nanoparticles from denatured whey protein by pH-cycling treatment. Food Hydrocol. 24(4): 341–346.

Gómez-Estaca, J., M. P. Balaguer, R. Gavara et al. 2012. Formation of zein nanoparticles by electrohydrodynamic atomization: Effect of the main processing variables and suitability for encapsulating the food coloring and active ingredient curcumin. Food Hydrocol. 28(1): 82–91.

Gortzi, O., S. Lalas, I. Chinou et al. 2006. Reevaluation of antimicrobial and antioxidant activity of *Thymus* spp. extracts before and after encapsulation in liposomes. J. Food Prot. 69(12): 2998–3005.

Gu, B., Q. Jiang, B. Luo et al. 2021. A sandwich-like chitosan-based antibacterial nanocomposite film with reduced graphene oxide immobilized silver nanoparticles. Carbohydr. Polym. 260: 117835.

Gülseren, İ., Y. Fang and M. Corredig. 2012. Zinc incorporation capacity of whey protein nanoparticles prepared with desolvation with ethanol. Food Chem. 135(2): 770–774.

Gulzar, S. and S. Benjakul. 2020. Characteristics and storage stability of nanoliposomes loaded with shrimp oil as affected by ultrasonication and microfluidization. Food Chem. 310: 125916.

Gupta, A., H. B. Eral, T. A. Hatton et al. 2016. Nanoemulsions: formation, properties and applications. Soft Matter. 12(11): 2826–2841.

Hanani, Z. N., Y. Roos and J. Kerry. 2014. Use and application of gelatin as potential biodegradable packaging materials for food products. Int. J. Biol. Macromol. 71: 94–102.

He, M., M. Chen, Y. Dou et al. 2020. Electrospun silver nanoparticles-embedded feather keratin/poly (vinyl alcohol)/poly (ethylene oxide) antibacterial composite nanofibers. Polym. 12(2): 305.

Hemmati, F., A. Bahrami, A. F. Esfanjani et al. 2021. Electrospun antimicrobial materials: Advanced packaging materials for food applications. Trends Food Sci. Tech.

Hosseini, S. F., L. Ramezanzade and M. Nikkhah. 2017b. Nano-liposomal entrapment of bioactive peptidic fraction from fish gelatin hydrolysate. Int. J. Biol. Macromol. 105: 1455–1463.

Hosseini S.M., F. Giasi and M. Jahromi 2017a. Nanocapsule formation by complexation of biopolymers in Nanoencapsulation technologies for the food and nutraceutical industries: pp. 447–492, Elsevier.

Hu, D., C. Lin, L. Liu et al. 2012. Preparation, characterization, and *in vitro* release investigation of lutein/zein nanoparticles via solution enhanced dispersion by supercritical fluids. J. Food Eng. 109(3): 545–552.

İlyasoğlu, H., M. Nadzieja and Z. Guo. 2019. Caffeic acid grafted chitosan as a novel dual-functional stabilizer for food-grade emulsions and additive antioxidant property. Food Hydrocol. 95: 168–176.

Islam, M. S. and J. H. Yeum. 2013. Electrospun pullulan/poly (vinyl alcohol)/silver hybrid nanofibers: preparation and property characterization for antibacterial activity. Colloids Surf A: Physicochem. Eng. Asp. 436: 279–286.

Jambeck, J. R., R. Geyer, C. Wilcox et al. 2015. Plastic waste inputs from land into the ocean. Sci. 347(6223):768–771.

Jangde, R. and D. Singh. 2016. Preparation and optimization of quercetin-loaded liposomes for wound healing, using response surface methodology. Artif Cells Nanomed. Biotechnol. 44(2): 635–641.

Janjarasskul, T., K. Tananuwong, T. Phupoksakul et al. 2020. Fast dissolving, hermetically sealable, edible whey protein isolate-based films for instant food and/or dry ingredient pouches. LWT 134: 110102.

Jantrawut, P., T. Chaiwarit, K. Jantanasakulwong et al. 2017. Effect of plasticizer type on tensile property and *in vitro* indomethacin release of thin films based on low-methoxyl pectin. Polym. 9(7): 289.

Jayeoye, T. J., F. N. Eze, O. O. Olatunde et al. 2021b. Synthesis of silver and silver@ zero valent iron nanoparticles using *Chromolaena odorata* phenolic extract for antibacterial activity and hydrogen peroxide detection. J. Environ. Chem. Eng. 9(3): 105224..

Jayeoye, T. J., F. N. Eze, S. Singh et al. 2021a. Synthesis of gold nanoparticles/polyaniline boronic acid/sodium alginate aqueous nanocomposite based on chemical oxidative polymerization for biological applications. Int. J. Biol. Macromol. 179: 196–205.

Jayeoye, T. J., O. F. Nwabor and T. Rujiralai. 2020a. Synthesis of highly stable and dispersed silver nanoparticles/poly (vinyl alcohol-co-ethylene glycol)/poly (3-aminophenyl boronic acid) nanocomposite: Characterization and antibacterial, hemolytic and cytotoxicity studies. J. Ind. Eng. Chem. 288–300.

Jayeoye, T. J., O. O. Olatunde, S. Benjakul et al. 2020b. Synthesis and characterization of novel poly (3-aminophenyl boronic acid-co-vinyl alcohol) nanocomposite polymer stabilized silver nanoparticles with antibacterial and antioxidant applications. Colloids Surf B Biointerfaces 193: 111112.

Jayeoye, T. J. and T. Rujiralai. 2020. Green, in situ fabrication of silver/poly (3-aminophenyl boronic acid)/sodium alginate nanogel and hydrogen peroxide sensing capacity. Carbohydr. Polym. 246: 116657.

Jiang, L., F. Jia, Y. Han et al. 2021. Development and characterization of zein edible films incorporated with catechin/β-cyclodextrin inclusion complex nanoparticles. Carbohydr. Polym. 261: 117877.

Jiménez, M., J. A. Domínguez, L. A. Pascual-Pineda et al. 2018. Elaboration and characterization of O/W cinnamon (*Cinnamomum zeylanicum*) and black pepper (*Piper nigrum*) emulsions. Food Hydrocol. 77: 902–910.

Kanmani, P. and J.-W. Rhim. 2014. Properties and characterization of bionanocomposite films prepared with various biopolymers and ZnO nanoparticles. Carbohydr. polym. 106: 190–199.

Karimi, A., G. Askari, M. S. Yarmand et al. 2020. Development, modification and characterization of ursolic acid-loaded gelatin nanoparticles through electrospraying technique. Food Bioprod. Process 124: 329–341.

Karn, P. R., W. Cho, H.-J. Park et al. 2013. Characterization and stability studies of a novel liposomal cyclosporin A prepared using the supercritical fluid method: comparison with the modified conventional Bangham method. Int. J. Nanomed. 8: 365.

Karthikeyan, A., N. Senthil and T. Min. 2020. Nanocurcumin: A promising candidate for therapeutic applications. Front Pharmacol. 11.

Kumar, S., I. B. Basumatary, H. P. Sudhani et al. 2021. Plant extract mediated silver nanoparticles and their applications as antimicrobials and in sustainable food packaging: A state-of-the-art review. Trends Food Sci. Tech.

Kumar, S., A. Mudai, B. Roy et al. 2020. Biodegradable hybrid nanocomposite of chitosan/gelatin and green synthesized zinc oxide nanoparticles for food packaging. Foods 9(9): 1143.

Kumar, S., A. Shukla, P. P. Baul et al. 2018. Biodegradable hybrid nanocomposites of chitosan/gelatin and silver nanoparticles for active food packaging applications. Food Packag Shelf Life 16: 178–184.

Lasic, D., R. Joannic, B. Keller et al. 2001. Spontaneous vesiculation. Adv. Colloid Interface Sci. 89: 337–349.

Li, K.-K., S.-W. Yin, Y.-C. Yin et al. 2013. Preparation of water-soluble antimicrobial zein nanoparticles by a modified antisolvent approach and their characterization. Journal Food Eng. 119(2): 343–352.

Li, S., H. Chen, W. Cheng et al. 2021. Impact of arabinoxylan on characteristics, stability and lipid oxidation of oil-in-water emulsions: arabinoxylan from wheat bran, corn bran, rice bran, and rye bran. Food Chem. 129813.

Liang, J., H. Yan, X. Wang et al. 2017. Encapsulation of epigallocatechin gallate in zein/chitosan nanoparticles for controlled applications in food systems. Food Chem. 231: 19–24.

Liu, J., C.-g. Meng, Y.-h. Yan et al. 2016. Protocatechuic acid grafted onto chitosan: Characterization and antioxidant activity. Int. J. Biol. Macromol. 89: 518–526.

Liu, W., A. Ye, W. Liu et al. 2015. Behaviour of liposomes loaded with bovine serum albumin during in vitro digestion. Food Chem. 175: 16–24.

Lv, S., H. Zhou, L. Bai et al. 2021. Development of food-grade Pickering emulsions stabilized by a mixture of cellulose nanofibrils and nanochitin. Food Hydrocol. 113: 106451.

Madureira, A. R., C. I. Pereira, A. M. Gomes et al. 2007. Bovine whey proteins–Overview on their main biological properties. Food Res. Int. 40(10): 1197–1211.

Mazandrani, H. A., S. Javadian and S. Bahram. 2016. The effect of encapsulated fennel extracts on the quality of silver carp fillets during refrigerated storage. Food Sci. Nut. 4 (2): 298–304.

Mo, X., X. Peng, X. Liang et al. 2021. Development of antifungal gelatin-based nanocomposite films functionalized with natamycin-loaded zein/casein nanoparticles. Food Hydrocol. 113: 106506.

Mohanty, B., V. Aswal, J. Kohlbrecher et al. 2005. Synthesis of gelatin nanoparticles via simple coacervation. J. Surf. Sci. Technol. 21(3/4): 149.

Moon, S. and J. I. Lee. 2013. Electrospun poly (ethylene oxide)/hydrated iron oxide/sodium alginate composite nanofibers with multi-functional properties. Polym. Eng. Sci. 53(6): 1321–1326.

Mozafari, M. 2006. Bioactive entrapment and targeting using nanocarrier technologies: An introduction. pp 1–16. *In*: Mozafari, M. R. (Ed.). Nanocarrier technologies: Frontiers of nanotherapy. The Netherlands: Springer.

Mozafari, M., K. Khosravi-Darani, G. G. Borazan et al. 2008. Encapsulation of food ingredients using nanoliposome technology. Int. J. Food Prop. 11(4): 833–844.

Mozafari, M. and S. Mortazavi. 2005. Nanoliposomes: from fundamentals to recent developments. Trafford Pub. Ltd. Oxford, UK.

Mozafari, M. R. 2007. Nanomaterials and nanosystems for biomedical applications: Springer Science & Business Media.

Muzzarelli, R. A. 2009. Genipin-crosslinked chitosan hydrogels as biomedical and pharmaceutical aids. Carbohydr. Polym. 77(1): 1–9.

Naskar, S., S. Sharma and K. Kuotsu. 2019. A smart gelatin nanoparticle for delivery of metoprolol succinate: A strategy for enhancing the therapeutic efficacy by improving bioavailability. J. Drug Deliv. Sci. Technol. 53: 101214.

Nasrollahzadeha, M. N. S., M. Mahamb and Z. Nezafata. 2021. Protein and polypeptide biopolymer chemistry. Biopolymer-Based Metal Nanoparticle Chemistry for Sustainable Applications: Volume 1: Classification, Properties Synthesis 107.

Olatunde, O. O., S. Benjakul and K. Vongkamjan. 2020a. Olatunde, O. O., S. Benjakul and K. Vongkamjan. 2020. Microbial diversity, shelf-life and sensory properties of Asian sea bass slices with combined treatment of liposomal encapsulated ethanolic coconut husk extract and high voltage cold plasma. LWT 134: 110232.

Olatunde, O. O., S. Benjakul and K. Vongkamjan. 2020b. Cold plasma combined with liposomal ethanolic coconut husk extract: A potential hurdle technology for shelf-life extension of Asian sea bass slices packaged under modified atmosphere. Innov. Food Sci. Emerg. Technol. 65: 102448.

Olatunde, O. O., S. Benjakul, K. Vongkamjan and T. Amnuaikit. 2019. Liposomal encapsulated ethanolic coconut husk extract: Antioxidant and antibacterial properties. J. Food Sci. 84(12): 3664–3673.

Olatunde, O. O., S. Benjakul, K. Vongkamjan and T. Amnuaikit. 2020c. Influence of stabilising agents on the properties of liposomal encapsulated ethanolic coconut husk extract. Int. J. Food Sci. Technol. 55(2): 702–711.

Ortega, F., V. B. Arce and M. A. Garcia. 2021. Nanocomposite starch-based films containing silver nanoparticles synthesized with lemon juice as reducing and stabilizing agent. Carbohydr. Polym. 252: 117208.

Pandit, R. S., S. C. Gaikwad, G. A. Agarkar et al. 2015. Curcumin nanoparticles: physico-chemical fabrication and its in vitro efficacy against human pathogens. 3 Biotech. 5(6): 991–997.

Park, H. J., Y. J. Byun, Y. T. Kim et al. 2014. Processes and applications for edible coating and film materials from agropolymers. In Innovations in food packaging: Elsevier.

Patel, A., Y. Hu, J. K. Tiwari and K. P. Velikov. 2010. Synthesis and characterisation of zein–curcumin colloidal particles. Soft Matter 6(24): 6192–6199.

Patel, R., S. Singh, S. Singh et al. 2009. Development and characterization of curcumin loaded transfersome for transdermal delivery. J. Pharm. Sci. Res. 1(4): 71.

Peles, Z., I. Binderman, I. Berdicevsky et al. 2013. Soy protein films for wound-healing applications: antibiotic release, bacterial inhibition and cellular response. J. Tissue Eng. Regen. Med. 7(5): 401–412.

Penalva, R., I. Esparza, M. Agüeros et al. 2015. Casein nanoparticles as carriers for the oral delivery of folic acid. Food Hydrocol. 44: 399–406.

Peng, S., L. Zhou, Q. Cai et al. 2020. Utilization of biopolymers to stabilize curcumin nanoparticles prepared by the pH-shift method: Caseinate, whey protein, soy protein and gum Arabic. Food Hydrocol. 107: 105963.

Ribeiro, E. F., T. T. de Barros-Alexandrino, O. B. G. Assis et al. 2020. Chitosan and crosslinked chitosan nanoparticles: Synthesis, characterization and their role as Pickering emulsifiers. Carbohydr. Polym. 250: 116878.

Ribes, S., A. Fuentes, P. Talens et al. 2017. Influence of emulsifier type on the antifungal activity of cinnamon leaf, lemon and bergamot oil nanoemulsions against *Aspergillus niger.* Food Control 73: 784–795.

Roy, S., J.-W. Rhim and L. Jaiswal. 2019. Bioactive agar-based functional composite film incorporated with copper sulfide nanoparticles. Food Hydrocol. 93: 156–166.

Sarwar, M. S., M. B. K. Niazi, Z. Jahan et al. 2018. Preparation and characterization of PVA/nanocellulose/ Ag nanocomposite films for antimicrobial food packaging. Carbohydr. Polym. 184: 453–464.

Savaghebi, D., M. Barzegar and M. R. Mozafari. 2020. Manufacturing of nanoliposomal extract from Sargassum boveanum algae and investigating its release behavior and antioxidant activity. Food Sci. Nut. 8(1): 299–310.

Savaghebi, D., M. Ghaderi-Ghahfarokhi and M. Barzegar. 2021. Encapsulation of *Sargassum boveanum* Algae extract in nano-liposomes: Application in functional mayonnaise production. Food Bioprocess Technol. 1–15.

Schmitt, C., L. Aberkane and C. Sanchez. 2009. Protein–polysaccharide complexes and coacervates. In Handbook of hydrocolloids: Elsevier.

Sekhon, B. S. 2010. Food nanotechnology–an overview. Nanotechnol. Sci. Appl. 3: 1.

Shaikh, S., N. Nazam, S. M. D. Rizvi et al. 2019. Mechanistic insights into the antimicrobial actions of metallic nanoparticles and their implications for multidrug resistance. Int. J. Mol. Sci. 20(10): 2468.

Shankar, S., L.-F. Wang and J.-W. Rhim. 2017. Preparation and properties of carbohydrate-based composite films incorporated with CuO nanoparticles. Carbohydr. Polym. 169: 264–271.

Shoji, Y. and H. Nakashima. 2004. Nutraceutics and delivery systems. J. Drug Target 12(6): 385–391.

Siegel, D. and B. Tenchov. 2008. Influence of the lamellar phase unbinding energy on the relative stability of lamellar and inverted cubic phases. Biophys. J. 94(10): 3987–3995.

Singh, A., A. Mittal and S. Benjakul. 2021a. Chitosan nanoparticles: Preparation, food applications and health benefits. Science Asia 47: 1–10.

Singh, S., O. F. Nwabor, D. M. Syukri et al. 2021b. Chitosan-poly(vinyl alcohol) intelligent films fortified with anthocyanins isolated from *Clitoria ternatea* and *Carissa carandas* for monitoring beverage freshness. Int. J. Biol. Macromol. 182: 1015–1025.

Slavin, Y. N., J. Asnis, U. O. Häfeli et al. 2017. Metal nanoparticles: understanding the mechanisms behind antibacterial activity. J. Nanobiotechnology 15(1): 1–20.

Suryani, N. H. A. Halid, N. I. Akib, Rahmanpiu and N. Mutmainnah. 2017. Preparation of curcumin nanoparticle by using reinforcement ionic gelation technique. Paper read at AIP Conference Proceedings.

Syahida, S. N., M. Ismail-Fitry, Z. Ainun et al. 2021. Effects of gelatin/palm wax/lemongrass essential oil (GPL)-coated Kraft paper on the quality and shelf life of ground beef stored at 4°C. Food Packag Shelf Life 28: 100640.

Tan, H., L. Han and C. Yang. 2021. Effect of oil type and β-carotene incorporation on the properties of gelatin nanoparticle-stabilized pickering emulsions. LWT 141: 110903.

Tang, J., N. Lin, Z. Zhang et al. 2019. Nanopolysaccharides in emulsion stabilization. In Advanced Functional Materials from Nanopolysaccharides: Springer.

Tavares, L., H. K. Souza, M. P. Gonçalves et al. 2021. Physicochemical and microstructural properties of composite edible film obtained by complex coacervation between chitosan and whey protein isolate. Food Hydrocoll. 113: 106471.

Teng, Z., Y. Li and Q. Wang. 2014. Insight into curcumin-loaded β-lactoglobulin nanoparticles: Incorporation, particle disintegration, and releasing profiles. J. Agric. Food Chem. 62(35): 8837–8847.

Teng, Z., Y. Luo and Q. Wang. 2012. Nanoparticles synthesized from soy protein: preparation, characterization, and application for nutraceutical encapsulation. J. Agric. Food Chem. 60(10): 2712–2720.

Teng, Z., Y. Luo and Q. Wang. 2013. Carboxymethyl chitosan–soy protein complex nanoparticles for the encapsulation and controlled release of vitamin D3. Food Chem. 141(1): 524–532.

Thompson, R. C., C. J. Moore, F. S. Vom Saal et al. 2009. Plastics, the environment and human health: current consensus and future trends. Philos. Trans. R Soc. Lond B Biol. Sci. 364(1526): 2153–2166.

Trucillo, P., R. Campardelli and E. Reverchon. 2019. A versatile supercritical assisted process for the one-shot production of liposomes. J. Supercrit. Fluids 146: 136–143.

Wang, X.-L., K.-K. Yang and Y.-Z. Wang. 2003. Properties of starch blends with biodegradable polymers. J. Macro Sci. C Polym. Rev. 43(3): 385–409.

Wang, Y., L. Yu, Q. Sun et al. 2021. Hydroxypropyl methylcellulose and hydroxypropyl starch: Rheological and gelation effects on the phase structure of their mixed hydrocolloid system. Food Hydrocol. 115: 106598.

Warnakulasuriya, S. N. and M. T. Nickerson. 2018. Review on plant protein–polysaccharide complex coacervation, and the functionality and applicability of formed complexes. J. Sci. Food Agric. 98(15): 5559–5571.

Wei, Y., C. Wang, X. Liu et al. 2020. Impact of microfluidization and thermal treatment on the structure, stability and in vitro digestion of curcumin loaded zein-propylene glycol alginate complex nanoparticles. Food Res. Int. 138: 109817.

Wongphan, P. and N. Harnkarnsujarit. 2020. Characterization of starch, agar and maltodextrin blends for controlled dissolution of edible films. Int. J. Biol. Macromol. 156: 80–93.

Xiao, N., W. He, Y. Zhao et al. 2021. Effect of pH and xanthan gum on emulsifying property of ovalbumin stabilized oil-in water emulsions. LWT: 111621.

Xu, C., S. Guan, J. Xu et al. 2021. Preparation, characterization and antioxidant activity of protocatechuic acid grafted carboxymethyl chitosan and its hydrogel. Carbohydr. Polym. 252: 117210.

Yanat, M. and K. Schroën. 2021. Preparation methods and applications of chitosan nanoparticles; with an outlook toward reinforcement of biodegradable packaging. React. Funct. Polym. 104849.

Yuan, D., F. Zhou, P. Shen et al. 2021. Self-Assembled soy protein nanoparticles by partial enzymatic hydrolysis for pH-driven encapsulation and delivery of hydrophobic cargo curcumin. Food Hydrocol. 106759.

Yuan, W., H. W. Lee and H.-G. Yuk. 2017. Antimicrobial efficacy of *Cinnamomum javanicum* plant extract against *Listeria monocytogenes* and its application potential with smoked salmon. Int. J. Food Microbiol. 260: 42–50.

Zagana, P., M. Haikou, P. Klepetsanis et al. 2008. *In vivo* distribution of arsonoliposomes: effect of vesicle lipid composition. Int. J. Pharm. 347(1-2): 86–92.

Zhang, C.-L. and S.-H. Yu. 2014. Nanoparticles meet electrospinning: recent advances and future prospects. Chem. Soc. Rev. 43(13): 4423–4448.

Zhao, C. and T. J. Ashaolu. 2020. Bioactivity and safety of whey peptides. LWT: 109935.

Zhou, P., L. Zhou, C. Zhu et al. 2016. Nanogel-electrospinning for controlling the release of water-soluble drugs. J. Mat. Chem. B 4(12): 2171–2178.

Zhu, W. and Z. Zhang. 2014. Preparation and characterization of catechin-grafted chitosan with antioxidant and antidiabetic potential. Int. J. Biol. Macromol. 70: 150–155.

Zou, L., J. Wang, Y. Gao et al. 2018. Synergistic antibacterial activity of silver with antibiotics correlating with the upregulation of the ROS production. Sci. Rep. 8(1): 1–11.

Zou, Y., C. Zhang, P. Wang et al. 2020. Electrospun chitosan/polycaprolactone nanofibers containing chlorogenic acid-loaded halloysite nanotube for active food packaging. Carbohydr. Polym. 247: 116711.

7

Biopolymers, Hydrogel and Their Effect on Drug Development

Bikash Ranjan Jena,[1] *Bhisma Narayan Ratha,*[1,2,*] *Gurudutta Pattnaik,*[1] *Naga Jogayya Kothakota,*[3] *Biswajit Puhan,*[1] *Sk Sahirul Islam*[1] and *Anshuman Panda*[1,2]

1. Introduction

Biopolymers are an intriguing category of polymers derived from organic sources, such as plants, animals, and microbes. The aforementioned polymers possess numerous advantages in comparison to their synthetic equivalents due to their enhanced sustainability, reduced environmental impact, and frequent biodegradability (Carter and Grove 2019). Biopolymers play a pivotal role in addressing the escalating concerns surrounding plastic pollution and the increasing need for more environmentally sustainable materials across various industries (McClements 2018). Biopolymers, which are derived from living organisms such as plants, animals, and microbes, constitute a distinct subset within the broader category of polymers (Sikkema et al. 2020). Distinguishing between natural polymers, synthetic polymers, and semi-synthetic hydrogels, which are derived from modified natural polymers and incorporate synthetic linkers, is possible. However, while considering the chemical aspect, it can be shown that synthetic polymers exhibit greater durability compared to natural polymers.

[1] School of Pharmacy and Life Sciences, Centurion University of Technology and Management, Bhubaneswar –752050, Odisha, India.
[2] School of Agricultural and Bio-Engineering, Centurion University of Technology and Management, Odisha, India.
[3] School of Forensic Sciences, Centurion University of Technology and Management, Vizianagaram, Andhra Pradesh, India.
* Corresponding author: bhisma.n.ratha@gmail.com

Due to their exceptional mechanical durability, these entities exhibit an extended lifespan and significantly sluggish rate of decay. The necessity to reconcile these two contradictory characteristics necessitates the implementation of an optimal design (Sikkema et al. 2020). Chemical gels utilize covalent bonding as a means to provide the necessary structural integrity for gel formation, whereas biological gels employ substances such as enzymes or amino acids to facilitate the gel-forming process. Physical gels have the ability to undergo a transition from a liquid state to a gel state in response to external stimuli. Due to its renewable nature and positive environmental impact, there is a growing trend of extensive research and implementation of these technologies across several industries. The utilization of biopolymers such as polylactic acid (PLA), polyhydroxyalkanoates (PHA), and starch-based polymers in the production of packaging materials offers a potential solution for reducing the environmental consequences associated with traditional plastic packaging (Wen et al. 2018). Figure 1 represents a schematic diagram that provides clarification on the many applications of biopolymers.

The primary applications for biopolymers are in the fields of tissue engineering, drug delivery systems, and medical device manufacturing. Wound dressings, tissue regeneration scaffolds, and controlled drug release systems are all examples of applications for chitosan and alginate, respectively. In the food sector, biopolymers have a number of applications, including those of thickeners, stabilizers, and emulsifiers (Zueva et al. 2019).

The various categories of biopolymers in drug delivery systems are elucidated in Figure 2.

Several biopolymers are used to improve, including pectin, alginate, and carrageenan (Zueva et al. 2019). These biopolymers have various characteristic features, which are demonstrated in Table 1.

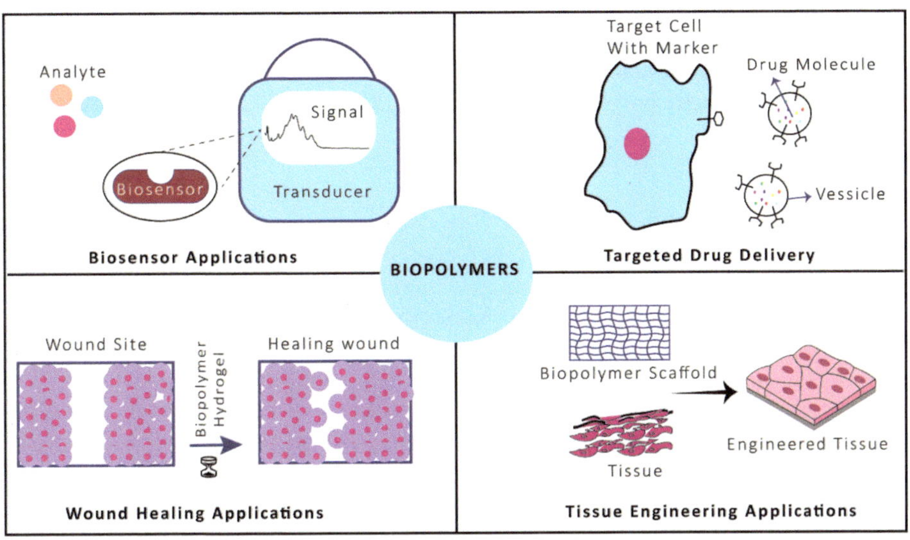

Figure 1: Schematic illustration of various applications of biopolymer (Modified from Biswas et al. 2021).

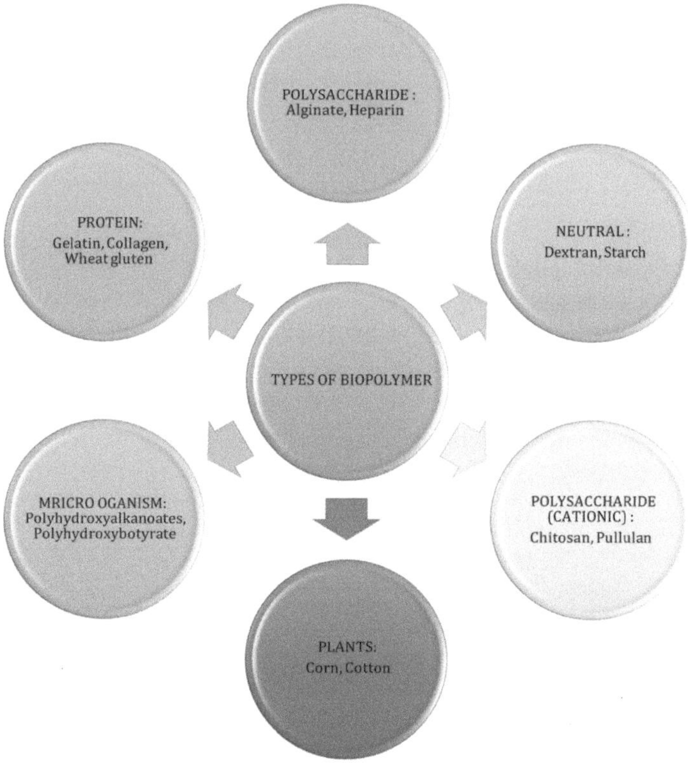

Figure 2: Categories of biopolymers.

In inference, biopolymers constitute a realistic option for creating more sustainable materials in a society that struggles with plastic waste and environmental damage. Because of their renewable status, biodegradability, and versatility, they are a valuable resource for many industries seeking environmentally benign alternatives to traditional synthetic polymers (Rostami et al. 2019).

2. Hydrogels

A hydrogel is defined as a 3D framework consisting of hydrophilic polymers that possess the capacity to imbibe water and retain it inside the interconnected structure. The preservation of the structure is accomplished by means of the chemical or physical crosslinking of the polymer chains (Denizli et al. 2004). For a substance to be classified as a hydrogel, it is required that it constitutes at least 10% of the total weight of the material (Chai et al. 2017). A multitude of theories has been created over time to enable the anticipation of structural attributes of hydrogels, encompassing elasticity, porosity, and pore dimensions within the hydrogel framework. These theories include many thermodynamic parameters, such as enthalpy and entropy, to determine the structure and pore sizes of hydrogels. Furthermore, computational modeling methodologies can be employed to construct the desired hydrogel design effectively. Hydrogels can be classified into two primary categories: those comprised

Table 1: Characteristic features of biopolymers and its sources.

S. No	Characteristic Features	Sources/Explanations	References
01	Renewable source	Typically, renewable resources like starch, cellulose, proteins, and other types of carbohydrates are used to create biopolymers. These resources can replace fossil fuels since they are abundant and can be regenerated throughout time.	(Zhang et al. 2016; Tirrell and Tirrell 1996)
02	Biodegradability	One of the biggest benefits of biopolymers is that they decay spontaneously in the environment, as opposed to many synthetic polymers that last for decades or even centuries. This quality considerably lessens the harm that plastic trash does to ecosystems and lessens environmental pollution.	(Makam and Gazit 2018)
03	Less carbon footprint	The production of biopolymers usually involves less energy use and greenhouse gas emissions than the fabrication of synthetic polymers. This is because biopolymers can be made from materials that are renewable and utilizing processes that are more environmentally friendly.	(Pina et al. 2022)
04	Diversity	Because biopolymers come in a number of forms and possess a range of properties, they can be tailored for specific uses. Polymers made from starch, cellulose, proteins, and polyhydroxyalkanoates (PHA) are a few examples of common types. Each type has unique properties that make it suited for different industries.	(Joye and McClements 2016)
05	Biocompatibility	A lot of biopolymers are biocompatible, which means that employing them in pharmaceutical and medical applications will not have any negative effects on people. Examples include biopolymers utilized in tissue engineering scaffolds, medication delivery systems, and medical sutures.	(Makshakova et al. 2022)
06	Functional characteristics	It is possible to create biopolymers with certain functional characteristics, such as flexibility, mechanical strength, and thermal stability. Researchers are always working to improve the functionality of biopolymers so that they can match or even outperform the properties of traditional synthetic polymers.	(Hilal et al. 2023)
07	Versatile applications	Biopolymers are used in a wide range of industries, including packaging, textiles, agriculture, food, medicine, and cosmetics. For instance, bio-based packaging materials can replace single-use plastics, and bio-based textiles can offer environmentally acceptable alternatives to conventional fabrics.	(Hilal et al. 2023)
08	Obstacles	While there are many benefits to using biopolymers, there are also disadvantages, such as higher manufacturing costs than regular plastics, limitations on some features, and competition for land and other resources that are used as their raw materials. Research and innovation are essential to overcoming these challenges and improving biopolymers' competitiveness in many applications.	(Hwang and Damodaran 1996)
09	Regulatory Support	By putting in place policies that encourage the reduction of plastic waste and the use of sustainable materials, numerous governments and organizations worldwide are promoting the use of biopolymers. The adoption and continued development of biopolymers may be influenced by this regulatory assistance. This regulatory support could hasten biopolymers' development and acceptability.	(Refojo and Holly 1977)

of natural polymers and those comprised of synthetic polymers. Moreover, it is worth noting the existence of a distinct classification referred to as semi-synthetic hydrogels. These hydrogels are produced by modifying natural polymers in conjunction with synthetic linkers (Silva et al. 2009). Synthetic polymers demonstrate enhanced chemical strength as compared to their natural counterparts. The remarkable durability of these materials can be ascribed to their robust mechanical properties, which concurrently facilitate a moderate rate of biodegradation. The need to harmonize these two divergent characteristics requires the execution of an idealized structure (Berkovitch and Seliktar 2017). Hydrogels demonstrate many gelation mechanisms, encompassing physical, chemical, and biological processes. Physical gels exhibit a transformation from a fluidic phase to a gel phase in response to external stimuli. In contrast, chemical gels depend on creating covalent connections to provide the requisite structural integrity required for the process of gelation. Biochemical gels utilize various substances, such as enzymes or amino acids, to aid the process of gelation (Tabata 2009). The gels possess cationic, anionic, or neutral characteristics depending on the ionic charges of the bound groups (Chai et al. 2017).

Hydrogels have the ability to undergo structural alterations in response to many stimuli, encompassing both chemical and physical impacts. The synthesis of hydrogels with shape-changing properties can be accomplished through the stratification of distinct hydrogel layers, each exhibiting a unique response to a particular stimulus (Tabata 2009). Hydrogels can be categorized into two primary classifications, specifically homopolymers and copolymers, which are identified using diverse polymerization techniques, including suspension, block, solution, and emulsion (Chai et al. 2017; Tabata 2009). Homopolymers consist only of a single type of monomer in their molecular structure. The observation of a crosslinked arrangement in homopolymers is contingent upon the characteristics of the monomer and the particular polymerization process employed. Copolymeric hydrogels consist

Table 2: Biomolecules as biopolymers and their biomedical applications.

Nature of Biopolymer	Biomedical Application	Examples	References
Polypeptide	Drug delivery	Collagen/chitosan gel	(Li et al. 2019)
	Tissue engineering	IKVAV, and RGD peptide	(Liyanage et al. 2017)
	Tissue implants	IEIK13	(Dufour et al. 2019)
Polysaccharides	Drug delivery	Hyaluronic acid	(Zhang et al. 2021)
	Tissue engineering	Alginic acid	(Rastogi and Kandasubramanian 2019)
	Biomedical devices	Polysaccharides	(Fois et al. 2019)
Nucleic acid	Drug delivery	DNA	(Mo et al. 2021)
	Tissue engineering	DNA	(Chen et al. 2022a)
	Cancer medicine	DNA; siRNA	(Mo et al. 2021; Ramakrishnan 2011)
	Biosensor development	Stimuli responsive DNA hydrogel	(Chen et al. 2022b)

of a combination of two distinct monomers, with the prerequisite that at least one of these monomers demonstrates hydrophilic properties. The interpenetrating network (IPN) can be formed by combining two polymers, given that one of the polymers is already present in the solution (Chai et al. 2017; Tabata 2009).

This investigation pertains to the identification of hydrogel biopolymers that are commonly seen in scientific literature and research. The characteristic properties of the typical hydrogel are described as follows. The detail biomedical Applications of biomolecules as biopolymers has been depicted in Table 2 (Li et al. 2019; Liyenage et al. 2017; Doufour et al. 2019; Zhang et al. 2021; Rastogi and Kandasubramanian 2019; Fois et al. 2019; Mo et al. 2022; Chen et al. 2022a; Mo et al. 2021; Ramakrishnan 2011; Chen et al. 2022b).

2.1 Physical and Chemical Properties of Hydrogels on Biopolymers

Hydrogels are a specific type of polymeric material with a three-dimensional network architecture that enables them to encapsulate and retain substantial amounts of water effectively. Due to their discernible physical attributes, they are extensively utilized across various industries, including environmental engineering, agriculture, and medicine. The following are key physical attributes of hydrogels (Nele et al. 2020). The phenomenon of swelling capacity absorption and retention are two fundamental properties of hydrogels that hold significant importance. In an aqueous milieu, these entities have the capacity to undergo significant expansion, often surpassing their weight in a desiccated state. Medicine is one of the domains where applications can be found. One of the most essential characteristics of hydrogels is their ability to absorb and retain water effectively. When subjected to an aqueous environment, they have the ability to increase their mass by several orders of magnitude compared to their weight in a dehydrated state. This property is crucial for various applications, such as medicine delivery techniques, wound dressings, and soil moisture retention.

The hydrogels' porosity is crucial in determining their water absorption capacity, permeability, and the rate of material diffusion inside the gel matrix. The features of hydrogel materials can be influenced by adjusting the pore size throughout the manufacturing process. The reference provided is in the form of a numerical citation, indicating that there is a corresponding. The mechanical properties of hydrogels can exhibit significant variations depending on factors such as crosslinking density and polymer composition, resulting in a wide range of characteristics ranging from soft and flexible to stiff and rigid. Certain hydrogels have been found to be suitable for tissue engineering and biomedical applications due to their ability to replicate the mechanical properties of actual tissues (Raghavan et al. 2018). Numerous hydrogels exhibit biocompatibility, indicating their capacity to be tolerated by living organisms. The interaction of hydrogel with biological systems is a critical characteristic that renders it suitable for applications in tissue engineering, medicine delivery, and medical devices. Hydrogels often exhibit a reduced thermal conductivity compared to a majority of other materials. This characteristic can be advantageous in specific contexts, such as the utilization of hydrogels in temperature-sensitive drug delivery systems. The electrical conductivity of hydrogels is contingent upon their specific structure and composition. Conductive hydrogels are utilized in several electronic applications, including biosensors and wearable electronics (Macaya and Spector 2012).

Hydrogels are considered highly suitable for applications of limited duration, such as drug administration or the provision of scaffolds for tissue regeneration, owing to their potential for gradual degradation. The degradation rate can be changed based on the specific requirements of the application. Certain hydrogels can exhibit reversible changes in their swelling behavior or other features in response to variations in pH or ionic concentration. The utilization of these hydrogels that exhibit responsiveness to stimuli is employed in the fields of sensing and drug delivery (Hu et al. 2021). Hydrogels may exhibit adhesive properties to adhere to tissues or surfaces. Additionally, these entities exhibit a high degree of cohesion, hence contributing to the overall stability of their organizational framework. It is important to acknowledge that the selection of polymers, crosslinking agents, and manufacturing methods can modify the physical properties of hydrogels, hence enabling their utilization in many industries and fields (Table 3).

Table 3: Major properties of hydrogels.

Physical Properties	Chemical Properties	Mechanical Properties
Swelling capacity	Hydration and swelling	Viscoelasticity
Porosity and pore size	Hydrophilicity	Gelation and crosslinking
Mechanical strength and elasticity	Crosslinking	Storage (G') and loss (G") moduli
Biocompatibility	Biocompatibility	Frequency and strain dependence
Thermal conductivity	Permeability and diffusion	Thixotropic and self-healing
Electrical conductivity	pH sensitivity	Temperature and pH sensitivity
Degradation and biodegradability	Thermo-responsiveness	Inject ability and processing
PH and ionic Responsiveness	Electrical conductivity	Modeling and characterization
Adhesion and cohesion	Chemical modification	Swelling behavior of hydrogel

3. Investigation of the Swelling Behavior of Hydrogels on Biopolymers

Hydrogels undergo expansion due to the absorption of water molecules into their polymer matrix. The polymer chains inside the hydrogel network exhibit a hydrophilic characteristic, indicating a strong affinity toward water. The hydrogel's volume is increased due to water molecule diffusion, which leads to the stretching of polymer chains within the hydrogel. In an equilibrium swelling state, hydrogels have the ability to achieve a balance between the rate of water absorption and water release. The equation $Q = M_s/M_d$ can be employed to establish the equilibrium swelling ratio (Q), which represents the relationship between the swelled mass (M_s) and the dry mass (M_d) of the hydrogel. This statement highlights the hydrogel's capacity for water absorption. The phenomenon of swelling is characterized by an abnormal increase in the size or volume of a body. The study of kinetics involves the examination of the rate and factors at which chemical reactions occur. The rate of hydrogel expansion is influenced by two factors: the transportation of water molecules inside the polymer network and the influence of external stimuli, such as changes in pH or temperature. The rate at which hydrogels of different types undergo swelling can vary from rapid to slow.

The Phenomenon of Swelling Equilibrium in the Context of Thermodynamics: The equilibrium swelling behavior of hydrogels can be described using thermodynamic models such as the Flory-Rehner theory. These models establish a relationship between the extent of swelling and other factors such as temperature, solvent quality, crosslinking intensity, and pH sensitivity. Certain hydrogels have pH-responsive characteristics, wherein they undergo expansion or contraction in response to fluctuations in pH levels. The utilization of pH changes is commonly employed for the aim of regulated medicine administration or other applications. Temperature sensitivity is observed in certain hydrogels, where their swelling behavior is contingent upon changes in temperature. The ability to undergo expansion or contraction in response to temperature variations renders them suitable for applications such as intelligent textiles or temperature-responsive drug administration. The network topology of a hydrogel, encompassing the polymer chains' density and flexibility, significantly influences its swelling behavior. The reduced swelling observed in strongly crosslinked hydrogels can be attributed to the restricted chain mobility they exhibit (DeFrates et al. 2018).

Hydrogel Applications: Hydrogels find utility in various domains owing to their unique swelling properties. These applications encompass tissue engineering scaffolds, contact lenses, wound dressings, agricultural practices (for soil moisture retention), sensors, and drug delivery systems. Multiple cycles of swelling and deswelling can be conducted on hydrogels without causing substantial modifications to their swelling characteristics. This property can be advantageous for applications that need frequent swelling and shrinking, such as those seen in responsive materials (DeFrates et al. 2018)). Environmental factors, such as the presence of ions, solvents, or other substances, can influence the swelling behavior of hydrogels. Environmental factors can have an impact on the dynamics of swelling and its equilibrium. The user did not provide any text. In summary, the swelling behaviors of hydrogels are a complex phenomenon that is influenced by several factors. Through meticulous development of the polymer network and a comprehensive understanding of the influence of many variables on their swelling behaviors, researchers and practitioners have the ability to tailor the properties of hydrogels to specific applications (Macaya and Spector 2012; Hu et al. 2021; DeFrates et al. 2018).

3.1 Gelation Mechanism of Hydrogel

The hydrophilic polymer networks that make up hydrogels are three-dimensional networks with the capacity to absorb and hold large volumes of water or other biological fluids. Depending on the particular hydrogel being generated and the polymer system being employed, the gelation process of hydrogels might change. A broad summary of hydrogels' gelation mechanism is provided here.

Polymer Dissolution and Swelling: The process typically begins with dissolving the hydrophilic polymer(s) in a solvent, usually water. As the polymer chains dissolve, they begin to swell due to interactions between the polymer and the solvent molecules. The degree of swelling depends on factors such as polymer concentration, temperature, and solvent quality (Cikrikci et al. 2018).

Polymer Crosslinking: The hydrogel's structure and stability are provided by the creation of physical or chemical crosslinks between polymer chains during the gelation process. Crosslinking can happen through a number of mechanisms.

Chemical Crosslinking: Covalent bonds are created in this process between polymer chain segments. The polymer solution is supplemented with crosslinking agents, which are frequently molecules with two or more functions. The functional groups on the polymer chains react with these substances to generate a network. Michael addition, photopolymerization, and free radical polymerization are all typical chemical crosslinking processes (Denizli et al. 2004). Physical crosslinking can also be facilitated by physical interactions such as hydrogen bonds, hydrophobic interactions, and ionic interactions. Reversible external inputs like temperature, pH, or ionic strength can control these interactions. Polymers may self-assemble into a gel network as a result of physical crosslinking (Sundar et al. 2010).

Gelation and Network Formation: The polymer chains bond as the crosslinking process continues, creating a three-dimensional network that traps water or other fluids. The hydrogel behaves solid-like because of the network structure's ability to stop polymer chains from diffusing away while yet having a high-water content.

Equilibrium Swelling: The hydrogel eventually reaches an equilibrium swelling stage where the rate of water absorption and release are equal once the gel network has formed. The administration of drugs and the healing of wounds are two applications where this swelling characteristic is essential (Gačanin 2017). It is crucial to remember that the gelation mechanism might change depending on the hydrogel's specific use, the kind of polymer utilized (natural or synthetic), and the manner of crosslinking (chemical or physical). To obtain desired features, including mechanical strength, swelling behavior, and reactivity to environmental stimuli, the choice of polymer and crosslinking technique may be customized.

4. Methods for Characterizing the Structure of Hydrogel Networks for Making Biopolymers

Using Scanning Electron Microscopy (SEM) and Transmission Electron Microscopy (TEM) to characterize hydrogel network architectures can be a highly effective way to learn more about the morphology, porosity, and general organization of hydrogel materials at the micro- and nanoscale levels. Researchers can study the internal structure and surface characteristics of hydrogels using these approaches, which helps them better understand their characteristics and possible uses. An overview of the applications for TEM and SEM in this context is given below.

4.1 Scanning Electron Microscopy (SEM)

SEM is a popular method for seeing the surface morphology of materials up close. The general surface structure, porosity, and size distribution of pores within the hydrogel network may all be learned through SEM in the context of hydrogels. The following are the fundamental procedures for hydrogel characterization using SEM.

(A) Sample Setup: Hydrogel samples must be set up correctly for SEM imaging. In order to maintain its structure and prevent collapse during imaging, the hydrogel is frequently frozen or freeze-dried in this manner. Also, to avoid charging artifacts, the sample is frequently thinly coated with conductive material (such as gold or platinum) after freezing (Alavarse et al. 2022).

(B) Imaging: The prepared hydrogel sample is placed in the SEM chamber. A focused electron beam scans the surface of the sample, and as electrons interact with the sample, various signals (secondary electrons and backscattered electrons) are detected to create an image of the sample's surface morphology.

(C) Analysis: The surface texture, pore size, and distribution inside the hydrogel can all be seen in the SEM pictures. Pore size distribution, porosity, and other structural properties may be determined quantitatively (Zhu et al. 2022).

4.2 Transmission Electron Microscopy (TEM)

The more sophisticated TEM method produces pictures with better resolution and enables researchers to see interior structures and nanoscale specifics in hydrogel networks. Even though it is more difficult and necessitates smaller sample slices, TEM can provide useful insights into the most intricate aspects of hydrogel structures (Wan et al. 2015):

(A) Sample Preparation: When compared to SEM, TEM requires significantly thinner samples, frequently in the tens to hundreds of nanometre range. The hydrogel is often embedded in a resin before being cut into thin sections with a microtome. Then, sections are arranged on a TEM grid (Sáez-Orviz et al. 2022).

(B) Imaging: A concentrated electron beam is sent across the material in TEM. The sample's internal structures, such as the configuration of polymer chains, crosslinking sites, and nanoscale features, are revealed in a detailed picture that is produced by the interaction of electrons with the sample (Bogdanova et al. 2019).

(C) Analysis: Examining nanoscale aspects like polymer chain alignment, network structure, and component interactions is possible with TEM images.

Combining data from SEM and TEM may provide a thorough knowledge of hydrogel network topologies at all scales, from the macro to the nano. These methods are essential for developing and perfecting hydrogel products for a range of uses, including drug delivery, tissue engineering, and biomaterials (Shchipunov et al. 2005).

- Tissue Imaging: It uses an electron microscope, allowing researchers to examine the interior structure of hydrogels at the nanoscale using transmission electron microscopy (TEM), which has an even better resolution. It offers details on the configuration of polymer chains, crosslinking sites, and other nanoscale components (Kalia and Sharma 2022).

- Sample Preparation: Hydrogel samples require more intricate preparation for TEM. The hydrogel has to be cut into small sections (ultra-thin slices) utilizing methods such as embedding it in resin and microtome sectioning. Then, a TEM grid is placed with these portions on it (Vert 2011).
- Imaging: In TEM, a concentrated electron beam passes through the delicate hydrogel sections. This results in finely detailed pictures of the polymer network's nanoscale structures and configurations (Gun'ko et al. 2017).
- Analysis: Nanoscale aspects like polymer chain alignment, network structure, and component interactions may all be examined using TEM images (Heim et al. 2010).

4.3 Fourier Transform Infrared Spectroscopy (FTIR)

In order to gain a better understanding of the structure, content, and properties of protein-based hydrogels, characterization techniques such as Fourier Transform Infrared Spectroscopy (FTIR) and Matrix-Assisted Laser Desorption/Ionization Time-of-Flight Mass Spectrometry (MALDI-TOF MS) are being utilized. The following is a summary of the procedures that are followed by each method to make their contribution to the characterization process:

In order to gain a better understanding of the structure, content, and properties of protein-based hydrogels, the techniques of FTIR and MALDI-TOF MS are being utilized to characterize these hydrogels. The following table (Sabet Ghadam Haghighi 2021) provides a summary of the stages that each technique takes to contribute to the characterization process.

A Representative Preparation: In most cases, an examination requires the production of a thin film or powder from the protein-based hydrogel sample, which is accomplished by air-drying or freeze-drying the sample. After this, the sample is placed into the FTIR spectrometer (Guo et al. 2021).

4.4 Spectral Acquisition and Analysis

- FTIR examines how infrared light interacts with the sample to cause molecular vibrations that produce various spectral bands of absorption (Bealer et al. 2020).
- The absorbance (intensity of absorbed light)/wavenumber (the reciprocal of wavelength) is plotted to get the FTIR spectrum (Shishir et al. 2018).
- Particular peaks in the spectrum represent various chemical vibrations and functional groups (Zhou et al. 2016).
- Protein-based hydrogels can provide information on the secondary structure of proteins, such as random coils, beta sheets, and alpha helices, by FTIR analysis (Cortés-Morales et al. 2021).
- It can recognize particular functional groups that are exclusive to proteins (such as amide bonds) (Luo et al. 2015).

- The FTIR spectrum may be changed before and after gel formation to identify interactions and conformational changes within the hydrogel (Davidov-Pardo et al. 2015).

4.5 *MALDI-TOF MS*

By using MALDI-TOF MS, it is possible to recognize and examine the mass-to-charge ratio of biomolecules, such as proteins and peptides, to learn more about their molecular structure and sequence.

Sample Preparation: Proteases (such as trypsin) are commonly used to enzymatically degrade protein-based hydrogel samples into smaller peptides. A matrix solution, a substance that promotes ionization, is added to the peptide mixture before it is spotted onto a MALDI target plate (Liu et al. 2019).

Interpretation: Based on the peaks' m/z values, the resultant mass spectrum shows several peptides as distinct clusters (Devi et al. 2017). By comparing the reported peptide weights to known protein sequences in databases, researchers may determine which proteins are present in the hydrogel. MALDI-TOF MS can also shed light on changes in protein composition and post-translational modifications.

4.6 *FTIR and MALDI-TOF MS*

By combining the information obtained from FTIR and MALDI-TOF MS, researchers can:

- Recognize how protein-based hydrogels' molecular structure and conformational alterations evolve.
- Determine the amount of each protein component in the hydrogel.
- Learn about the changes, interactions, and possible pollutants of proteins.
- Improve the formulation and design of hydrogels for particular applications (An et al. 2022).

In conclusion, using FTIR and MALDI-TOF MS in combination gives a thorough method for characterizing protein-based hydrogels, allowing researchers to understand their molecular makeup, structure, and prospective functions (McClements 2018).

5. Applications of Hydrogel in Biomedicine

The topics of attention comprise tissue engineering, drug delivery, and biocompatibility. Hydrogels possess a diverse range of applications in the field of biomedicine owing to their notable characteristics, such as elevated water content, biocompatibility, and the ability to be tailored to certain physical attributes. Hydrogels have found extensive applications in the field of biomedicine, encompassing a diverse range of functions (He et al. 2020).

5.1 Tissue Engineering and Regenerative Medicine

Hydrogels are three-dimensional scaffold materials that may be used to construct tissues' extracellular matrix (ECM). They offer a favorable setting for cell development, expansion, and differentiation.

Cell Delivery: Hydrogels are capable of encasing and delivering cells to injured tissues, aiding in the regeneration of those tissues. They improve cell survival and give cells a regulated environment in which to behave.

Organ and Tissue healing: By offering mechanical support and directing tissue growth, hydrogels are utilized to heal injured tissues and organs, such as cartilage, bone, skin, and neural tissues (Chen et al. 2019).

5.2 Drug Delivery Systems

Localized Delivery: One potential application of hydrogels is their usage as localized drug delivery systems, wherein pharmaceuticals or therapeutic substances can be encapsulated within the hydrogel matrix. Pharmaceutical agents are administered in a regulated and prolonged manner, thereby reducing adverse reactions and enhancing the efficacy of treatment.

Targeted Therapy: Hydrogels possess the capability to be engineered in a manner that allows them to exhibit responsive behavior toward particular stimuli, such as variations in pH, temperature, or the presence of enzymes. This unique attribute empowers hydrogels to facilitate the precise delivery of therapeutic agents to predetermined locations within the human body (Kratochvil et al. 2019).

Oral Drug Delivery: Hydrogels possess the potential to serve as a viable medium for oral drug delivery since they offer the advantage of safeguarding delicate pharmaceutical compounds against degradation inside the gastrointestinal environment while concurrently augmenting their absorption capabilities.

5.3 In Vivo and In Vitro Testing of Biopolymers and Hydrogel using Animal Models and Cell Culture Techniques

In-vivo and *in-vitro* testing of biopolymers and hydrogels using animal models and cell culture techniques are essential steps in evaluating the biocompatibility, efficacy, and safety of these materials for various biomedical applications. Here is an overview of how these testing methods are used (Vanderburgh et al. 2017).

5.4 In Vitro Testing

In vitro testing involves conducting experiments in a controlled laboratory environment using isolated cells or tissues. It provides insights into the interactions between biopolymers, hydrogels, and biological systems at the cellular level. Common techniques include:

Cell Viability and Cytotoxicity Assays: Cell cultures are exposed to biopolymer or hydrogel samples to assess their effects on cell viability and proliferation. Assays like MTT, Alomar Blue, and LIVE/DEAD staining evaluate cell metabolic activity and membrane integrity.

Cell Adhesion and Migration Studies: Cells are cultured on surfaces coated with biopolymers or hydrogels to study their adhesion, migration, and interaction with the material (Chang et al. 2010).

Protein Adsorption and Release Studies: Hydrogels are incubated with proteins to assess their adsorption capacity and controlled release properties (Duval et al. 2017).

Mechanical Characterization: Biomechanical properties of hydrogels (e.g., stiffness, elasticity) are measured using techniques like atomic force microscopy (AFM) or rheology.

Gene Expression and Signaling Pathway Analysis: Changes in gene expression and signaling pathways are analyzed to understand cellular responses to biopolymers and hydrogels.

5.5 In Vivo Testing

In vivo testing involves studying the behavior and effects of biopolymers and hydrogels within living organisms, typically animal models. This step provides insights into the material's performance in a more complex biological environment. Common techniques include the following.

Implantation Studies: Biopolymers or hydrogels are implanted into animals to assess their biocompatibility, integration, and tissue response. Animal models may include rodents, rabbits, or larger animals, depending on the application (Andersen et al. 2015).

Tissue Regeneration and Healing Studies: Biopolymers and hydrogels are used to facilitate tissue regeneration in animal models, such as wound healing or bone repair studies.

Drug Delivery Evaluation: Hydrogels loaded with drugs are tested in animal models to assess their controlled release and therapeutic efficacy.

Imaging Techniques: Imaging modalities like MRI, CT, PET, or optical imaging are used to track the distribution and behavior of hydrogel-based materials *in vivo*.

Biomechanical Testing: The mechanical properties of hydrogel implants in living organisms can be evaluated using techniques like indentation tests or ultrasound electrography.

Immunological Responses: *In vivo* studies assess the immune response triggered by biopolymers or hydrogels, including inflammation and foreign body reactions (Favi et al. 2013).

Both *in vitro* and *in vivo* testing provide complementary information about the behavior of biopolymers and hydrogels in biological systems. *In vitro* studies offer controlled and detailed insights into cellular interactions, while *in vivo* studies provide a more holistic understanding of material performance within a living organism. Together, these approaches help researchers make informed decisions about the suitability of biopolymers and hydrogels for various biomedical applications (Pattanashetti et al. 2017).

6. Future Developments and Trends in Biopolymers and Hydrogel Research

Multiple hypotheses are employed to make predictions about the structure of hydrogels and to analyze the properties of the gels. By incorporating these derived theories into computational methodologies, it becomes feasible to forecast the composition and performance of hydrogels before their synthesis, therefore expediting the advancement of hydrogels for medicinal purposes. Numerous applications and innovative applications exist for the development of novel hydrogels. The testing efficiency and material design capabilities can be significantly enhanced using various types of gels that are presently accessible, including pH-sensitive, heat-sensitive, electro-sensitive, light-sensitive, and shape-changing hydrogels (Mahmood et al. 2022). Hydrogels can be produced using many manufacturing methods, from traditional approaches involving basic chemical reactions to cutting-edge concepts in 3D printing. The aforementioned technological advancements represent only the initial stages of hydrogel synthesis and are establishing the foundation for significant progress in production methods. Hydrogels have demonstrated efficacy in the realm of tissue regeneration by exerting control over the localization of encapsulated cells within specified tissue areas. To optimize results and mitigate potential drawbacks, drug- and molecule-dispensing hydrogels, such as hydrogels designed for insulin release and hydrogels designed for cancer drug delivery, can effectively transport insulin and cancer treatments to their intended locations. The utilization of hydrogels in various applications has proven to be highly efficient, particularly in the medical domain, since 2015. This progress has subsequently facilitated the advancement of hydrogels with enhanced capacity to retain more medications, cells, and other substances. Utilizing these more recent hydrogels would allow for less invasive tissue regeneration, potentially eliminating the need for surgery and more efficient drug delivery for enhanced patient outcomes. In the case of widespread back pain, for instance, new hydrogels may be able to alleviate people's daily distress without invasive procedures.

Potential improvements in the field of biopolymer-based hydrogels, including the utilization of natural polymers with distinct characteristics, such as elastin-cellulose and elastin-alginate combinations, hold promise for enhancing tissue regeneration in the realm of tissue engineering. Due to the considerable potential of hydrogels in drug delivery and tissue regeneration, developing novel biopolymer-based hydrogels with enhanced characteristics, such as prolonged and controlled drug release and reduced degradation rates, presents an intriguing avenue of research for other scholars. Hydrogel therapy is anticipated to significantly influence forthcoming medical interventions owing to its remarkable adaptability, biocompatibility, and tunability.

7. Conclusion

Hydrogels are polymeric materials characterized by the presence of hydrophilic molecules, which facilitate their ability to retain a significant amount of water inside their interstitial structures. By employing biopolymers, specifically, it becomes feasible to fabricate hydrogels that possess distinctive properties and find diverse uses. Hydrogels can be synthesized using a diverse range of biopolymers, such as peptides and saccharides, each possessing unique merits and potential applications. Hydrogels have been employed as a means of using polypeptides, including collagen, gelatin, and elastin. Polysaccharide hydrogels can be produced using several substances, including alginate, cellulose, glycosaminoglycan, and nucleic acids, among others. Hydrogels have been creatively integrated and stratified by innovative techniques, shown by the amalgamation of two distinct hydrogels that exhibit varied responsivity to stimuli. This resultant tubular architecture has been harnessed to address vascular injuries within the human body. The diverse range of network densities exhibited by hydrogels has facilitated the emergence of double-network hydrogels, which possess unique mechanical characteristics. The utilization of inner networks, which can tolerate tensional loads, in conjunction with outer networks, which can withstand shear forces, holds potential advantages in the treatment of ligaments.

References

Alavarse, Ana Carolina, Edicarla Cristina Frachini, Ricardo Luiz da Silva et al. 2022. Crosslinkers for polysaccharides and proteins: Synthesis conditions, mechanisms, and crosslinking efficiency, a review. International Journal of Biological Macromolecules 202: 558–569.

An, Dan, Qingan Ban, Huaye Du et al. 2022. Nanofibrils of food-grade proteins: Formation mechanism, delivery systems, and application evaluation. Comprehensive Reviews in Food Science and Food Safety 21(6): 4847–4871.

Andersen, Tim B., Pernille Auk-Emblem and Michael Dornish. 2015. 3D cell culture in alginate hydrogels. Microarrays 4 (2): 133–161.

Bealer, Erika J, Shermin Onissema-Karimu, Ariel Rivera-Galletti, et al. 2020. Protein–polysaccharide composite materials: Fabrication and applications. Polymers 12 (2):464.

Berkovitch, Yael, and Dror Seliktar. 2017. Semi-synthetic hydrogel composition and stiffness regulate neuronal morphogenesis. Int. J. Pharm. 523: 545–555.

Biswas, Manik Chandra, Bodiuzzaman Jony, Pranab Kumar Nandy et al. 2021. Recent advancement of biopolymers and their potential biomedical applications. Journal of Polymers and the Environment: 1–24.

Bogdanova, Lyudmila R., Alexey M. Rogov, Olga S. Zueva et al. 2019. Lipase enzymatic microreactor in polysaccharide hydrogel: Structure and properties. Russian Chemical Bulletin: 400–404.

Carter, Nathan A. and Tijana Z. Grove. 2019. Functional protein materials: Beyond elastomeric and structural proteins. Polymer Chemistry 10(23): 2952–2959.

Chai, Qiyuan, Yanpeng Jiao and Xiaojun Yu. 2017. Hydrogels for biomedical applications: Their characteristics and the mechanisms behind them. Gels 3: 6.

Chang, Rick, Kaveh Emami, Huanhuan Wu et al. 2010. Biofabrication of a three-dimensional liver micro-organ as an in vitro drug metabolism model. Biofabrication 2(4): 045004.

Chen, C., J. Zhou, J. Chen et al. 2022. Advances in DNA supramolecular hydrogels for tissue engineering. Macromol. Biosci. 22(12): e2200152.

Chen, Guoping, Wanwan Tang, Xue Wang et al. 2019. Applications of hydrogels with special physical properties in biomedicine. Polymers 11(9): 1420.

Chen, M., Y. Wang, J. Zhang et al. 2022. Stimuli-responsive DNA-based hydrogels for biosensing applications. J. Nanobiotechnology 20(1): 40.

Cikrikci, Sevde, Burak Mert and M. H. Oztop. 2018. Development of pH sensitive alginate/gum tragacanth based hydrogels for oral insulin delivery. J. Agric. Food Chem. 66: 11784–11796.

Cortés-Morales, E. A., G. Mendez-Montealvo and G. Velazquez. 2021. Interactions of the molecular assembly of polysaccharide-protein systems as encapsulation materials. A review. Advances in Colloid and Interface Science 295: 102398.

Davidov-Pardo, Gabriel, Iris J. Joye and David Julian McClements. 2015. Food-grade protein-based nanoparticles and microparticles for bioactive delivery: fabrication, characterization, and utilization. Advances in Protein Chemistry and Structural Biology 98: 293–325.

DeFrates, Kayla G., Ryan Moore, Joseph Borgesi et al. 2018. Protein-based fiber materials in medicine: A review. Nanomaterials 8: 457.

Denizli, B. K., H. K. Can, Z. M. Rzaev et al. 2004. Preparation conditions and swelling equilibria of dextran hydrogels prepared by some crosslinking agents. Polymer 45(19): 6431–6435.

Devi, N., Manash Khatun Sarmah and Tapan Kumar Maji. 2017. Encapsulation of active ingredients in polysaccharide–protein complex coacervates. Advances in Colloid and Interface Science 239: 136–145.

Dufour, A., M. Buffier, D. Vertu-Ciolino et al. 2019. Combination of bioactive factors and IEIK13 self-assembling peptide hydrogel promotes cartilage matrix production by human nasal chondrocytes. J. Biomed. Mater. Res. A. 107(4): 893–903.

Duval, Katherine, Hitesh Grover, Linheng Han et al. 2017. Modeling physiological events in 2D vs. 3D cell culture. Physiology 32(4): 266–277.

Favi, Pelin M., Roberto S. Benson, Nancy R. Neilsen et al. 2013. Cell proliferation, viability, and *in vitro* differentiation of equine mesenchymal stem cells seeded on bacterial cellulose hydrogel scaffolds. Materials Science and Engineering: C 33(4): 1935–1944.

Fois, M., P. Arrigo, A. Bacciu et al. 2019. The presence of polysaccharides, glycerol, and polyethyleneimine in hydrogel enhances the performance of the glucose biosensor. Biosensors (Basel) 9(3).

Gačanin, Jasna. 2017. Design and Synthesis of Multifunctional Hybrid Hydrogels for Medicinal Applications. PhD Thesis, Universität Ulm.

Gun'ko, Vladimir M., Irina N. Savina and Sergey V. Mikhalovsky. 2017. Properties of water bound in hydrogels. Gels 3(4): 37.

Guo, Qian, Xin Shu, Yating Hu et al. 2021. Formulated protein-polysaccharide-surfactant ternary complexes for co-encapsulation of curcumin and resveratrol: Characterization, stability and *in vitro* digestibility. Food Hydrocolloids 111: 106265.

He, Jie, Rui Wang, Weiyi Feng et al. 2020. Design of novel edible hydrocolloids by structural interplays between wheat gluten proteins and soy protein isolates. Food Hydrocolloids 100: 105395.

Heim, Markus, Lin Römer and Thomas Scheibel. 2010. Hierarchical structures made of proteins. The complex architecture of spider webs and their constituent silk proteins. Chemical Society Reviews 39(1): 156–164.

Hilal, Ahmad, Agnieszka Florowska and Małgorzata Wroniak. 2023. Binary hydrogels: induction methods and recent application progress as food matrices for bioactive compounds delivery-a bibliometric review. Gels 9(1): 68.

Hu, Xun, Stephen Ricci, Sebastian Naranjo et al. 2021. Protein and polysaccharide-based electroactive and conductive materials for biomedical applications. Molecules 26: 4499.

Hwang, Dong Chul and Srinivasan Damodaran. 1996. Chemical modification strategies for the synthesis of protein-based hydrogel. Journal of Agricultural and Food Chemistry 44(3): 751–758.

Joye, Iris J. and David Julian McClements. 2016. Biopolymer-based delivery systems: challenges and opportunities. Current Topics in Medicinal Chemistry 16(9): 1026–1039.

Kalia, Susheel and Swati Sharma. 2022. Protein-based biopolymers: From Source to Biomedical Applications: Woodhead Publishing.

Kratochvil, Matthew J., Andrew J. Seymour, Ting Lu Li et al. 2019. Engineered materials for organoid systems. Nature Reviews Materials 4(9): 606–622.

Li, M., M. Han, Y. Sun et al. 2019. Oligoarginine mediated collagen/chitosan gel composite for cutaneous wound healing. Int. J. Biol. Macromol. 122: 1120–1127.

Liu, Lulu, Xiang-Tian Li, Ning Zhang et al. 2019. Novel soy β-conglycinin nanoparticles by ethanol-assisted disassembly and reassembly: Outstanding nanocarriers for hydrophobic nutraceuticals. Food Hydrocolloids 91: 246–255.

Liyanage, W., H. A. Ardona, H. Q. Mao et al. 2017. Cross-linking approaches to tuning the mechanical properties of peptide pi-electron hydrogels. Bioconjug. Chem. 28(3): 751–759.

Luo, Yang, Ke Pan and Qixin Zhong. 2015. Casein/pectin nanocomplexes as potential oral delivery vehicles. International Journal of Pharmaceutics 486(1-2): 59–68.

Macaya, Daniel and Myron Spector. 2012. Injectable hydrogel materials for spinal cord regeneration: A review. Biomed. Mater. 7: 012001.

Mahmood, Aneesa, Dhruv Patel, Bethany Hickson et al. 2022. Recent progress in biopolymer-based hydrogel materials for biomedical applications. Int. J. Mol. Sci. 23(3): 1415.

Makam, Pandeeswar and Ehud Gazit. 2018. Minimalistic peptide supramolecular co-assembly: expanding the conformational space for nanotechnology. Chemical Society Reviews 47(10): 3406–3420.

Makshakova, Olga N., Lyudmila R. Bogdanova, Anna O. Makarova et al. 2022. κ-Carrageenan hydrogel as a matrix for therapeutic enzyme immobilization. Polymers 14(19): 4071.

McClements, David Julian. 2018. Encapsulation, protection, and delivery of bioactive proteins and peptides using nanoparticle and microparticle systems: A review. Advances in Colloid and Interface Science 253: 1–22.

Mo, F., K. Jiang, D. Zhao et al. 2021. DNA hydrogel-based gene editing and drug delivery systems. Adv. Drug Deliv. Rev. 168: 79–98.

Nele, Valeria, Jonathan P. Wojciechowski, James P. K. Armstrong et al. 2020. Tailoring gelation mechanisms for advanced hydrogel applications. Advanced Functional Materials 30(42): 2002759.

Pattanashetti, N. A., G. B. Heggannavar and M. Y. Kariduraganavar. 2017. Smart biopolymers and their biomedical applications. Procedia Manufacturing 12: 263–279.

Pina, Sandra, Rui L. Reis and Joaquim M. Oliveira. 2022. Natural polymeric biomaterials for tissue engineering. In Tissue Engineering Using Ceramics and Polymers.

Raghavan, Shyni R., Nikhil J. Fernandes and Bhawneshwar H. Cipriano. 2018. Shape-changing tubular hydrogels. Gels 4: 18.

Ramakrishnan, S. 2011. Hydrogel-siRNA for cancer therapy. Cancer Biol. Ther. 11(9): 849–51.

Rastogi, P. and B. Kandasubramanian. 2019. Review of alginate-based hydrogel bioprinting for application in tissue engineering. Biofabrication 11(4): 042001.

Refojo, M. F. and F. J. Holly. 1977. Tear protein adsorption on hydrogels: a possible cause of contact lens allergy. Eye & Contact Lens 3(1): 23–36.

Rostami, Mahshid, Mohammad Yousefi, Ahmad Khezerlou et al. 2019. Application of different biopolymers for nanoencapsulation of antioxidants via electrohydrodynamic processes. Food Hydrocolloids 97: 105170.

Sabet Ghadam Haghighi, Saman. 2021. Bilayer emulsion filled in hydrogel as a novel bioactive delivery system, ResearchSpace@ Auckland.

Sáez-Orviz, Sergio, Manuel Rendueles and Mario Diaz. 2022. Impact of adding prebiotics and probiotics on the characteristics of edible films and coatings-a review. Food Research International: 112381.

Shchipunov, Yakov A., Tatiana Y. Karpenko and Anna V. Krekoten. 2005. Hybrid organic–inorganic nanocomposites fabricated with a novel biocompatible precursor using sol-gel processing. Composite Interfaces 11(8-9): 587–607.

Shishir, Mohammad Rezaul Islam, Linglin Xie, Cuixia Sun et al. 2018. Advances in micro and nano-encapsulation of bioactive compounds using biopolymer and lipid-based transporters. Trends in Food Science & Technology 78: 34–60.

Sikkema, Raymond, Kevin Baker and Igor Zhitomirsky. 2020. Electrophoretic deposition of polymers and proteins for biomedical applications. Advances in Colloid and Interface Science 284: 102272.

Silva, Ana, Cédric Richard, Michel Bessodes et al. 2009. Growth Factor Delivery Approaches in Hydrogels. Biomacromolecules 10: 9–18.

Sundar, S., J. Kundu and S. C. Kundu. 2010. Biopolymeric nanoparticles. Science and Technology of Advanced Materials.

Tabata, Yasuhiko. 2009. Biomaterial technology for tissue engineering applications. J. R. Soc. Interface 6: S311–S324.

Tirrell, Judith G. and David A. Tirrell. 1996. Synthesis of biopolymers: proteins, polyesters, polysaccharides and polynucleotides. Current Opinion in Solid State and Materials Science 1(3): 407–411.

Vanderburgh, Joshua, Julie A. Sterling and Scott A. Guelcher. 2017. 3D printing of tissue engineered constructs for in vitro modeling of disease progression and drug screening. Annals of Biomedical Engineering: 164–179.

Vert, Michel. 2011. Not any new functional polymer can be for medicine: what about artificial biopolymers? Macromolecular Bioscience 11(12): 1653–1661.

Wan, Zhenlong, Jie Guo and Xiaoxi Yang. 2015. Plant protein-based delivery systems for bioactive ingredients in foods. Food & Function 6(9): 2876–2889.

Wen, Jie, Maram Al Gailani and Ning Yin. 2018. Filled hydrogel particles emulsion-based systems for delivery of food active compounds. Formation, Application, Health and Safety: 161–180.

Zhang, X., D. Wei, Y. Xu et al. 2021. Hyaluronic acid in ocular drug delivery. Carbohydr. Polym. 264: 118006.

Zhang, Zhen, Ruojie Zhang, Liqiang Zou et al. 2016. Protein encapsulation in alginate hydrogel beads: Effect of pH on microgel stability, protein retention and protein release. Food Hydrocolloids 58: 308–315.

Zhou, Meng, Tao Wang, Qiaoling Hu et al. 2016. Low density lipoprotein/pectin complex nanogels as potential oral delivery vehicles for curcumin. Food Hydrocolloids 57: 20–29.

Zhu, Yaqi, Yujie Han, Shifeng Peng et al. 2022. Hydrogels assembled from hybrid of whey protein amyloid fibrils and gliadin nanoparticles for curcumin loading: Microstructure, tunable viscoelasticity, and stability. Frontiers in Nutrition 9: 994740.

Zueva, Olga S., Anna O. Makarova and Yuri F. Zuev. 2019. Carbon nanotubes in composite hydrogels based on plant carbohydrates. Materials Science Forum 945: 522–527.

8

Application of Biomaterials in Tissue Engineering

Khadija Shahab Turabi,[1] *Saakshi Bangera,*[1] *Nikita Shinde,*[1] *Pranavi Mishra,*[1] *Shine Devarajan,*[1] *Soumya Basu*[2] and *Jyotirmoi Aich*[1,*]

1. Tissue Engineering and Its Collaboration with Biomaterials

1.1 Tissue Engineering and Its Applications

Tissue engineering (TE) is an integrative of biological science aimed at fabricating a three-dimensional (3D) functional network that combines scaffolding bioactive cellular components (Griffith and Swartz 2006). This field involves the scientific area of various biotechnology-related fields such as biological and material science, chemistry techniques, and medicinal aspects. Langer and Vacanti initially proposed and publicized the term "TE" to the scientific world in 1993 (Langer and Vacanti 1993). The production of biological alternatives to maintain, enhance, or restore the functions of the tissue is the ultimate goal of tissue engineering, according to their definition, which is still in use today (Langer and Vacanti 1993). As a result, TE can avoid tissue injury issues that are now handled by transplantation, mechanical implants, or surgery.

Medical treatments have prolonged and improved the lives of numerous individuals, but they often come with associated issues. For instance, organ transplants have significant drawbacks, such as immune rejection and donor scarcity, to accommodate all demands worldwide. Mechanical devices are unable to finish all tissue-related tasks and are unable to stop patients' bodies from deteriorating

1 School of Biotechnology and Bioinformatics, DY Patil Deemed-to-Be University, CBD Belapur, Navi Mumbai 400 614, Maharashtra, India.
2 Cancer and Translational Research Centre, Dr. D.Y. Patil Biotechnology & Bioinformatics Institute, Dr. D. Y. Patil Vidyapeeth, Pune 411 033, Maharashtra, India.
* Corresponding author: jyoti.aich@gmail.com

over time. Surgical reconstruction can sometimes result in prolonged complications (Lalan et al. 2001). As a result, there is a need to offer a more reliable option for network improvement in the clinic. This need will be met by creating *in vitro* gadgets that can restore damaged *in vivo* networks. Building an operational structure that repairs, maintains, or improves injured tissue or whole organs is the goal of network engineering. The FDA has authorized some engineering systems, such as artificial skin and cartilage, but their use in human patients is relatively minimal. TE additionally generates engineered tissues that enable studies pertaining to physiological processes (Griffith and Swartz 2006; Pampaloni et al. 2007). In the organism, cells develop in a well-organized three-dimensional ECM encircled by other cells. It is true that a cell's ability to proliferate, differentiate, undergo apoptosis, or invade depends on its interactions with other cells and the ECM. However, research on cellular biology has frequently been done on 2D cultures, wherein cells develop under non-physiological circumstances.

Tissue engineering has been found to have a wide variety of applications. One major use of TE has been found in regenerating various tissues, including bone, vascular, cardiac, and pancreatic tissues (Castells-Sala et al. 2013). Regenerative medicine integrates TE with other therapeutic strategies, such as cellular, genetic, and immunoregulatory, to elicit regeneration. TE utilizes cells, scaffolds, and growth factors to replenish tissues or substitute injured or unhealthy tissues (Han et al. 2020).

The development of *in vitro* human models that enable us to recognize and comprehend the elements that support cell functions is another emerging use. The goal is to analyze the function of various biochemical and physiological elements in a simplified environment by breaking down an intricate microenvironment into its component parts (Castells-Sala et al. 2013). These models effectively meet the requirements for reductive techniques to comprehend the biological mechanisms, enabling them to more accurately forecast the effects of medications and therapeutic treatment (Castells-Sala et al. 2013).

1.2 Biomaterials and Their Uses

Any substance, surface, or design interacting with living systems is considered a biomaterial. Biomaterials can be manufactured in the lab using metal, polymer, ceramic, or composites, or they can be naturally obtained (Detsch et al. 2018). Biomaterials are frequently employed in medical devices to supplement or substitute a biological function. Cardiac valves, hip replacements, and surgical and dental components are among the examples (Ng et al. 2019).

In modern medicine, biomaterials are crucial for improving recovery after damage or illness and regaining functionality in patients. In medical applications, biomaterials—either natural or artificial—are utilized to support, enhance, or substitute impaired tissue or biological function (Binyamin et al. 2006). Ancient Egyptians utilized sutures derived from animal musculature as the first known application of biomaterials (Ige et al. 2012). Combining medical, biological, physical, and chemical sciences and more recent contributions from TE approaches are the fields of biomaterials in modern times (Ratner et al. 2013). With advancements made in the field of biomaterials, innovative strategies are formulated to enhance the drug delivery system. For more than six decades, biomaterials have been developed in conjunction with numerous scientific disciplines, including engineering, chemistry, physics, biology, and medicine, for use in complex pharmaceutical administration systems (Yun et al. 2015).

Biomaterials today have a plethora of applications in the medical and healthcare industry. From implantation devices and regenerating tissues to biosensors, their uses are numerous. Biomaterials are now synthesized to target specific biological components, tissues or organs; this is a useful application in terms of targeting specific cells in diseases (Huebsch and Mooney 2009). Similarly, biomaterials are also employed in cancer therapeutics, where they are designed to target tumor cells. They have also been employed to manufacture diagnostic equipment that can acquire atypical cell masses (Huebsch and Mooney 2009).

1.3 Application of Biomaterials in Tissue Engineering

Using proper cell implantation techniques, donor tissue that has been separated for the purpose, and tissue-compatible scaffolding materials, tissue engineering is the process of creating bioengineered tissue and resulting alterations *in vivo* from cell development and function. To encourage the optimum cell-cell attachment, movement, and accumulation of endogenous extracellular matrix components by cells, biomaterials for TE should have surface chemistry, permeability, and regulated degradability (Berthiaume 2011). To create a large cell mass capable of performing the various roles required for TE approaches to shift cells from growth to differentiation and vice versa—which frequently are mutually exclusive—are used. Depending on how well the various cell types present in TE adhere to one another and to substrates, the composition of cells and components can reorganize. Finally, for effective nutritional intake and waste removal, network development should be carefully incorporated into the target hosts (Berthiaume 2011).

2. Types of Biomaterials Based on Types of Tissue Engineering

As mentioned earlier, biomaterials have a diverse set of applications in the field of TE; this section highlights the various kinds of biomaterials used for different kinds of TE. Table 1 gives an overview of the types of biomaterials used for different kinds of TE.

2.1 Bone Tissue Engineering

Bone tissues are highly diverse, structurally ranging from long to short, flat and irregular and functionally ranging from movement to protecting organs. Bone TE aims to develop alternative treatment options to eventually eliminate the drawbacks of current treatments, such as immune rejection, donor site morbidity, and restricted availability. Understanding the physiology of the bone, its mechanics, and how bone tissue forms the basis of bone tissue engineering (Amini et al. 2012). The various types of bone TE include calcium phosphate ceramics (CaP), bioactive glass, metal-based materials, polymers, and composites.

Table 1. List of biomaterials based on the type of tissue engineering.

Type of TE	Biomaterials Used	References
Bone	CaP ceramics, bioactive glasses, metal-based materials, biopolymers, and composites	(Amini et al. 2012; Díaz-cuenca et al. 2022; Lee et al. 2014; Kokubo and Takadama 2006; Chen et al. 2019; Guo et al. 2021; Mieszawska et al. 2010)
Osteochondral	Single-phase scaffold, biphasic scaffold, and gradient scaffold	(Martin et al. 2007; Lee et al. 2014; Jiang et al. 2010; Mohan et al. 2011)
Cardiac	Extracellular matrix proteins and nanomaterials	(Mousavi et al. 2021; Kc et al. 2019; Esmaeili et al. 2022)
Pancreas	Extracellular matrix proteins	(Goh et al. 2013; Baidal et al. 2019)
Stem cell-based	Mesenchymal stem cells	(Sharma et al. 2019; Augustine et al. 2021)

2.1.1 Calcium Phosphate Ceramics

CaP ceramics are a type of biomaterial that can stimulate the creation of ectopic bone by signaling its surroundings to do so (Amini et al. 2012). They are weakly crystalline and naturally produced in the body. They form very tiny crystals in close proximity to the protein's developing matrix. Consequently, the ideal biochemical and physiological attributes for synthesized CaP materials should be maintained (Díaz-cuenca et al. 2022). The nontoxicity, tissue compatibility, and biochemical and structural resemblance of CaPs are combined with the features of inert tissue compatibility in these chemical formulations. This combination, in turn, increases the bioactivity, porosity, elasticity, and mechanical strength of the ceramic (Díaz-cuenca et al. 2022). These bioceramics, however, remain inherently rigid and have slow degradability. In order to create more effective matrixes, they are frequently coupled with biopolymers that are easily degradable. The most popular CaP bioceramic formulations are hydroxyapatite (HA), β-tricalcium phosphate, and a combination of both (Lee et al. 2014).

2.1.2 Bioactive Glasses

Materials that can undergo certain surface responses when inserted into the body are referred to as bioactive materials. They are in charge of creating a strong connection with both hard and soft tissues (Kokubo and Takadama 2006). Silicate bioactive glasses are built on a SiO_2 matrix that forms a 3D glass and contains a coordinate link between the Si and O atoms. A carbonate-substituted HA-like layer forms on the glass surface of 45S5 glass, a commonly used binding glass, when it comes into contact with body fluids. This later tightly adheres to the live bone and tissue because it resembles the mineral composition of bone (Hench 1998). Borate bioactive glasses break down more rapidly and entirely into hydroxyapatite-like material than silicate glasses (Huang et al. 2006). Due to the presence of their ions in the organic mineral component of the bone, phosphate bioactive glasses and bone share chemical properties (Rahaman et al. 2011).

2.1.3 Metal-Based Materials

Porous metallic scaffolds have been researched for bone TE because they have excellent physical qualities and tissue growth-promoting properties. Out of the many biomaterials, titanium (Ti) is the most commonly used metal for clinical purposes, such as dental implants and bone implants. Ti is widely used for implants because its porous structure mimics the human bone, making it excellent for bone grafting (Chen et al. 2019). Apart from that, Ti and its alloys show outstanding mechanical properties, biocompatibility and corrosion resistance (Kent et al. 2011). Surface modification of titanium implants with new Ti coatings may enhance the functioning of the implant for a favorable bone-implant interaction (Simon et al. 2005). Apart from Ti, tantalum is also used for bone tissue engineering. Tantalum scaffolds possess titanium's favorable properties and demonstrate superior attachment and proliferation capacity as compared to Ti (Balla et al. 2010).

2.1.4 Polymers

Polymers are highly versatile materials for tissue engineering. They can be engineered in the form of fibers or gel-based materials and can be chemically modified to improve the bioactivity of the material in the body (Lee et al. 2014). The clinical practice uses both natural and synthetic biopolymeric substances; however, biopolymers are more compatible and easily degradable than synthetic ones. Chitosan, collagen, silk fibroin, cellulose, and gelatin are a few examples of natural polymers used for bone TE (Guo et al. 2021). Conversely, there are certain downsides to biopolymers, including processing challenges, immunogenicity, and source-based unpredictability (Lee and Shin 2007). An artificial polymer with outstanding mechanical characteristics that is thermostable and has a slow rate of biodegradation is called polycaprolactone (PCL), which has received FDA approval (Rao et al. 2017).

2.1.5 Composites

Composites are made to improve at least two materials' distinctive features. Frequently, bioactive glasses or calcium phosphate ceramics are mixed with biopolymers or artificial polymers (Lee et al. 2014). One example of a composite is one made from silk fibroin and silica-based bioglass. The bioactivity of the bioglass was combined with the silk fibroin for its excellent mechanical strength and biodegradability (Mieszawska et al. 2010). Calcium phosphate scaffolds were combined with collagen hydrogel to enhance bone synthesis (Yu et al. 2012). One of the most popular biopolymers for bone TE is alginate, which is a biomaterial. The polysaccharide is non-toxic, non-immunogenic, biocompatible, and biodegradable due to the guluronic and mannuronic acids that make up its makeup. Furthermore, its usage as a scaffold is facilitated by its accessibility and affordability (Venkatesan et al. 2014).

2.2 Osteochondral Tissue Engineering

Due to the joint's mechanical brittleness, osteochondral abnormalities frequently trigger degenerative processes. An osteochondral graft typically comprises an underlying calcified tissue and a surface cartilaginous layer. The mechanical and biological viability of the joint can be restored with this design, which is a promising approach. However, the grafts have some drawbacks, such as material scarcity and donor site morbidity. These limitations may be overcome by tissue engineering of osteochondral composites. Three-dimensional grafts of a pre-defined shape can be developed by merging the patient's own cells with porous biomaterials. (Martin et al. 2007).

2.2.1 Single-Phase Scaffold

Single-phase scaffolds are commonly used for osteochondral TE. They are made up of a single biomaterial or a composite of several. Irrespective of the number of biomaterials employed, they must all be distributed evenly throughout the scaffold (Lee et al. 2014). Bone morphogenic protein (BMP-2) chemically conjugated PCL scaffolds were developed to evaluate the potential for neocartilage production (Jeong et al. 2012). Hydroxyapatite microparticles influence the hypertrophic chondrocyte culture (Zhou et al. 2011).

2.2.2 Biphasic/Multiphasic Scaffold

Biphasic scaffolds and multiphasic scaffolds are better equipped to summarize the subchondral bone and the articular cartilage (Lee et al. 2014). In order to simulate the chondro-phase, agar hydrogel containing chondrocytes was created, and bioglass with osteoblasts was created to simulate the bone layer (Jiang et al. 2010). Bilayer scaffolds made of porous tantalum metal and rMSC-seeded polyethylene glycol hydrogel were inserted into the rabbit's osteochondral deformations. The metal induced the regeneration of subchondral bone and aided in the production of hyaline cartilage (Bal et al. 2010).

2.2.3 Gradient Scaffold

Gradient scaffolds, as compared to traditional scaffolds, improve the continuity between the various layers of the scaffold. Gradients can be integrated by modifying the composition of the scaffold or by adjusting the distribution of bioactive molecules (Lee et al. 2014). A gradient scaffold for implantation in a rabbit femoral condyle showed excellent histological points for the distribution and amount of cartilage thickness and bone mineralization around the osteochondral defect site. This gradient was created by changing the flow rates at which TGF-1 or BMP-2/HA-containing PLGA microparticles were delivered to a cylindrical mold. Chondrocyte-forming particles made up the top layer of this scaffold, and a combination of chondrogenic and osteogenic particles made up the basal layer (Mohan et al. 2011).

2.3 Cardiac Tissue Engineering

Lack of self-regenerative capacity, abnormal distribution of stress and problems with the ability to contract. Cardiac muscles require vigorous mechanical strength to contract efficiently and continuously throughout the individual's lifetime. Numerous strategies have been developed to repair the functions of the muscle using various scaffolds, manufacturing techniques and cell sources (Sharma et al. 2019). The biomaterial utilized in cardiac tissue engineering must be able to endure the contraction of the heart muscle and permit cell penetration inside the matrixes (Pope et al. 2008). Extracellular proteins, natural biomaterials, and artificial biomaterials are the most commonly used biomaterials for cardiac tissue engineering (Sharma et al. 2019).

2.3.1 Extracellular Matrix Proteins

Cardiac scaffolds in the form of injectable hydrogels are capable of mimicking the myocardial ECM. Injectable hydrogels are water-insoluble biomaterials that can absorb water, become swollen, and eventually retain their shape. They are crosslinked and viscoelastic in nature. Due to the high water content of hydrogels, they are easy to administer and provide a simple microenvironment for cell growth (Mousavi et al. 2021). Polymers such as polyethylene glycol and polycaprolactone are used to develop injectable hydrogels (Bearzi et al. 2014; Wang et al. 2015). Natural biopolymers such as gelatin, collagen, alginate, hyaluronic acid, and chitosan are also used for cardiac TE (Mousavi et al. 2021). Decellularized cardiac ECM, which consists of proteins, proteoglycans, growth factors, and cytokines, is considered to be highly suitable to mimic the myocardium ECM. The decellularized ECM provides a complex blend of biochemical and physical signals that support proliferation, cell attachment and differentiation (Kc et al. 2019).

2.3.2 Nanomaterials

First-generation biomaterials such as collagen, fibrin, gelatin, and decellularized matrix are extensively used for cardiac tissue regeneration. However, these biomaterials are poorly electroconductive, which hinders the proper functioning of the tissue. In order to compensate for this disadvantage, electroconductive biomaterials, such as nanoparticles and electroconductive polymers, are used. Materials such as silicon-derived nanoparticles, gold nanoparticles, and carbon-based nanoparticles are used for designing electroconductive biomaterials (Esmaeili et al. 2022). The performance of traditional biomaterials can also be enhanced by incorporating nanomaterials and developing a combined blend of the same (Ahadian et al. 2017). Nanoparticles are advantageous in tissue engineering because of their large proportion of surface area to volume. Nanoparticles are useful for cellular absorption because they can transport large numbers of biomolecules from one place to another. Because they have the capacity to increase cardiomyocyte activity, carbon nanotubes are frequently used in the engineering of heart tissue. The electrical and mechanical properties of carbon nanotubes also allow them to induce stem-cell differentiation (R Amin et al. 2020). Gelatin methacryloyl gel with carbon nanotubes has also been used to differentiate embryoid bodies into cardiomyocetes (Ahadian et al. 2016).

2.4 Pancreas Tissue Engineering

Developing an engineered pancreas using the appropriate biomaterials can provide an alternative to traditional diabetes treatment (hormonal therapy). Pancreas tissue engineering can potentially overcome the long-term complications of hormonal therapy during the treatment of diabetes (Salg et al. 2019). The idea of whole organ perfusion decellularization has been pondered upon to generate ECM scaffolds from complex organs such as the heart, lungs, or pancreas. Organ matrixes offer a convenient scaffold because of their physiological similarity with the target tissue. Bioengineered scaffolds like this provide preserved gaps of the ECM components, intact 3D structure and a vascular network that can be easily repopulated with new cells. Perfusion decellularization of a whole pancreas can also be used to develop a natural scaffold (Goh et al. 2013). The greater omentum, which is an extensively vascularized structure of connective, adipose and vascular cells, has been considered an ideal site for islet transplantation (Baidal et al. 2019).

2.5 Stem Cell-Based Tissue Engineering

Pluripotent stem cells can be used in tissue engineering to enhance cell survival, tissue repair and immunological pathways. Stem-cell therapies have a strong potential to repair musculoskeletal and periodontal conditions. Mesenchymal stem cells have been extensively used for direct differentiation into forming a cell and the creation of regenerative pathways by producing growth factors and hormones (Sharma et al. 2019). Under certain environmental circumstances, stem cells can be made to develop into smooth muscle cells or fibroblasts (Augustine et al. 2021). Induced pluripotent stem cells (iPSCs) are considered the finest source of stem cells for tissue engineering because of their pluripotency and flexibility (Liu et al. 2013).

3. Types of Biomaterials in Tissue Engineering

3.1 Polymers (Synthetic and Natural)

Polymeric biomaterials are materials that can be used in physiological environme nts over a long or short duration of time. Biopolymers, including collagens, alginate, proteoglycans, chitosans, and chitins, have all been employed as scaffolds. These polymers support cell-cell attachment and development on their surface and have both biocompatibility and bioactivity. Due to their significant biologically active components, collagen scaffolds provide good cell adherence to their surfaces (Dolcimascolo 2019). They are thoroughly tested and widely used as a remedy for recovering functionality in impaired brain tissue. They are extremely versatile in terms of shape and biomechanical properties, and alternative biomaterials like ceramics cannot equal them in terms of being biocompatible (Boni et al. 2018). Polymers are either natural or synthetic in origin. Combinations of materials with unique attributes are utilized as scaffolds to increase the utilization of biopolymers in regenerating tissues. The main reason behind using a hybrid biomaterial system with different biological and physiological relevance is to obtain the synergy of one component with both features. For instance, integrating synthetic and natural polymers with microstructures, chemically and structurally specified components, and biosynthetic features might provide a useful link between the two techniques (Lee et al. 2018).

3.1.1 Natural Polymers

As the name suggests, natural polymers are derived naturally. They are primarily created by animals, plants, and microorganisms and fall into three basic categories: polypeptides, polysaccharides, and polyesters. These polymers have applications in the biomedical sector in cosmetics and pharmaceuticals for regenerative medicine (Abbasian et al. 2019). In contrast to artificial polymers, natural polymers have higher bioactivity and biocompatibility. They are frequently comparable, if not exact replicas, of compounds that are already found in the human body, reducing the likelihood of cytotoxic activity and immunogenicity after insertion in the body (Gomes et al. 2017). In tissue engineering, biobased polymers can function as matrix precursors, gel-forming reagents, or drug release regulators. They can also be quickly changed to address a physiological morphological deficit (Madigan et al. 2009). Despite these benefits, biobased polymers have a number of disadvantages that may limit their use in tissue engineering, including the high cost of manufacture in certain instances, variations within batches that are primarily caused by their complex structural, morphological and chemical properties, an unregulated hydration rate, inadequate resources, and the possibility of microbiological contamination. These polymers are also not easily processable (Abbasian et al. 2019).

3.1.2 Synthetic Polymers

Synthetic polymers have attracted a lot of attention from the scientific community for the purposes of nerve TE and regenerative medicine due to their ability to modify the physical, chemical, and mechanical characteristics for the replication of biological tissues. These polymers are excellent for TE since they are hydrolyzed in a predictable manner across all hosts (Amani et al. 2019). Modification of polymeric structure is quite difficult with natural polymers as compared to synthetic polymers.

Artificial polymers like PCL, PLGA and PLA are extensively utilized in bone TE (Figure 1 a, b, and c). Due to their processability, regulatable properties, and biodegradability, they are authorized by the FDA for human utilization in numerous medical uses, such as suture material, implants, and pharmaceutical delivery systems (Lee et al. 2018). They lack inherent biological characteristics, and the

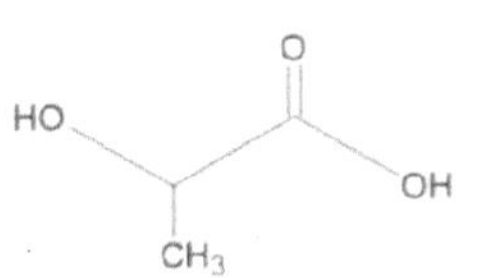

(a) Polylactic acid (PLA)

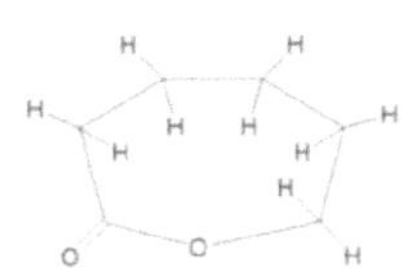

(b) Polycaprolactone (PCL)

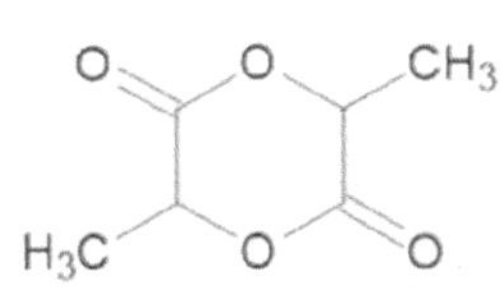

(c) Poly Lactic-co-glycolic acid (PLGA)

Figure 1: Chemical structure of (a) polylactic acid (PLA), (b) polycaprolactone (PCL) and (c) poly Lactic-co-glycolic acid (PLGA).

Table 2: Comparison between natural and synthetic polymer .

Types of Polymers	Advantages	Disadvantages
Natural polymer	Biological properties; imitate content and natural structure of ECM; hydrophilic in nature	Difficult to control the biodegradation rate; weak mechanical stability; temperature sensitive
Synthetic polymer	Easy to process; controllable degradation rate; controlled surface characteristics	Less biological properties; degrade to form a by-product; hydrophobic in nature

byproducts they produce *in vivo* may have negative effects or affect the surrounding environment. Additionally, because synthetic polymers are not hydrophobic, they may cause fibrous encapsulation and protein degradation in the implant's domain (Kim and Mooney 1998).

The various advantages and disadvantages of biopolymers and artificial polymers have been discussed in Table 2.

3.2 Hydrogels

Hydrogels are hydrophilic polymers that absorb water and retain a characteristic 3D conformation (Kopecek 2007). With the help of a tissue-engineered matrix, they can successfully maintain proliferating cells without experiencing biochemical stressors that could result from harmful interactions with the matrixes. They have demonstrated use as sealants, pharmaceutical delivery systems, and TE matrices due to their high absorption and adsorption capabilities (Love 2017). Due to their action of modifying the structural properties of scaffolds based on hydrogels, they closely resemble those of real tissues, making them a significant subclass of scaffolds. Copolymer hydrogels' functional qualities are highly reliant on the amount of water present in the structure, and precise regulation is required to obtain consistent data about the mechanical behavior within the systems. As demonstrated by hydrogel foams, the capacity to control the copolymer volume relative to the fluid in which it is closed offers the potential to build soft structures (Chen et al. 1999). Due to this characteristic, hydrogels—especially natural hydrogels—have numerous applications in TE. The scaffolds serve as the extracellular matrix and can bind different growth factors tailored to the tissue application to promote the most efficient regeneration and mimic the microenvironment of the tissues. Biohydrogels are intrinsically biocompatible and non-toxic and also show increased cell-cell attachment and biodegradation due to proteolysis. By simulating tissue organization and improving cell integration within the scaffolds, using biofabrication methodologies for biohydrogels could produce useful structures (Elkhoury et al. 2021).

3.3 Injectables

The importance of injectable biomaterials is significant because they enable accurate insertion into tightly contained anatomical areas and the healing of irregular lesions (Kretlow et al. 2009). Injectable biomaterials have evolved from structural fillers to multipurpose systems as a result of changing clinical requisites and developments in material science. As a result of being effectively transported, encased, maintained, and supplied to the target site of injection and the biomaterials, alone or in combination with therapeutic agents, which are vital in tissue restoration and regeneration, these biomaterials combined with a wide variety of cells, genetic material, and growth factors can also produce better results (Zhou et al. 2019). Due to their ability to be injected into tiny, abnormally structured spaces, injectables are utilized as fillers or binders to physically fill tissue spaces during plastic, orthopedic, and orthodontic surgical procedures (Zhou et al. 2019). Hydrogels can be administered intravenously. For cancer and other disease therapeutics, tissue restoration and regeneration, and as carriers for therapeutics, including medications, cellular components, biologically active compounds, and proteins, these injectable hydrogels have undergone substantial research. As traditional hydrogels that can be injected into target areas with little to no invasive activity and can be used for areas with irregularity in conformation, injectables have an advantage over the disadvantages of preformed hydrogels in medical applications (Tavakol et al. 2020).

Table 3 highlights the different biomaterials and their applications in tissue engineering.

Table 3: Biomaterials and their applications.

Types of Biomaterials	Role	Applications	References
Polymers (natural and synthetic)	Are used in modulating cell adhesion, expansion and enhancement of electrical properties of the adhered cells	To improve mechanical and physical properties by the inclusion of nanoparticles and the method of preparation	(Boccaccini and Gough 2007; Hussain and Thomas 2021)
Hydrogels	Hydrophilic polymers sustaining a unique 3-dimensional structure	Pharmaceutical delivery systems, cellular carriers for TE and biosensors	(Kalia 2015)
Injectables	By filling irregular flaws and possessing manageable chemical and physical qualities, they improve native tissue integration	Wide range of applications in medical implants	(Chun et al. 2018)

4. Biomaterials Enhancing the Drug Delivery System in Tissues

With advancements made in the field of biomaterials, innovative strategies are formulated to enhance the drug delivery system (Yun et al. 2015). Specifically, polymeric and lipid-based substances for drug administration have been pushed by breakthroughs in a wide variety of industries. Biomaterials offer a diverse variety of efficacy over pharmaceutical substances (Langer 1990). Several of them are made to deliver a certain quantity of therapeutic for a predetermined duration and may be changed to function in targeted areas, lowering the quantity of drugs needed to accomplish the intended action and reducing the patient's toxicity (Hoare and Kohane 2008). In order to achieve maximum therapeutic action, biomaterials' physical and chemical properties and delivery methods can be changed. Biomaterials are most progressed for injected and oral drug delivery techniques; however, novel new drug delivery options such as pulmonary, transdermal, ophthalmic, and nasal routes have also been increasingly adopted (Anselmo and Mitragotri 2014).

4.1 Biocompatibility

Biocompatibility is the biomaterial's ability to carry out its function without giving rise to toxic or harmful effects on biological systems but produces appropriate host responses in certain cases. The ISO 10993 series is an internationally recognized standard for testing the end point of biocompatibility for various medical devices. The aim of biocompatibility is to ensure that the drug will not pose any physiological harm to the patient or the consumer (Kohane et al. 2003). A variety of pharmaceutical delivery techniques have been used to provide delayed release of bupivacaine, including polymeric particles, spray-dried lipids, proteins, sugar particles, liposomes, crosslinked hyaluronic acid gels, and polysaccharide blends (Kohane et al. 2002). These compositions generate varied inflammation in the absence of medication but little to no actual damage. But when combined with bupivacaine, all result in varying degrees of muscular damage (Kohane et al. 2010).

4.2 Biodegradability

Biodegradability is the ability of living creatures to biologically break down organic material into simplified molecules (Kellie 2016). Certain biodegradable polymers that occur naturally are expressed in a way that can be used in the implementation system of drug administration due to their widespread occurrence naturally, their biocompatible nature and their inclusion of essential polysaccharides (Pillai and Panchagnula 2001). Biologically degradable polymers can occur naturally and can be synthesized. In some ways, synthetic polymers are superior to naturally occurring ones because they can be produced using a repeatable process that results in the same polymeric structure with identical content repeatedly. Unlike naturally occurring polymers, which are available limitedly, they may be created with a variety of properties that can be altered according to the uses.

The majority of contemporary drug development methods are made with polymeric substances that have sufficient biodegradability. The most crucial aspect of TE is managing the discharge of bioactive molecules. Improving and regenerating the network is a challenging process that calls for numerous growth factors and associated hormones. For network engineering to be successful, a distribution network that can enable the controlled release of these components is crucial (Bae et al. 2010).

Using biodegradable polymers provides certain benefits compared to other materials, mostly because of their ability to change mechanically. The ability of polymers to be combined to lessen the need for additional surgery removal is their main benefit, as it is cost and time-efficient (Dhaliwal and Dosanjh 2018).

4.3 Biomaterials Act as a Catalyst to the Drug Delivery System

To avoid or reduce side effects caused by drug delivery in cells and to increase curative efficiency and rehabilitation, it is important to consider the use of biocompatible and biodegradable biomaterials. Thus, it can be said that biomaterials act as catalysts in the pharmaceutical administration system.

The need for components that can deliver medications to specific areas with no issues has arisen as a result of the quest for the discharge of controlled medications. resulting from complications linked with the administration and distribution of traditional medications. In order to quickly remediate and recover patients during the therapy duration, medications are typically sent periodically on the body's prescription. By doing this, the medication concentration will rise, and when it exceeds the body's tolerance limit, complications with excessive medication levels may arise, which may result in varied extent of toxicity in the patient's body (Schneider 2017). As a result of the limited drug concentration at the target site, which can happen as a result of metabolic and degradative activity, as well as its movement away from the site, it is also possible that the release rate of the drug acting on the patient is so rapid that curative measures do not function adequately (Schneider 2017). As a result, this phenomenon will squander pharmaceuticals and cause a moderate amount of transportation loss, as well as high-risk adverse effects on the nearby body cells, tissues, and organs. A pharmaceutical carrier with the ability to deliver a sustained discharge flow to the specific site of action and enable rehabilitation of overall therapy prior to breakdown and removal of the concentrated surplus of medicines and carrier media is the answer to this issue (Park 2014). Capsule-based medications and carriers can be prepared for parenteral or oral administration (Yun et al. 2015).

There is a need for materials that enable the discharge of the medication without any alteration or decomposition over time, as well as those that have a prolonged duration of action. In order to improve shipping effectiveness, these subordinates are designed for direct injection or implantation into the sick tissue or cell target (Bertrand et al. 2014). Target medications have been administered using ligands that have adequate binding affinity and that have been applied to the surface of biomaterials to enable uptake and usage by cells and weak tissue (Kamaly et al. 2012). Drug vehicles must be designed with biomaterials that prevent rupture or degeneration at the specific site and allow surface alteration with ligands.

5. Modifying Biomaterials for Tissue Engineering

Interactions between tissue and scaffolds take place at the surface of the scaffold, therefore the biomaterials' surface is crucial in successful implantation. The scaffold interface is crucial in determining how cells behave since it serves as a framework for cellular growth, differentiation, and adherence (Vasita et al. 2008).

Developing a new biomaterial for tissue engineering is a tedious task as compared to modifying the existing ones to perform better. Most modification techniques involve the improvement of the surface properties of already researched biomaterials (Jiao and Cui 2007). Surface modification is a convenient approach to change the biochemical interactions of a biomaterial to develop desired scaffolds (Jiao and Cui 2007). The idea of biomaterial surface reformation has been accepted over time. Biomaterials (both natural and synthetic), such as polyglycolic acid, chitosan, alginate, polyvinyl alcohol, collagen, gelatin, etc., have been investigated for surface modification. On the basis of the place of implantation, distinct surface alteration techniques for every natural biomaterial are being investigated. At the same time, the physicochemical characteristics of synthetic biomaterials can be altered merely by changing their molecular structure. Most biological scaffolds are biodegradable by hydrolysis, enzymatic activity, radiation or chemical treatments (solvents); therefore, the process of surface modification should be free of high temperatures, enzymes, solvents and irradiation conditions. Chemical modification of a biomaterial provides better surface stability than physical modification due to covalent bonds. Chemical modification also provides better biocompatibility (Vasita et al. 2008). Interactions of the scaffold with the tissues are determined by the scaffold's structure, porosity, distribution, texture, and other chemical properties. Cell behaviors, such as differentiation, proliferation, attachment and other biochemical reactions at the surface, strongly depend on the physicochemical properties of the biomaterials' interface. The chemical and molecular conformation, ionicity, and other factors influence the interaction between cells and scaffolds (Jiao and Cui 2007).

Physical modification of a biomaterial involves covering a biomaterial with cell adhesion proteins like collagen, fibronectin and chitosan (Lotfi et al. 2013). These methods improve cell adherence and cell proliferation on biomaterials with protein coatings. Chemically-immobilized ECM proteins are also used for coating scaffolds (Tallawi et al. 2015). Chemically altering biomaterials by alkaline breakdown of aliphatic polyester surfaces is a widespread practice. Chemical alterations like this create active functionalities that allow covalent attachment between the implant and the tissue (Vasita et al. 2008). The production of hydroxyl and carboxyl groups as a result of these active functionalities is used to immobilize ligands like extracellular matrix proteins. Plasma treatment is also used to modify the properties of the surface of a biomaterial and enhance the affinity of cells toward the scaffold. Photochemical techniques can be used to design biomaterial surfaces and promote polymer cross-linking, photopolymerization and grafting of molecules on the polymers (Vasita et al. 2008).

6. Conclusion and Future Perspectives

In conclusion, there are numerous uses for biomaterials in TE. It has achieved sizable strides in medicine and will have a substantial influence on upcoming clinical applications. The biomimetic surface topography and regulated distribution of bioactive signals to promote restoration are the main areas of study. More extensive research is being conducted on natural and synthetic polymers for the production of medical devices. They can be further utilized in the biofabrication of three-dimension porous biomimetic bioactive scaffolds that are a viable alternative for structural and functional restoration strategies in TE. Natural and synthetic polymers, hydrogels, ceramics, and metals are the primary classifications used in producing scaffolds. In order to promote cell development and, ultimately, the regeneration of tissues, the use of biomaterials in medicine has a significant impact on tissue regeneration. They offer a structural foundation resembling native ECM. Different types of biomaterials are being studied for different methods of tissue repair and to improve the growth of various damaged tissues or organs. Apart from polymers, other biomaterials are also covered in this chapter. Then, they have been used singly or in combination with other materials as composites. Additionally, by processing and characterizing biomaterials from their biological origin, new biomaterials can be investigated to improve activity and lessen side effects related to the medication.

Defining boundaries with tissue components, reestablishing post-injury, and preserving them are the biggest challenges in creating the interfaces. The correlation between the biomolecules' and receptors' structure and function at the area of insertion, as well as the mechanism behind the creation of the interface, is better understood as a result. The principle underlying the differentiating, proliferating, and migrating cells of the novel creations is complex. With the development of a new range of stem-cell treatments, emerging techniques in TE can offer functional tissue replacements. Development of this field pivots to some extent on the advancement of new materials and further developed platform handling procedures. Furthermore, a profound cognisance of the fundamental elements of the growth of bones and ligaments would considerably help frame properties fit in a way that is more reminiscent of the natural environment. Advance momentum examination and backing novel therapeutics, which have phenomenal potential in restoring cardiovascular pathologies as well. These strategies and procedures work with better protein adsorption, cell connection, expansion, cell connection, and separation. A more profound understanding of the fundamental systems of bone and ligament development will considerably help frame features that are highly reminiscent of the natural environment. Right now, the emphasis has been on tending to biomimetic surface geology for impacting cellular conduct, controlled conveyance of bioactive signs to animate recovery, vascularizing the bone, articular ligament zonal design, and osteochondral interface mix. Advancements in innovative therapeutic research and support have great prospects for treating heart diseases as well. Better protein adsorption, cell connection, expansion, cell connection, and separation are the results of the following strategies and techniques. Developing tissue recovery and regeneration medicine can also be aided by understanding surface cooperation, system, and their impact on holding with viable tissues.

References

Abbasian, M., B. Massoumi, R. Mohammad-Rezaei et al. 2019. Scaffolding polymeric biomaterials: Are naturally occurring biological macromolecules more appropriate for tissue engineering? Int. J. Biol. Macromol. 134: 673–694. https://doi.org/10.1016/j.ijbiomac.2019.04.197.

Ahadian, S., S. Yamada, J. Ramón-Azcón et al. 2016. Hybrid hydrogel-aligned carbon nanotube scaffolds to enhance cardiac differentiation of embryoid bodies. Acta Biomater. 31: 134–143. https://doi.org/10.1016/j.actbio.2015.11.047.

Ahadian, S., L. Davenport Huyer, M. Estili et al. 2017. Moldable elastomeric polyester-carbon nanotube scaffolds for cardiac tissue engineering. Acta Biomater. 52: 81–91. https://doi.org/10.1016/j.actbio.2016.12.009.

Amin, R., D., E. Sink, S. P. Narayan et al. 2020. Nanomaterials for cardiac tissue engineering. Molecules 25: 5189. https://doi.org/10.3390/molecules25215189.

Amani, H., H. Kazerooni, H. Hassanpoor et al. 2019. Tailoring synthetic polymeric biomaterials towards nerve tissue engineering: a review. Artif. Cells Nanomed. Biotechnol. 47: 3524–3539. https://doi.org/10.1080/21691401.2019.1639723.

Amini, A. R., C. T. Laurencin and S. P. Nukavarapu. 2012. Bone tissue engineering: Recent advances and challenges. Crit. Rev. Biomed Eng. 40(5): 363–408. https://doi.org/10.1615/CritRevBiomedEng.v40.i5.10.

Anselmo, A. C. and S. Mitragotri. 2014. An overview of clinical and commercial impact of drug delivery systems. J. Control. Release 190: 15–28. https://doi.org/10.1016/j.jconrel.2014.03.053.

Augustine, R., P. Dan, A. Hasan et al. 2021. Stem cell-based approaches in cardiac tissue engineering: controlling the microenvironment for autologous cells. Biomed. Pharmacother. 138: 111425. https://doi.org/10.1016/j.biopha.2021.111425.

Bae, S. E., D. H. Choi, D. K. Han et al. 2010. Effect of temporally controlled release of dexamethasone on in vivo chondrogenic differentiation of mesenchymal stromal cells. J. Control. Release 143: 23–30. https://doi.org/10.1016/j.jconrel.2009.12.024.

Baidal, D. A., M. Infante, V. Fuenmayor et al. 2020. A biologic resorbable scaffold for tissue engineering of the endocrine pancreas: Clinical experience of islet transplantation on the omentum. In Transplantation, Bioengineering, and Regeneration of the Endocrine Pancreas. pp. 269–276. https://doi.org/10.1016/B978-0-12-814831-0.00019-1.

Bal, B. S., M. N. Rahaman, P. Jayabalan et al. 2010. *In vivo* outcomes of tissue-engineered osteochondral grafts. J. Biomed. Mater. Res. B Appl. Biomater. 93: 164–174. https://doi.org/10.1002/jbm.b.31571.

Balla, V. K., S. Bodhak, S. Bose et al. 2010. Porous tantalum structures for bone implants: fabrication, mechanical and in vitro biological properties. Acta Biomater. 6: 3349–3359. https://doi.org/10.1016/j.actbio.2010.01.046.

Bearzi, C., C. Gargioli, D. Baci et al. 2014. PlGF-MMP9-engineered iPS cells supported on a PEG-fibrinogen hydrogel scaffold possess an enhanced capacity to repair damaged myocardium. Cell Death Dis. 5: e1053. https://doi.org/10.1038/cddis.2014.12.

Berthiaume, F., T. J. Maguire and M. L. Yarmush. 2011. Tissue engineering and regenerative medicine: history, progress, and challenges. Annu. Rev. Chem. Biomol. Eng. 2: 403–30. https://doi.org/10.1146/annurev-chembioeng-061010-114257.

Bertrand, N., J. Wu, X. Xu et al. 2014. Cancer nanotechnology: the impact of passive and active targeting in the era of modern cancer biology. Adv. Drug Deliv. Rev. 66: 2–25. https://doi.org/10.1016/j.addr.2013.11.009.

Binyamin, G., B. M. Shafi and C. M. Mery. 2006. Biomaterials: a primer for surgeons. Semin. Pediatr. Surg. 15:276–283. https://doi.org/10.1053/j.sempedsurg.2006.07.007.

Boccaccini, A. R. and J. Gough (Eds.). 2007. Tissue engineering using ceramics and polymers. Woodhead Publishing, Cambridge, England.

Boni, R., Ali, A., Shavandi, A., and Clarkson, A.N. 2018. Current and novel polymeric biomaterials for neural tissue engineering. J. Biomed. Sci. 25. https://doi.org/10.1186/s12929-018-0491-8.

Castells-Sala, C., M. Alemany-Ribes, T. Fernández-Muiños et al. 2013. Current Applications of Tissue Engineering in Biomedicine. Journal of Biochips & Tissue chips S2: 004. https://doi.org/10.4172/2153-0777.S2-004.

Chen, J., H. Park and K. Park. 1999. Synthesis of superporous hydrogels: Hydrogels with fast swelling and superabsorbent properties. J. Biomed. Mater. Res. 44: 53–62. https://doi.org/10.1002/(sici)1097-4636(199901)44:1<53::aid-jbm6>3.0.co;2-w.

Chen, Y., P. Han, L.-J. Vandi et al. 2019. A biocompatible thermoset polymer binder for Direct Ink Writing of porous titanium scaffolds for bone tissue engineering. Mater. Sci. Eng. C Mater. Biol. Appl. 95: 160–165. https://doi.org/10.1016/j.msec.2018.10.033.

Chun, H. J., K. Park, C.-H. Kim et al. (Eds.). 2018. Novel biomaterials for regenerative medicine. Springer, Singapore, Singapore., 1st ed. https://doi.org/10.1007/978-3-319-74854-2_6.

Detsch, R., J. Will, J. Hum et al. 2018. Biomaterials. pp. 91–105. *In:* Kasper, C., V. Charwat, A. Lavrentieva (eds.). Cell Culture Technology. Springer International Publishing, Cham. https://doi.org/10.1007/978-3-319-74854-2_6.

Dhaliwal, K. and P. Dosanjh. 2018. Biodegradable polymers and their role in drug delivery systems. Biomed. J. Sci. Tech. Res. 11. https://doi.org/10.26717/bjstr.2018.11.002056.

Díaz-Cuenca, A., D. Rabadjieva, K. Sezanova et al. 2022. Biocompatible calcium phosphate-based ceramics and composites. Mater. Today 61: 1217–1225. https://doi.org/10.1016/j.matpr.2022.01.329.

Dolcimascolo, A., G. Calabrese, S. Conoci et al. 2019. Innovative Biomaterials for Tissue Engineering. *In* Biomaterial-supported Tissue Reconstruction or Regeneration. IntechOpen. https://doi.org/10.5772/intechopen.83839.

Elkhoury, K., M. Morsink, L. Sanchez-Gonzalez et al. 2021. Biofabrication of natural hydrogels for cardiac, neural, and bone Tissue engineering Applications. Bioact. Mater. 6: 3904–3923. https://doi.org/10.1016/j.bioactmat.2021.03.040.

Esmaeili, H., A. Patino-Guerrero, M. Hasany et al. 2022. Electroconductive biomaterials for cardiac tissue engineering. Acta Biomater. 139: 118–140. https://doi.org/10.1016/j.actbio.2021.08.031.

Goh, S.-K., S. Bertera, P. Olsen et al. 2013. Perfusion-decellularized pancreas as a natural 3D scaffold for pancreatic tissue and whole organ engineering. Biomaterials 34: 6760–6772. https://doi.org/10.1016/j.biomaterials.2013.05.066.

Gomes, M. E., M. T. Rodrigues, R. M. A. Domingues et al. 2017. Tissue engineering and regenerative medicine: New trends and directions-A year in review. Tissue Eng. Part B Rev. 23: 211–224. https://doi.org/10.1089/ten.TEB.2017.0081.

Guo, L., Z. Liang, L. Yang et al. 2021. The role of natural polymers in bone tissue engineering. J. Control. Release 338: 571–582. https://doi.org/10.1016/j.jconrel.2021.08.055.

Griffith, L. G. and M. A. Swartz. 2006. Capturing complex 3D tissue physiology *in vitro*. Nat. Rev. Mol. Cell Biol. 7(3): 211–24. https://doi.org/10.1038/nrm1858.

Han, F., J. Wang, L. Ding et al. 2020. Tissue engineering and regenerative medicine: Achievements, future, and sustainability in Asia. Front. Bioeng. Biotechnol. 8: 83. https://doi.org/10.3389/fbioe.2020.00083.

Hench, L. L. 2005. Bioceramics. J. Am. Ceram. Soc. 81: 1705–1728. https://doi.org/10.1111/j.1151-2916.1998.tb02540.x.

Hoare, T. R. and D. S. Kohane. 2008. Hydrogels in drug delivery: Progress and challenges. Polymer (Guildf.) 49: 1993–2007. https://doi.org/10.1016/j.polymer.2008.01.027.

Huang, W., D. E. Day, K. Kittiratanapiboon et al. 2006. Kinetics and mechanisms of the conversion of silicate (45S5), borate, and borosilicate glasses to hydroxyapatite in dilute phosphate solutions. J. Mater. Sci. Mater. Med. 17: 583–596. https://doi.org/10.1007/s10856-006-9220-z.

Huebsch, N. and D. J. Mooney. 2009. Inspiration and application in the evolution of biomaterials. Nature 462: 426–432. https://doi.org/10.1038/nature08601.

Hussain, C. M. and S. Thomas (Eds.). 2021. Handbook of polymer and ceramic nanotechnology. Springer Nature, Cham., Switzerland, 1st ed. https://doi.org/10.1007/978-3-030-40513-7.

Ige, O. O., L. E. Umoru and S. Aribo. 2012. Natural products: A minefield of biomaterials. ISRN Mater. Sci. 2012: 1–20. https://doi.org/10.5402/2012/983062.

Jeong, C. G., H. Zhang and S. J. Hollister. 2012. Three-dimensional polycaprolactone scaffold-conjugated bone morphogenetic protein-2 promotes cartilage regeneration from primary chondrocytes in vitro and in vivo without accelerated endochondral ossification. J. Biomed. Mater. Res. A 100: 2088–2096. https://doi.org/10.1002/jbm.a.33249

Jiang, J., Tang, A., Ateshian, G.A., Guo, X.E., Hung, C.T., and Lu, H.H. 2010. Bioactive stratified polymer ceramic-hydrogel scaffold for integrative osteochondral repair. Ann. Biomed. Eng. 38: 2183–2196. https://doi.org/10.1007/s10439-010-0038-y.

Jiao, Y.-P. and F.-Z. Cui. 2007. Surface modification of polyester biomaterials for tissue engineering. Biomed. Mater. 2: R24–37. https://doi.org/10.1088/1748-6041/2/4/R02.

Kalia, S. (Ed.). 2015. Polymeric Hydrogels as Smart Biomaterials. Springer International Publishing, Cham, Switzerland, 1st ed. https://doi.org/10.1007/978-3-319-25322-0.

Kamaly, N., Z. Xiao, P. M. Valencia et al. 2012. Targeted polymeric therapeutic nanoparticles: design, development and clinical translation. Chem. Soc. Rev. 41: 2971–3010. https://doi.org/10.1039/c2cs15344k.

Kc, P., Y. Hong and G. Zhang. 2019. Cardiac tissue-derived extracellular matrix scaffolds for myocardial repair: advantages and challenges. Regen. Biomater. 6: 185–199. https://doi.org/10.1093/rb/rbz017.

Kellie, G. (Ed.). 2016. Advances in Technical Nonwovens. Woodhead Publishing, Cambridge, England.

Kent, D., G. Wang, Z. Yu et al. 2011. Strength enhancement of a biomedical titanium alloy through a modified accumulative roll bonding technique. J. Mech. Behav. Biomed. Mater. 4: 405–416 https://doi.org/10.1016/j.jmbbm.2010.11.013.

Kim, B. S. and D. J. Mooney. 1998. Development of biocompatible synthetic extracellular matrices for tissue engineering. Trends Biotechnol. 16: 224–230. https://doi.org/10.1016/s0167-7799(98)01191-3

Kohane, D. S., M. Lipp, R. C. Kinney et al. 2002. Biocompatibility of lipid-protein-sugar particles containing bupivacaine in the epineurium. J. Biomed. Mater. Res. 59: 450–459. https://doi.org/10.1002/jbm.1261.

Kohane, D. S., S. E. Smith, D. N. Louis et al. 2003. Prolonged duration local anesthesia from tetrodotoxin-enhanced local anesthetic microspheres. Pain 104: 415–421. https://doi.org/10.1016/s0304-3959(03)00049-6.

Kohane, D. S. and R. Langer. 2010. Biocompatibility and drug delivery systems. Chem. Sci. 1: 441–446. https://doi.org/10.1039/c0sc00203h.

Kokubo, T. and H. Takadama. 2006. How useful is SBF in predicting in vivo bone bioactivity? Biomaterials 27: 2907–2915. https://doi.org/10.1016/j.biomaterials.2006.01.017.

Kopecek, J. 2007. Hydrogel biomaterials: a smart future? Biomaterials 28: 5185–5192. https://doi.org/10.1016/j.biomaterials.2007.07.044.

Kretlow, J. D., S. Young, L. Klouda et al. 2009. Injectable biomaterials for regenerating complex craniofacial tissues. Adv. Mater. 21: 3368–3393. https://doi.org/10.1002/adma.200802009.

Lalan, S., I. Pomerantseva and J. P. Vacanti. 2001. Tissue engineering and its potential impact on surgery. World Journal of Surgery 25(11): 1458–1466. https://doi.org/10.1007/s00268-001-0131-3.

Langer, R. 1990. New methods of drug delivery. Science 249: 1527–1533. https://doi.org/10.1126/science.2218494.

Langer, R. and J. P. Vacanti. 1993. Tissue engineering. Science. 260(5110): 920–6. https://doi.org/10.1126/science.8493529.

Lee, E. J., F. K. Kasper and A. G. Mikos. 2014. Biomaterials for tissue engineering. Ann. Biomed. Eng. 42: 323–337. https://doi.org/10.1007/s10439-013-0859-6.

Lee, S.-H. and H. Shin. 2007. Matrices and scaffolds for delivery of bioactive molecules in bone and cartilage tissue engineering. Adv. Drug Deliv. Rev. 59: 339–359. https://doi.org/10.1016/j.addr.2007.03.016.

Lee, S. J., J. J. Yoo and A. Atala. 2018. Biomaterials and tissue engineering. pp. 17–51. *In:* Kim, B. (eds). Clinical Regenerative Medicine in Urology. Springer Singapore, Singapore. https://doi.org/10.1007/978-981-10-2723-9_2.

Liu, J., W. Chen, Z. Zhao et al. 2013. Reprogramming of mesenchymal stem cells derived from iPSCs seeded on biofunctionalized calcium phosphate scaffold for bone engineering. Biomaterials 34: 7862–7872. https://doi.org/10.1016/j.biomaterials.2013.07.029.

Lotfi, M., M. Nejib and M. Naceur. 2013. Cell adhesion to biomaterials: Concept of biocompatibility. In Advances in Biomaterials Science and Biomedical Applications. InTech. https://doi.org/10.5772/53542.

Love, B. J. 2017. Biomaterials: A systems approach to engineering concepts. Academic Press, San Diego, CA.

Madigan, N. N., S. McMahon, T. O'Brien et al. 2009. Current tissue engineering and novel therapeutic approaches to axonal regeneration following spinal cord injury using polymer scaffolds. Respir. Physiol. Neurobiol. 169: 183–199. https://doi.org/10.1016/j.resp.2009.08.015.

Martin, I., S. Miot, A. Barbero et al. 2007. Osteochondral tissue engineering. J. Biomech. 40: 750–765. https://doi.org/10.1016/j.jbiomech.2006.03.008.

Mieszawska, A. J., N. Fourligas, I. Georgakoudi et al. 2010. Osteoinductive silk-silica composite biomaterials for bone regeneration. Biomaterials 31: 8902–8910. https://doi.org/10.1016/j.biomaterials.2010.07.109.

Mohan, N., N. H. Dormer, K. L. Caldwell et al. 2011. Continuous gradients of material composition and growth factors for effective regeneration of the osteochondral interface. Tissue Eng. Part A 17: 2845–2855. https://doi.org/10.1089/ten.tea.2011.0135.

Mousavi, A., S. Mashayekhan, N. Baheiraei et al. 2021. Biohybrid oxidized alginate/myocardial extracellular matrix injectable hydrogels with improved electromechanical properties for cardiac tissue engineering. Int. J. Biol. Macromol. 180: 692–708. https://doi.org/10.1016/j.ijbiomac.2021.03.097.

Ng, I. C., P. Pawijit, J. Tan et al. 2019. Anatomy and physiology for biomaterials research and development. pp. 225–236. *In:* Narayan, R. (eds.). Encyclopedia of Biomedical Engineering. Elsevier. https://doi.org/10.1016/B978-0-12-801238-3.99876-3.

Pampaloni, F., E. G. Reynaud and E. H. Stelzer. 2007. The third dimension bridges the gap between cell culture and live tissue. Nature reviews. Molecular Cell Biology. 8(10): 839–845. https://doi.org/10.1038/nrm2236.

Park, K. 2014. Controlled drug delivery systems: past forward and future back. J. Control. Release 190: 3–8. https://doi.org/10.1016/j.jconrel.2014.03.054.

Pillai, O. and R. Panchagnula. 2001. Polymers in drug delivery. Curr. Opin. Chem. Biol. 5: 447–451. https://doi.org/10.1016/s1367-5931(00)00227-1.

Pope, A. J., G. B. Sands, B. H. Smaill et al. 2008. Three-dimensional transmural organization of perimysial collagen in the heart. Am. J. Physiol. Heart Circ. Physiol. 295: H1243–H1252. https://doi.org/10.1152/ajpheart.00484.2008.

Rahaman, M. N., D. E. Day, B. S. Bal et al. 2011. Bioactive glass in tissue engineering. Acta Biomater. 7: 2355–2373. https://doi.org/10.1016/j.actbio.2011.03.016.

Rao, S. H., B. Harini, R. P. K. Shadamarshan et al. 2018. Natural and synthetic polymers/bioceramics/bioactive compounds-mediated cell signalling in bone tissue engineering. Int. J. Biol. Macromol. 110: 88–96. https://doi.org/10.1016/j.ijbiomac.2017.09.029.

Ratner, B. D., A. S. Hoffman, F. J. Schoen et al. 2013. Biomaterials science: An evolving, multidisciplinary endeavor. pp. xxv–xxxix. *In:* Ratner, B. D., A. S. Hoffman, F. J. Schoen et al. (eds.). Biomaterials Science. Elsevier. https://doi.org/10.1016/B978-0-08-087780-8.00153-4.

Salg, G. A., N. A. Giese, M. Schenk et al. 2019. The emerging field of pancreatic tissue engineering: A systematic review and evidence map of scaffold materials and scaffolding techniques for insulin-secreting cells. J. Tissue Eng. 10: 2041731419884708. https://doi.org/10.1177/2041731419884708.

Schneider, C., R. Langer, D. Loveday et al. 2017. Applications of ethylene vinyl acetate copolymers (EVA) in drug delivery systems. J. Control. Release 262: 284–295. https://doi.org/10.1016/j.jconrel.2017.08.004.

Sharma, K., M. A. Mujawar and A. Kaushik. 2019. State-of-art functional biomaterials for tissue engineering. Front. Mater. 6. https://doi.org/10.3389/fmats.2019.00172.

Simon, M., C. Lagneau, J. Moreno et al. 2005. Corrosion resistance and biocompatibility of a new porous surface for titanium implants. Eur. J. Oral Sci. 113: 537–545. https://doi.org/10.1111/j.1600-0722.2005.00247.x.

Tallawi, M., E. Rosellini, N. Barbani et al. 2015. Strategies for the chemical and biological functionalization of scaffolds for cardiac tissue engineering: a review. J. R. Soc. Interface 12: 20150254. https://doi.org/10.1098/rsif.2015.0254.

Tavakol, D. N., J. Tratwal, F. Bonini et al. 2020. Injectable, scalable 3D tissue-engineered model of marrow hematopoiesis. Biomaterials 232: 119665. https://doi.org/10.1016/j.biomaterials.2019.119665.

Vasita, R., I. K. Shanmugam and D. S. Katt. 2008. Improved biomaterials for tissue engineering applications: surface modification of polymers. Curr. Top. Med. Chem. 8: 341–353. https://doi.org/10.2174/156802608783790893.

Venkatesan, J., I. Bhatnagar, P. Manivasagan et al. 2015. Alginate composites for bone tissue engineering: A review. Int. J. Biol. Macromol. 72: 269–281. https://doi.org/10.1016/j.ijbiomac.2014.07.008.

Wang, X., Y. W. Chun, L. Zhong et al. 2015. A temperature-sensitive, self-adhesive hydrogel to deliver iPSC-derived cardiomyocytes for heart repair. Int. J. Cardiol. 190: 177–180. https://doi.org/10.1016/j.ijcard.2015.04.139.

Yu, H.-S., G.-Z. Jin, J.-E. Won et al. 2012. Macrochanneled bioactive ceramic scaffolds in combination with collagen hydrogel: a new tool for bone tissue engineering. J. Biomed. Mater. Res. A 100: 2431–2440. https://doi.org/10.1002/jbm.a.34163.

Yun, Y. H., B. K. Lee and K. Park. 2015. Controlled Drug Delivery: Historical perspective for the next generation. J Control Release. 219: 2–7. https://doi.org/10.1016/j.jconrel.2015.10.005.

Zhou, H., C. Liang, Z. Wei et al. 2019. Injectable biomaterials for translational medicine. Mater. Today (Kidlington) 28: 81–97. https://doi.org/10.1016/j.mattod.2019.04.020.

Zhou, J., C. Xu, G. Wu et al. 2011. *In vitro* generation of osteochondral differentiation of human marrow mesenchymal stem cells in novel collagen-hydroxyapatite layered scaffolds. Acta Biomater. 7: 3999–4006. https://doi.org/10.1016/j.actbio.2011.06.040.

9

Strategies for Development of Biofilms

Ambikesh Soni,[1,*] *Pratiksha Singh*[2] and *Gagan Kant Tripathi*[2]

1. Introduction

Forming biofilms enables single-celled organisms to adopt a multicellular lifestyle temporarily, promoting collective survival in challenging conditions. What was once considered a simple attachment of microbes to a surface has now revealed itself as a complex and dynamic developmental process (Kostakioti et al. 2013). The shift from individual planktonic growth to biofilm formation is triggered by environmental changes orchestrated by various regulatory networks. These networks translate signals into synchronized changes in gene expression, allowing bacterial cells to reorganize both spatially and temporally. This cellular reprogramming alters the expression of surface molecules, nutrient utilization, and virulence factors, equipping bacteria with various traits that enable them to endure harsh conditions (Böttcher et al. 2013).

The biofilm matrix serves as a protective shield for bacteria within it, guarding them against the body's natural immune defenses like opsonization and phagocytosis, as well as antibiotic treatments when they are inside the host. This protection enables the exchange of drug-resistance traits and other virulence factors among bacteria, contributing to the persistence of biofilm-forming pathogens. These resilient pathogens are responsible for causing chronic and challenging infections, including urinary tract infections (like those caused by uropathogenic *Escherichia coli* and *Klebsiella pneumoniae*), upper respiratory infections, catheter-related infections,

1 Regenerative Medicine and Stem Cell (RMS) Lab, Department of Biomedical Engineering, Indian Institute of Technology, Hyderabad, Telangana, India.
2 School of Nanotechnology, Rajiv Gandhi Proudyogiki Vishwavidyalaya (RGPV), India.
* Corresponding author: ambikeshsoni14@gmail.com

periodontitis (involving mixed biofilms of Streptococcus mutants and other bacteria), and more (involving bacteria like *E. coli* and *Enterococcus faecalis*). In individuals with compromised immune systems, infections by opportunistic biofilm-forming organisms can be especially dangerous, leading to severe symptoms and, sadly, often resulting in fatalities (Tanzil et al. 2016).

Bacteria are enveloped inside the biofilm by a self-produced extracellular matrix, which makes up 90% of the biofilm's mass. The primary composition of this matrix consists of extracellular polymeric substances (EPS), which function as a stabilizing framework for the three-dimensional structure of the biofilm. Furthermore, it includes carbohydrate-binding proteins and extracellular DNA (eDNA). Hydrophilic polysaccharides in the matrix form hydrogen bonds with water, effectively retaining it, while nutrients are trapped within for the resident bacteria to use for their metabolism (Divakar et al. 2019). In response to the availability of nutrients, bacteria release enzymes that alter the composition of EPS, enabling the biofilm's architecture to adjust to its particular surroundings (Veerubhotla et al. 2018). The structural elements within the matrix create a well-hydrated, sturdy framework with impressive tensile strength. This framework keeps bacteria closely packed together, allowing for intimate cell-to-cell interactions and DNA exchange, and shields the biomass from drying out, predators, oxidative compounds, radiation, and other harmful agents. Moreover, environmental variations within the biomass play a role in enhancing biofilm resilience. Distinct bacterial groups display diverse gene expression patterns based on local nutrient levels and oxygen availability, effectively resulting in a "division of labor" within the community (Nira Rabin et al. 2017).

The process encompasses reversible and irreversible stages of bacterial aggregation, followed by subsequent biofilm development, along with other elements that are conserved and unique to specific species. The initial phase involves placing bacteria onto a somewhat random surface. It occurs due to factors like Brownian motion and gravity, with additional influences from the surrounding hydrodynamic forces. Identifying the surface extracellular appendages, which are sticky, mediate the adhesions to release at the surface (Mishra et al. 2020). The effectiveness of these appendages in binding is not absolute due to their reversible and dynamic nature. This allows bacteria to reattach even after being detached by factors like hydrodynamic forces (which can dislodge bacteria from the surface), repulsive forces, or in response to nutrient availability (Dumaru et al. 2019). Pathogens like *P. aeruginosa*, known for their tendency to form biofilms, employ various attachment structures to adhere irreversibly to surfaces. *P. aeruginosa* utilizes a type IV pili-mediated twitching motility to navigate the interface between liquid and solid surfaces. This allows it to move along the surface and maintain its adhesion. The bacterium also expresses several CUP fimbriae, including CupA, which plays a role in surface adhesion and the self-aggregation of bacterial cells (Toole et al. 2000).

A promising approach to nanoparticle (NPs) production involves utilizing biofilms, which have recently emerged as the predominant growth mechanism for bacteria. Biofilms created by metal-reducing bacteria can serve as a highly effective framework for the reduction and formation of NPs, surpassing the capabilities of planktonic cells. Electrochemical redox processes are facilitated inside biofilms, establishing a favorable environment that includes natural reducing agents such as proteins, peptides, and heterocyclic compounds. This leads to a remarkable enhancement in the reduction of metals into nanoparticles (NPs), achieving efficiency levels of up to 600 times greater than what is observed in planktonic cells (Yu et al. 2019). It is crucial to acknowledge that research about nanoparticle (NP) production within biofilms has been constrained, and the mechanisms accountable for stabilizing NPs in biofilms are still predominantly uncharted. Consequently, acquiring a more profound comprehension of how biofilms generate NPs compared to their planktonic counterparts holds significant importance. Such insights will aid forthcoming researchers in crafting more resilient microbial systems for the efficient and precise biosynthesis of NPs with particular sizes and shapes. A substantial portion of research concentrates on characterizing the synthesized nanoparticles (NPs) rather than delving into the fundamental processes at play. For example, in bacterial sludge-forming biofilms, these biofilms exhibit a notable capacity to reduce Ag^+ ions and externally generate, isolate, and sustain AgNPs efficiently (Mishra et al. 2020).

2. Structure of Biofilms

The extracellular polymeric substance (EPS) matrix comprises various components within the biofilm, including extracellular polysaccharides, DNA, proteins, and more. This matrix forms a network of channels that facilitate the passage of water, air, and nutrients throughout the entire biofilm structure. Exopolysaccharides, a key component of EPS, are produced within or outside cells and are then released into the environment (Zhang et al. 2020). They appear as long linear or branching strands that attach to the cell surfaces and extend outward, forming extensive networks when observed through electron microscopy. Exopolysaccharides function as frameworks for various molecules, including carbohydrates, proteins, nucleic acids, and lipids. It is important to emphasize that exopolysaccharides' constituents, configurations, and attributes may differ among distinct biofilms. The compositions and linkages of the EPS matrix within biofilms of *K. pneumoniae, P. aeruginosa, S. aureus, E. faecalis, A. baumannii*, and *E.* spp. were investigated. The most frequently encountered carbohydrates identified in these biofilms comprise mannose, galactose, and glucose. N-acetyl-glucosamine, galacturonic acid, arabinose, fucose, rhamnose, and xylose follow closely in abundance. It is essential to highlight that while many exopolysaccharides are not exclusive to biofilms, specific ones, like colanic acid in *E. coli* and alginate in *P. aeruginosa*, are generated in response to challenging or stressful conditions (Veerubhotla et al. 2018).

The M antigen, also referred to as colanic acid, is an extracellular heteropolysaccharide observed in *Enterobacteriaceae*. It is composed of repeat units comprising L-fucose, D-galactose, D-glucuronate, and D-glucose, incorporating O-acetyl and pyruvate side chains enhancing its structure. Within the cell, glycosyltransferases collaborate on the inner cell membrane to construct these repetitive units. After that, Wzx flips the repeat units from cytoplasm to periplasm. Initial polymerization is handled by the Wzy system, while Wzc and Wzb regulate high-level polymerization (Zhang et al. 2020). Colonic acid polymers are transported into the extracellular space by the outer membrane protein Wza. In *E. coli* K-12 strains where colanic acid synthesis is absent, the rate of initial attachment is reduced but not completely abolished. In *P. aeruginosa* biofilms, there are three prominent exopolysaccharides – Pel, Psl, and alginate – each fulfilling crucial functions (Guzmán-Soto et al. 2021).

Alginate consists of alternating D-mannuronic acid and L-guluronic acid residues in its chemical structure. In the presence of alginate, *P. aeruginosa* develops increased resistance to antibiotics like ciprofloxacin, gentamicin, ticarcillin, and ceftazidime. Alginate forms a protective barrier around bacterial cells, impeding the penetration and efficacy of antibiotics. Furthermore, alginate can absorb antibiotics, further diminishing their effectiveness. Additionally, alginate disrupts the host's immune response to *P. aeruginosa* infection. It can hinder the activities of neutrophils and macrophages, essential white blood cells responsible for combating diseases. Alginate can also suppress the production of antibodies and other immune components. These combined effects make alginate a significant contributor to the persistence of *P. aeruginosa* infections, particularly in cystic fibrosis patients. Strains of *P. aeruginosa* that produce alginate are associated with more extensive lung damage and unfavorable clinical outcomes. The alginate and lipopolysaccharide production enzyme AlgC was discovered to be implicated (LPS) (Coenye et al. 2020). The enzyme encoded by the algC gene possesses the activities of both phosphomannomutase and phosphoglucomutase. It holds a vital function in converting Man-6-P into Man-1-P and glucose-6-P into glucose-1-P. Man-1-P is pivotal as an intermediate in the synthesis of alginate and LPS (Dumaru et al. 2019).

Within mature *Pseudomonas aeruginosa* biofilms, the biofilm matrix comprises two pivotal exopolysaccharides, Psl and Pel, and alginate and LPS. Both Pel and Psl hold indispensable functions in these biofilms. The Psl polysaccharide consists of repeating pentamers, which facilitate the initial surface attachment process by incorporating D-mannose, L-rhamnose, and D-glucose residues. In the case of strain PA14, the Pel genes are accountable for producing exopolysaccharides rich in glucose. Conversely, in strains ZK2870 and PAO1, the Psl genes participate in the formation of exopolysaccharides rich in mannose (Guzmán-Soto et al. 2021). In instances where Psl genes are absent, a deficient PAO1 strain can still establish mature biofilms by elevating the expression of Pel genes. The AlgC enzyme occupies a pivotal position in synthesizing both Psl and Pel. Notably, the generation of Psl, alginate, and LPS all hinges on a shared intermediate, Man-1-P. Recent research findings have unveiled the intricate interplay among these biofilm components; an increase in the production of one part corresponds to a reduction in the production of the others (Verderosa et al. 2019). Notably, *Pseudomonas aeruginosa* possesses 24 genes involved in the synthesis and secretion of alginate.

Eight are in charge of alginic acid export, and 12 are in order of polysaccharide biosynthesis. The remaining four genes control the synthesis. Polysaccharide intercellular adhesin (PIA), initially identified by Mack and colleagues in *Staphylococcus*, is a significant exopolysaccharide associated with biofilm formation. It comprises a linear polymer consisting of glucosamine residues linked at the 1,6 positions, and it is also referred to as poly-N-acetyl glucosamine (Kartin Schilcher 2020). In *Staphylococcus epidermidis,* a notable feature of PIA is that more than 80% of its residues are N-acetylated, while the remaining 20% carry a

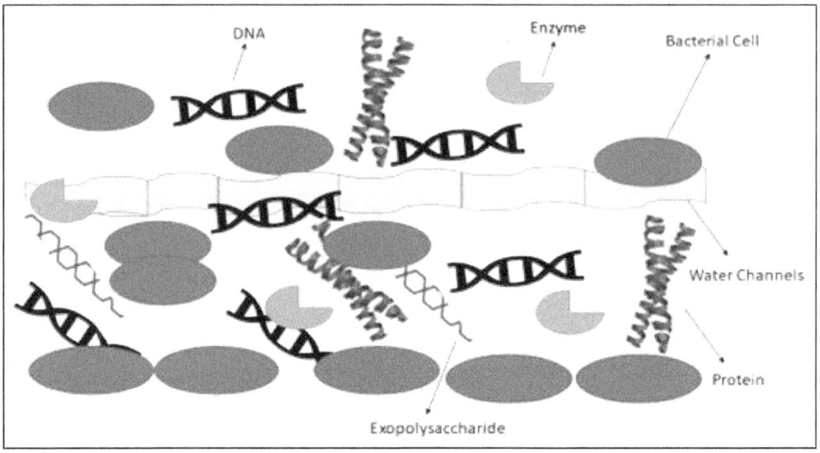

Figure 1. Structure of the biofilm (source: self-created).

positive charge. "The synthesis of PIA is coordinated by four essential proteins: IcaA, IcaB, IcaC, and IcaD" (Kartin Schilcher 2020). IcaA and IcaD operate as N-acetylglucosaminyltransferases, collaborating with UDP-GlcNAc to construct 20-residue oligomers. In contrast, IcaC is crucial in crafting extended oligomers exceeding 130 residues and facilitates their transportation to the extracellular space. Meanwhile, IcaB contributes by deacetylating, introducing positive charges essential for interactions with cell membranes. In the case of *Bacillus subtilis*, the composition of the biofilm-associated exopolysaccharide can vary, presenting either as EPS or polyglutamate. This variability depends on the specific strain and the prevailing environmental conditions (Kokare et al. 2009).

2.1 Extracellular Proteins

Extracellular proteins represent another crucial component of the EPS matrix, contributing significantly to the formation and stability of biofilms. These proteins establish connections between cell surfaces and polysaccharides, promoting the preservation of biofilm architecture. For instance, in *S. mutans* biofilms, glucan-binding proteins (Gbps) serve as an example. Gbps plays a pivotal role by bridging bacteria and exopolysaccharides, ensuring the integrity of biofilm structure. When Gbps are mutated, the resulting biofilms exhibit a substantial reduction in height. Additionally, amyloids, a specific type of insoluble fibrous protein, also aid biofilm formation (Bi et al. 2021). This phenomenon is illustrated in *Pseudomonas spp.*, particularly in the case of Fap amyloids. When Fap amyloids are overexpressed, it leads to the expedited processes of cell aggregation and biofilm development. In *B. subtilis* biofilms, one of the critical components is the amyloid protein TasA. TasA is responsible for creating robust fibers that bind biofilm cells together and exhibit impressive resistance to external pressures, ensuring biofilm integrity even under harsh conditions. Another notable group to consider is the biofilm-associated protein (bap) family, which includes proteins like Bap in *S. aureus* and Esp in

E. faecalis. Members of the bap family serve a dual purpose, actively participating in both biofilm development and the facilitation of infections. There are additional instances, such as the outer-membrane lectins discovered in Azospirillum brasiliense, along with the galactophilic lectin le-cA and the L-fucose binding lectin lecB in *P. aeruginosa*, all of which contribute significantly to the process of biofilm formation. Additionally, certain enzymes within biofilms are engaged in the breakdown process (Taraszkiewicz et al. 2013).

Within the EPS matrix, various substrates like cellulose, proteins, lipids, polysaccharides, nucleic acids, and other components, along with substances trapped within the matrix, all serve as targets for enzymes. These enzymes play a critical role in breaking down biopolymers, thus supplying biofilm cells with essential carbon and energy. This becomes especially vital when the biofilm cells face nutrient scarcity conditions. Enzymatic actions are also required for biofilm separation and dissemination (Zhao et al. 2017). The commencement of a new biofilm lifecycle occurs as biofilm cells are released when enzymes internally degrade the EPS matrix. An instance of this is the DspB protein, which plays a role in the detachment of *Actinobacillus pleuropneumoniae* biofilms from surfaces. Additionally, the understanding of extracellular DNAs (eDNAs) shifted when Mattick and colleagues revealed DNase I could effectively inhibit the development of *P. aeruginosa* biofilms. Before this discovery, eDNAs were commonly considered remnants resulting from cell lysis. eDNA is actively released and derived from lyzed cells, which suggests that eDNA plays an essential role in biofilm development (Kokare et al. 2009). It has been determined that eDNA plays a crucial role in biofilm adhesion. Initially, its negative charge serves as a repellent factor. However, when the cell and the surface gap narrow to just a few nanometers, eDNA engages with receptors on the substrate surface, facilitating adhesion. Furthermore, researchers have found that eDNA also coordinates cell movement, particularly in the expansion of *P. aeruginosa* biofilms through twitching motility (Yu et al. 2019).

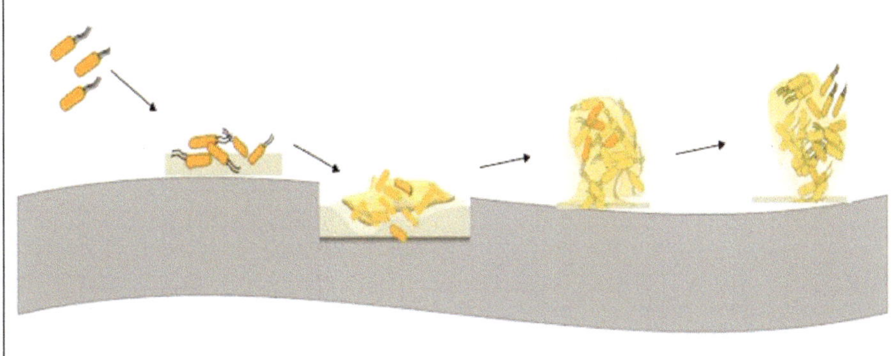

Figure 2: Steps showing the synthesis of biofilm (source: self-created) biofilms development methods.

Due to its negative charge, eDNA can bind with metal cations and certain antibiotics that carry positive charges. In Gram-negative bacteria like *P. aeruginosa, Salmonella Typhimurium*, and others, eDNA can specifically chelate Mg^{2+} and activate the PhoPQ/PmrAB two-component systems, enhancing resistance to antimicrobial peptides. In the case of *S. epidermidis*, eDNA has been found to obstruct the transport of vancomycin within biofilms, thus safeguarding the bacteria trapped within the biofilm structure. However, the current understanding of eDNA's precise roles in biofilm maturation and persistence remains limited, emphasizing the need for further research to understand its significance in biofilm development better (Costerton et al. 2003).

2.1.1 Dental Plaque

Biofilms play a significant role in dentistry, and researchers have conducted numerous studies on dental biofilms. Many bacteria and archaea, totaling over 700 species, have been identified in dental plaque. Dental plaque consists of various types of biofilms, and it is important to note that the composition of a healthy plaque biofilm significantly differs from that of plaque biofilms associated with disease. The unique microenvironment within the oral cavity, influenced by water, temperature fluctuations, solid surfaces, and the availability of carbon and nitrogen sources, contributes to this variation (De la Fuente-Núñez et al. 2013). Moreover, alterations in the local environment can impact the bacterial composition of biofilms. In healthy biofilms, innocuous bacterial species like *Streptococcus oralis, Actinomyces species, Streptococcus gordonii, Streptococcus sanguis,* and others are present but exhibit low acid tolerance.

Bacteria responsible for dental caries, such as *S. sobrinus* and *S. mutans*, notably tolerate acidic conditions. These acid-tolerant bacteria play a significant role in various oral disorders, including dental caries, periodontitis, gingivitis, and more, given their capacity to establish robust biofilms. In order to initiate colonization and the subsequent formation of biofilms on enamel surfaces, bacteria must communicate with one another (Abdallah et al. 2014). The bacteria are flushed away with the oral saliva if they do not linger on the tooth surface. When bacteria successfully adhere to surfaces, they form dental plaque, comprising many species. Oral microbes attach themselves to immobilized bacteria on various surfaces, including teeth, epithelial tissues, and other bacteria, a phenomenon known as co-adhesion. Moreover, genetically diverse cells can engage in cell-to-cell interactions in a suspended state, termed coaggregation. At least one coaggregation partner typically exists within the realm of oral bacteria. A schematic representation illustrates the dynamic interplay of early and late colonizers competing and cooperating during tooth colonization, with Streptococci and actinomycetes being the initial colonizing species (Bi et al. 2021).

The primary colonizers of dental plaque recognize receptors. Phosphate-rich proteins like statherin and enzymes like -amylase are among the receptors. Streptococci detect salivary pellicle receptors (coats the enamel). Proline-rich proteins and statherin bind to actinomyces. Fusobacteria, anaerobic Gram-negative bacteria, also bind statherin. Bacteria in the oral biofilm communicate by secreting and accumulating tiny chemicals. Autoinducer (AI)-2 is a ubiquitous intergeneric signaling molecule involved in forming biofilms. *S. oralis* and *Actinomyces oris* T14V aggregate in response to AI-2 (Alotaibi and Bukhari 2021). Small doses of AI-2 are adequate for oral bacteria coaggregation but not oral pathogens. The concentration of AI-2 rises as the commensal bacteria's cell density increases. Under conducive conditions, these microorganisms multiply and establish the biofilm at optimal concentrations. However, the circumstances are not favorable for their growth at lower concentrations. Fusobacteria, for instance, are prolific producers of AI-2, which they utilize to co-aggregate with various colonizing species. This strategic move allows Fusobacteria to emerge as the predominant bacteria in biofilms associated with healthy gums and plaque (Jiang et al. 2020).

2.1.2 Wounds

Chronic wounds often exhibit the presence of biofilms. Unlike acute injuries, which typically do not involve biofilms and tend to heal promptly, chronic wound cases are characterized by a slower and prolonged healing process. Biofilms usually develop on the surface of wounds. Still, in some instances, like *P. aeruginosa* biofilms, they can be deeply embedded within the wound's layers, making them challenging to detect using conventional wound swab cultures (De la Fuente-Núñez et al. 2013).

2.1.3 Prosthetic Joint Infection

Prosthetic joint infections frequently involve Gram-positive bacteria, notably staphylococci. Shortly after surgery, bacteria from the bloodstream or lymphatic system adhere to the surfaces of prosthetic joints, initiating the development of biofilms. Unlike typical bacterial infections that trigger immediate symptoms like fever, biofilms on these implants, such as pain, may not become apparent until some time has passed (Abdallah et al. 2014).

2.1.4 Persister Cells

Persisters are a small group of cells found within biofilms, and they have a unique characteristic: their growth rate is prolonged, sometimes even non-existent. When treating infections, most antibiotics are designed to target processes associated with rapid cell growth and division. As a result, these antibiotics prove ineffective against persister cells. This leads to a concerning situation where persister cells essentially act as hiding spots for diseases, as they can resurface as infectious agents once the antibiotic treatment has waned (Costerton et al. 2003).

2.1.5 EPS Matrix Protection

The physical barrier of the EPS matrix shields the biofilm cells that clump together. Specifically, in *Pseudomonas aeruginosa* biofilms, the exopolysaccharide known as alginate is a protective shield, guarding the biofilm bacteria from being killed by human white blood cells.

2.1.6 Horizontal Gene Transfer

Bacteria can acquire antibiotic resistance through two primary mechanisms: random mutations in their genes and the presence of antibiotic-resistant genes on plasmids. Plasmids are small DNA molecules that can be easily shared between bacteria through horizontal gene transfer. Interestingly, the transfer of plasmids is much more frequent in biofilms compared to individual planktonic cells. This means biofilms play a significant role in disseminating antibiotic-resistance genes carried on plasmids, primarily through mechanisms like conjugation and mobilization. Studies focusing on Staphylococcus aureus biofilms have highlighted the role of biofilms in increasing the spread of antibiotic-resistance genes through these processes (Van Houdt and Michiels 2010).

2.1.7 Efflux Pumps

Efflux pumps serve as a defense mechanism for bacterial cells, allowing them to expel harmful substances like antibiotics. While these pumps are present in individual planktonic cells, there's an interesting observation in biofilms—several efflux pump genes are upregulated, indicating their potential involvement in antibiotic resistance. Among these genes, PA1874–1877 is a newly discovered efflux pump gene in *Pseudomonas aeruginosa*. It stands out because it exhibits significantly higher expression levels in biofilms than planktonic cells. This efflux pump, generated by PA1874–1877, has been linked to increased resistance to antibiotics like ciprofloxacin, gentamicin, and tobramycin (Jiang et al. 2020).

2.1.8 Cardiac Valve Infection

The effect of bacterial biofilm on artificial biofilm causes prosthetic valve endocarditis. *S. aureus, Corynebacterium* spp., *S. epidermidis, Candida* spp., and *Streptococcus* spp. are the bacteria causing endocarditis. Biofilm build-up can disturb or clog the prosthetic heart valve, causing reduced flow, turbulence, and leakage. Biofilm cells detached from their biofilm can travel via the bloodstream and infect other organs. Biofilm fragments in the bloodstream might obstruct blood flow to vital organs such as the brain and kidneys (Donlan 2001).

2.1.9 Biofilm Resistance to Antibiotics and Host Immune System

Unlike planktonic bacteria, biofilm bacteria exhibit more excellent resistance to antibiotics and the reactive substances generated by the host immune system. It has been estimated that biofilm cells are approximately 10,000 times more resilient to antibiotics than their planktonic counterparts. For instance, when subjected to tobramycin treatment, planktonic cells could not survive beyond a concentration of 50 g/ml. In contrast, biofilm cells could withstand as much as 1 mg/ml of tobramycin when *P. aeruginosa* biofilm and planktonic cells were treated with this antibiotic (Kokare et al. 2009). In inhibiting *Staphylococcus aureus* biofilms, two distinct disinfectants, benzalkonium chloride, and the oxidizing agent sodium hypochlorite, have shown exciting results. These disinfectants required concentrations that were 50 times and 600 times higher, respectively, compared to what is needed to inhibit individual planktonic cells of the same bacterium. As for the growing problem of antibiotic resistance observed in bacterial biofilms, the underlying causes remain largely elusive. A combination of multiple factors likely comes into play, working together to protect biofilm cells from the effects of antibiotic treatments. Several potential mechanisms for antibiotic resistance in bacterial biofilms have been proposed (Abdallah et al. 2014).

3. Mechanism of Biofilm Formation

Microbes form biofilms by congregating in communities on the surfaces of inert materials and binding themselves with a viscous substance known as exopolysaccharide (EPS); these polymers, produced by the microorganisms within the biofilm, anchor bacteria to the material's surface. Bacteria, fungi, algae, protozoa, and other microorganisms can all be found in natural biofilms. Biofilms' amazing tolerance to various toxins and nutrition deprivation has piqued scientists' interest in studying their properties and structure for decades (Toole et al. 2000). The metabolic burden is shared by microbes incorporated in the biofilm matrix as a community via quorum sensing. Biofilms have found applications in various fields, such as bioremediation, wastewater treatment, and metal leaching, to name a few. An example of a valuable biofilm is the electroactive biofilm, which is the cornerstone in bio-electrochemical systems. The steps of biofilm formation will be discussed in detail in the next section. There are multiple phases to the creation of a biofilm. Bacteria travel toward the surface in the first step, which can be done aggressively or passively. Some planktonic microbes actively swim toward a suitable abiotic surface location. Other species use weak forces, including Brownian motion, gravity, and van der Waals forces, to reach the surface (Sousa et al. 2020).

The constant flow situation can bring bacteria closer to the surface of the support in some cases. The surface characteristics of the support mostly determine bacterial adhesion. A reversible adhesion of planktonic bacteria to the surface starts the second stage of biofilm growth. Cells' slender appendages, including pili, fimbriae, and flagella, stimulate cell adherence to the surface, resulting in cell phenotypic alterations. Microbe attachment is further shaped by environmental parameters such as pH, temperature, shear stress, and strain motility (De la Fuente-Núñez et al. 2013). The change from reversible to irreversible attachment on the surface occurs following adhesion. The development of microorganisms as a biofilm has a new beginning with this adhesion. The secretion of EPS is the most noticeable alteration seen in the sessile biofilm. EPS differs from organism to organism in terms of composition.

There are multiple phases to the creation of a biofilm. Bacteria travel toward the surface in the first step, which can be done aggressively or passively. Some planktonic microbes actively swim toward a suitable abiotic surface location (Alotaibi an Bukhari, 2021). Various species employ weak forces such as Brownian motion, gravity, and van der Waals to facilitate their surface arrival. A constant flow situation can sometimes bring bacteria near the support's surface. The surface properties of the substrate play a crucial role in bacterial adhesion—the reversible adhesion of planktonic bacteria to the surface marks initiating the second phase of biofilm growth. Slender cell appendages like pili, fimbriae, and flagella are pivotal

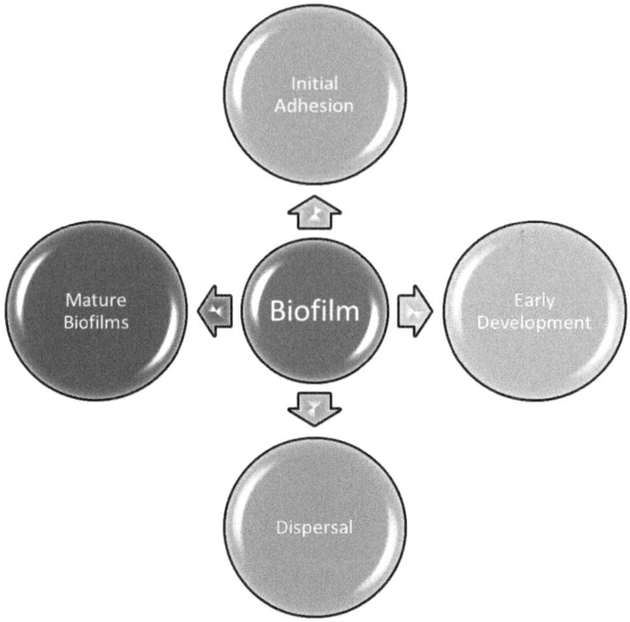

Figure 3: Synthesis and dispersal of biofilm—(a) initial adhesion, (b) early development, (c) mature biofilms, (d) dispersal (source: self-created).

in promoting cell adhesion to the surface, leading to noticeable changes in the cells' characteristics (Van Houdt and Michiels 2010).

Environmental factors like pH, temperature, shear stress, and motility patterns significantly influence bacterial attachment. The transition from reversible to irreversible attachment to the surface occurs once the initial adhesion occurs. The development of microorganisms as a biofilm has a new beginning with this adhesion. The secretion of EPS is the most noticeable alteration seen in the sessile biofilm. EPS differs from organism to organism in composition (Toole et al. 2000). For instance, *E. coli* bacteria utilize colonic acid as a crucial element in their biofilm formation, while *P. aeruginosa* bacteria rely on alginate. This biofilm is enveloped by an EPS matrix, protecting it against external pressures. Changes occurring within the biofilm involve modifications in metabolite production, downregulating genes responsible for flagella formation, and enhanced resistance to UV light exposure. The EPS matrix is believed to act as a barrier, hindering antibiotics from reaching the cell surfaces of microorganisms within the biofilm and shielding these bacteria from various environmental stresses (Zhao et al. 2017).

Once the EPS synthesis is complete, the biofilm enters a maturation phase primarily regulated by quorum sensing. Quorum-sensing molecules play a pivotal role in shaping the architecture of the biofilm, a characteristic unique to specific bacterial species. In its fully developed state, a biofilm exhibits a distinct three-dimensional structure, well-defined water channels, and significant thickness. The ultimate stage involves the detachment of bacteria from the biofilm. Detachment mechanism research is still in its early stages (De la Fuente-Núñez et al. 2013).

4. Methods to Detect the Biofilm Production

Standard antimicrobial susceptibility testing involves injecting an antimicrobial agent into planktonic (suspended) cells and will not reliably predict an agent's efficiency against biofilm-associated pathogens. The National Committee for Clinical Laboratory Standards (NCCLS) does not have a method for testing antimicrobial drugs' efficacy against biofilm-associated pathogens. Several techniques for evaluating antimicrobial drugs against biofilms have been proposed. An approach that can predict the susceptibility of biofilm-associated organisms *in vivo*, is reproducible, affordable, has a quick turnaround time, and is simple to use would be acceptable. The Calgary Biofilm Device was created to test numerous antimicrobial drugs against *E. coli*, *S. aureus*, and *P. aeruginosa* biofilms (Dumaru et al. 2019).

The gadget creates biofilms on duplicate "pegs", tested using a standard 96-well plate setup. The device's results demonstrated that biofilm formation on pegs was repeatable, simple to use, and affordable. The Modified Robbins Device is a tubular flow-through device that produces biofilms of specified organisms on duplicate membranes in flowing circumstances (Costerton et al. 2003). The membranes with biofilms are then withdrawn from the device and treated with various antimicrobial treatments. It was feasible to assess the susceptibility of biofilm and planktonic phenotypes of an organism by comparing findings from these tests with data acquired using the NCCLS protocols. However, any of these biofilm susceptibility testing techniques raises the question of how well they can mimic an *in vivo* or *in situ* biofilm. How well does the approach predict antimicrobial agent efficacy against biofilms on a catheter or a prosthetic heart valve, for example? Better estimations may be possible with model systems that closely imitate an implanted medical device. It created a technique for cultivating biofilms on central venous catheter needleless connectors that replicated the device's properties in operation, including flow rate, temperature, medium composition, the presence of a blood-conditioning coating on the device surface, and the organisms present (Coenye et al. 2020). Similar techniques could be used to test antimicrobial drugs' effectiveness against biofilms on these devices. Comparing results from an *in vitro* model with results from an *in vivo* animal model would be a natural next step.

Following CLSI recommendations, an antibiotic susceptibility test was done on Muller Hinton agar using the Kirby Bauer disk diffusion method. All the bacterial strains were subjected to a battery of common antibiotics, including ofloxacin (5 g), levofloxacin (5 g), ceftazidime (30 g), cefepime (30 g), amikacin (30 g), gentamicin (10 g), piperacillin (100 g), piperacillin-tazobactam (100/10 g), imipenem (10 g), and colistin (10 g). In the case of Pseudomonas species, the testing panel was expanded to include polymyxin B (300 units), tobramycin (10 g), and carbenicillin (100 g) (Veerubhotla et al. 2018).

4.1 ESBL and MBL Phenotypic Detection

A disk diffusion test was used to screen the ceftazidime-resistant bacteria for ESBL production. Between ceftazidime (30 g) and ceftazidime-clavulanate (30/10 g), a

5-mm rise in the zone's diameter was deemed ESBL positive. A combination disk diffusion assay was employed to assess the production of Metallo-β-lactamase (MBL) in imipenem-resistant bacteria. This assay involved using two imipenem disks, one of which had an additional 10 μl of 0.5 M EDTA applied to it. Compared to the zone of inhibition observed with the antibiotic disk on its own, the imipenem-EDTA disk exhibited a noticeably larger zone of inhibition, exceeding 7 mm. This enlargement indicated the presence of MBL production (Nira Rabin et al. 2017).

Two methods are commonly used for detecting biofilms: Congo red agar and the tube adherence method. Here is a description of the tube adherence method: In this method, the process begins with introducing a thriving microorganism into a solution of trypticase soy broth. Subsequently, the mixture is placed in an incubator at a temperature of 35°C for a duration of 24 hours. The tube was rinsed with phosphate buffer saline after discarding the supernatant. It was treated with 0.1% crystal violet for staining, then washed and dried. The presence of visible biofilm lining the tube's bottom and the wall was regarded as a favorable sign. The organisms were inoculated onto Congo red agar plates and incubated for 24 hours at 37°C (Tanzil et al. 2016).

5. Biofilm Growth

Cells that firmly attach to surfaces, resisting removal through gentle rinsing, will follow a sequence of events. They will initiate division, create microcolonies, and commence the production of extracellular polymers, which are responsible for defining the biofilm. These extracellular polymeric substances (EPSs), primarily polysaccharides, can be identified under a microscope and through chemical analysis. These EPSs play a crucial role in forming the matrix or structure of the biofilm. Remarkably, they have a high water content, comprising approximately 98 percent water, yet they firmly adhere to the underlying surface with remarkable tenacity (Verderosa et al. 2019). The structure of a biofilm is not just a straightforward, uniform layer of slime. Instead, it exhibits a dynamic and diverse composition that varies spatially and temporally. Within this structure, "water channels" serve as conduits, facilitating the essential flow of nutrients and oxygen to nourish the cells within the biofilm. Biofilms behave as filters, trapping diverse particles such as minerals and host components, including fibrin, RBCs, and platelets. Biofilm-associated organisms develop more slowly than planktonic organisms, owing to nutritional and oxygen deprivation in the cells. Cell development and division and the clearance of biofilm aggregates containing masses of cells cause cells to detach from the biofilm. Depending on several conditions, including the host immune system's response, these detachable cells may create a systemic infection (Chandel et al. 2018).

6. Factors Affecting Biofilm Formation

The development of biofilms in microbial fuel cells (MFCs) is noticeably shaped by three primary factors: the setup of the system, the operational conditions, and the biological aspects involved. The system setup encompasses various structural elements within MFCs. This includes aspects like the choice of electrode material, the spacing between electrodes, the type of membrane used, the electrical circuit's conductivity, and the overall working volume of the system. These components collectively define the system's configuration, influencing the formation of biofilms and the ultimate performance of MFCs. As a result, substantial research efforts have been dedicated to improving the design of MFC systems (Sousa et al. 2020). The selection of electrode material holds significant importance in biofilm development within microbial fuel cells (MFCs). This choice is pivotal because the electrode material serves as the foundation upon which the biofilm grows, and it also facilitates the exchange of electrons required for anaerobic respiration. In many studies involving MFCs, carbon-based electrodes have emerged as popular choices. These materials include carbon felt, glassy carbon electrodes, activated carbon fabric, graphite, and various other carbon-based alternatives. While specific metal-based electrodes like gold, platinum, or titanium have been employed in MFCs, their widespread use is constrained by their relatively high costs (Chandel et al. 2016). The properties of bioelectrode surfaces, which encompass characteristics such as porosity, hydrophobicity, conductivity, and charge, are primarily determined by the choice of material. These qualities play a substantial role in influencing how bacteria adhere to them. Porosity is an essential attribute of electrodes as it expands the available surface area, making it easier for bacteria to attach and establish colonies. In the case of electroactive bacteria, which are vital for forming conductive biofilms, they tend to favor electrode surfaces that are positively charged and hydrophilic. Since many of these electroactive bacteria are Gram-negative, the positive charge on the electrode surface enhances electrostatic interactions with these negatively charged bacterial cells. Moreover, hydrophilic characteristics encourage the formation of hydrogen bonds, further facilitating the attachment of bacteria to the electrode surface (Coenye et al. 2020).

6.1 Effect of the Type of Membrane

In microbial fuel cells (MFCs), the membrane maintains electroneutrality between the two chambers while facilitating selective cation or anion transfer between the anode and cathode. Proton exchange membranes (PEMs) are the most commonly employed membranes in MFCs, enabling the transfer of protons from the anode to the cathode. Research has demonstrated that the type and surface area of the membrane have a notable impact on the power generation efficiency of MFCs (Böttcher et al. 2013). The choice of membrane in use directly influences the transport of ions and, consequently, the pH of the solution. In some instances, other cations like Na+, K^+, Ca^{2+}, and Mg^{2+} from the anolyte are transported to the cathode instead of protons, leading to an increase in pH at the cathode and a corresponding decrease in pH at the anode when Na^+ ions are employed. This pH shift has implications for the growth and community structure of the biofilm, which will be discussed in greater detail in the subsequent sections (Puranik et al. 2021).

6.1.1 Operating Conditions

Physiological factors like pH, temperature, substrate availability, ion concentrations, and various other variables affect biofilm formation. This is due to the mutual influence of bacterial and material surfaces with their surrounding environments (Dumaru et al. 2019).

6.1.2 Effect of pH

pH stands out as the most critical factor influencing bacterial growth and the performance of microbial fuel cells (MFCs). Numerous research efforts are dedicated to unraveling the impact of anodic pH on MFC performance. In most MFC studies, the optimal pH has been determined to fall within the neutral (pH 7) or near-neutral (pH 6–8) range. Nevertheless, there have been limited reports on MFC operation at extreme pH values. Several parameters, such as ion transfer, substrate oxidation, and oxygen reduction, are pH-dependent during MFC operation. However, these effects can be mitigated by implementing appropriate catalysts, innovative ion-exchange membranes, and buffering agents. It is important to note that electrolyte pH directly influences bacterial populations' composition, development, and structure (Puranik et al. 2021).

6.1.3 Formation of Biofilm Microbial Cells

Bacteria are highly responsive to shifts in pH within their surroundings, and even slight alterations in pH levels can prompt significant adjustments in microbial metabolism. In contrast to enzymatic reactions, bacteria can adapt their metabolic processes in response to fluctuations in environmental pH. The pH of the immediate microenvironment notably impacts various physiological aspects, such as membrane potential, proton motive force, cytoplasmic electrochemical gradient, and the overall balance between proton production and consumption (Divakar et al. 2019).

6.1.4 Temperature

It affects bacterial biofilms' growth and electrocatalytic activity, as it does on pH. Bacteria are classified into three types based on the temperature range in which they grow: psychrophiles (40°C) and thermophiles (40°C). Most MFC research has been conducted in mesophilic temperatures, implying that most electroactive bacteria are active at these temperatures. Psychrophiles and thermophiles, on the other hand, have been reported to be electroactive. Temperature determines the type and distribution of microbial communities in MFCs with mixed cultures, as each species in the consortium has a different optimal temperature for growth. Furthermore, temperature variations affect the reaction rates and thermodynamics of anodic processes, indirectly impacting biofilm formation (Kartin Schilcher 2020).

6.1.5 Effect of Substrate Concentration

Within a mixed microbial community, the type of substrate present in the growth medium and its concentration play pivotal roles in determining which species predominates. In the context of microbial fuel cells (MFCs), a wide array of substrates has been employed, spanning from simple sugars and complex carbohydrates to simulated or real wastewater and lignocellulosic biomass, among other options. Notably, changes in the substrate used have been observed to lead to discernible shifts in the dominant microbial community. This phenomenon stems from the fact that the mixed consortia involved in MFCs, which extend beyond just electroactive bacteria, compete for the substrate as an energy source. Additionally, it is worth mentioning that the concentration of the substrate has a noticeable impact on the power output generated by the MFC system. Generally, higher concentrations of substrates have been observed to have detrimental effects on biofilms due to a phenomenon known as substrate inhibition (Chandel et al. 2018).

6.1.6 *Parameters of Biology*

The formation of biofilms is influenced by several critical factors, including the origin and type of inoculum, the growth rate of bacteria, the specific microorganism involved, electron transfer systems, and various other biological traits. Regarding the inoculum, it is important to highlight that most electroactive biofilms studied thus far have predominantly comprised Gram-negative bacteria. This preference stems from distinctions in the cell wall composition between Gram-negative and Gram-positive bacteria, which significantly affect the electrogenic behavior of these organisms. Gram-positive bacteria possess a peptidoglycan layer associated with teichoic acid, resulting in a relatively high Zeta potential. In contrast, Gram-negative bacteria feature lipopolysaccharides, leading to a lower Zeta potential than their Gram-positive counterparts. As explained earlier, the interaction between the surface charges of bacteria and those on the electrodes plays a pivotal role in determining the choice of material for the MFC system. As is commonly understood, bacterial biofilms can be formed using either a pure culture or a mixed culture (Toole et al. 2000). The composition of bacteria in the environment impacts the micro-colonization of the diverse species that constitute the bacterial biofilm matrix. When more electroactive species are present, substrate degradation occurs more rapidly, producing more protons and electrons. If all the electrons released during substrate degradation are efficiently channeled for energy generation, MFC outputs can be significantly enhanced. Mixed culture biofilms yield a higher electron flow than pure culture biofilms. However, the diverse culture's bacterial community exhibits variable diversity, which raises questions about the reproducibility of such biofilm ecosystems (Jiang et al. 2020).

6.1.7 *Inter- and Intra-species Interactions*

Scientists have proposed that a mix of direct and indirect (mediator-based) electron transfer mechanisms within mixed cultures can contribute to power generation. Within electroactive biofilms, researchers have detected different types of interactions between species. Quorum sensing (QS) is a mechanism that enables communication among bacteria of various species. It has been observed that certain QS compounds play a role in coordinating and controlling gene expression associated with the formation of biofilms. Enhanced power outputs in microbial fuel cells (MFCs) have been attributed to synergistic interactions among bacterial species. Additionally, in various studies involving conductive bacterial nanowires, there is evidence to suggest the presence of syntrophic interactions. Although we have observed symbiotic interactions among different bacterial groups in fermentative processes, our comprehension of cell-to-cell interactions within electroactive biofilms remains constrained. Consequently, there is a necessity for additional research to delve deeper into the specific synergistic relationships and other biological processes that occur within electroactive communities (Zhao et al. 2017).

7. Reason for Biofilm Formation

Many bacteria possess the ability to transition between two distinct states: planktonic and biofilm. In the planktonic state, bacteria exhibit a more rapid rate of cell growth and reproduction. However, it is noteworthy that the biofilm state appears to be the natural and prevailing condition for bacteria. Several factors can lead bacteria to adopt a biofilm state. Biofilms offer numerous advantages, notably bolstering a bacterium's ability to withstand harsh environmental conditions. When bacteria adhere to surfaces or tissues, they can avoid being swept away by water or the bloodstream movement. An illustrative example is seen in oral biofilms, which exhibit exceptional resilience against repetitive, high-shear forces. Biofilm cells display a remarkable resistance to external stresses, estimated to be approximately 1,000 times greater than planktonic cells. Moreover, the protective extracellular polymeric substance (EPS) matrix shields bacterial cells in deeper layers. This shielding primarily works by limiting the diffusion of antimicrobial agents, thereby safeguarding the biofilm community from the detrimental effects of these agents. Biofilms are more than just a physical barrier; they also promote increased bacterial cell densities, creating an ideal setting for exchanging extracellular DNA, including plasmids, through mechanisms like conjugation. This genetic exchange has the potential to bestow antibiotic resistance upon certain bacteria. Significantly, the rate of horizontal gene transfer within biofilms significantly surpasses that observed in planktonic cells (Van Houdt and Michiels 2010).

8. Conclusion

The protective layers formed by cells within the biofilm render antibiotics ineffective in combating biofilm-related infections. Consequently, antibiotic penetration is severely restricted, enabling the stationary cell community to persist even when antibiotics eliminate their motile counterparts effectively. Disrupting the extracellular polymeric substance (EPS) matrix may be necessary to achieve the dispersal of mature biofilms. Notably, components like alginate, Pel, and Psl, found within the EPS matrix of *P. aeruginosa* biofilms, represent potential targets for disrupting biofilm formation. Subsequently, once exposed, the drug molecule could exhibit bactericidal properties or function by dispersing the cells. Conversely, preventing initial processes such as cell adhesion to surfaces could serve as a strategy to impede the formation of new biofilms. This aspect is of utmost importance in advancing medical and dental implants, which are susceptible to colonization by harmful bacteria, leading to biofilm formation. However, a notable challenge lies in the disparity between data generated from *in vitro* and *in vivo* models. The selection of an appropriate animal model would profoundly impact the development of any antibiofilm treatment, contingent upon the model's similarity to human conditions.

Acknowledgments

The authors are thankful to the Faculty of Medicine, The University of Tokyo (Japan), CSIR-AMPRI, Indian Institute of Technology Hyderabad, and RGPV Bhopal (India) for providing the necessary institutional facilities and encouragement.

References

Abdallah, M., C. Benoliel, D. Drider et al. 2014. Biofilm formation and persistence on abiotic surfaces in the context of food and medical environments. Archives of Microbiology 196(7): 453–472. https://doi.org/10.1007/s00203-014-0983-1.

Alotaibi, G. F. and M. A. Bukhari. 2021. Factors influencing bacterial biofilm formation and development. Am. J. Biomed. Sci. & Res. 6: 617–626. https://doi.org/10.34297/AJBSR.2021.12.001820.

Bi, Y., G. Xia, C. Shi et al. 2021. Therapeutic strategies against bacterial biofilms. Fundamental Research 1(2): 193–212. https://doi.org/10.1016/j.fmre.2021.02.003.

Böttcher, T., I. Kolodkin-Gal, R. Kolter et al. 2013. Synthesis and activity of biomimetic biofilm disruptors. Journal of the American Chemical Society 135(8): 2927–2930. https://doi.org/10.1021/ja3120955.

Chandel, A. K. S., A. Bera, B. Nutan et al. 2016. Reactive compatibilizer mediated precise synthesis and application of stimuli-responsive polysaccharides-polycaprolactone amphiphilic co-network gels. Polymer. https://doi.org/10.1016/j.polymer.2016.07.033.

Chandel, A. K. S., B. Nutan, I. H. Raval et al. 2018. Self-Assembly of partially Alkylated Dextran-graft-Poly [(2-dimethylamino) ethyl methacrylate] Copolymer Facilitating Hydrophobic/Hydrophilic Drug Delivery and Improving Conetwork Hydrogel Properties. Biomacromolecules. https://doi.org/10.1021/acs.biomac.8b00015.

Coenye, T., B. Kjellerup, P. Stoodley et al. 2020. The future of biofilm research – Report on the 2019 Biofilm Bash. Biofilm, 2: 100012. https://doi.org/10.1016/j.bioflm.2019.100012.

Costerton, W., R. Veeh, M. Shirtliff et al. 2003. The application of biofilm science to the study and control of chronic bacterial infections. Journal of Clinical Investigation 112(10): 1466–1477. https://doi.org/10.1172/JCI200320365.

De la Fuente-Núñez, C., F. Reffuveille, L. Fernández et al. 2013. Bacterial biofilm development as a multicellular adaptation: Antibiotic resistance and new therapeutic strategies. Current Opinion in Microbiology, 16(5): 580–589. https://doi.org/10.1016/j.mib.2013.06.013.

Divakar, S., M. Lama and U. K. Asad. 2019. Antibiotics versus biofilm: an emerging battleground in microbial communities | Enhanced Reader. Antimicrobial Resistance and Infection Control, 8(76): 1–10.

Donlan, R. M. 2001. Biofilm formation: A clinically relevant microbiological process. Healthcare Epidemiology 33(8): 1387–1392. https://doi.org/10.1086/322972.

Dumaru, R., R. Baral and L. B. Shrestha. 2019. Study of biofilm formation and antibiotic resistance pattern of gram-negative Bacilli among the clinical isolates at BPKIHS, Dharan. BMC Research Notes, 12(1): 1–6. https://doi.org/10.1186/s13104-019-4084-8

Guzmán-Soto, I., C. McTiernan, M. Gonzalez-Gomez et al. 2021. Mimicking biofilm formation and development: Recent progress in *in vitro* and *in vivo* biofilm models. IScience 24(5). https://doi.org/10.1016/j.isci.2021.102443.

Jiang, Y., M. Geng and L. Bai. 2020. Targeting biofilms therapy: Current research strategies and development hurdles. Microorganisms 8(8): 1–34. https://doi.org/10.3390/microorganisms8081222.

Kartin Schilcher, A. R. H. 2020. Staphylococcal biofilm development: structure, regulation, and treatment strategies. Microbiology and Molecular Biology Reviews 84(3): 1–36.

Kokare, C. R., S. Chakraborty, A. N. Khopade et al. 2009. Biofilm: Importance and applications. Indian Journal of Biotechnology 8(2): 159–168.

Kostakioti, M., M. Hadjifrangiskou and S. J. Hultgren. 2013. Bacterial biofilms: Development, dispersal, and therapeutic strategies in the dawn of the postantibiotic era. Cold Spring Harbor Perspectives in Medicine 3(4). https://doi.org/10.1101/cshperspect.a010306.

Mishra, R., Panda, A. K., De Mandal, S., Shakeel, M., Bisht, S. S., & Khan, J. (2020). Natural Anti-biofilm Agents: Strategies to Control Biofilm-Forming Pathogens. Frontiers in Microbiology, 11(October). https://doi.org/10.3389/fmicb.2020.566325

Nira Rabin, Y. Z., Opoku-Temeng, C., Yixuan Du, E. B., & Sintim, H. O. (2017). Biofilm formation mechanisms and targets for developing antibiofilm agents. Future Medicinal Chemistry, 71(10), 493–512. https://doi.org/10.2307/j.ctvnwc0d0.18

Puranik, N., Parihar, A., & Khandia, R. (2021). Lactate Dehydrogenase a Potential Diagnostic Biomarker for Cancer : A Review Lactate Dehydrogenase a Potential Diagnostic Biomarker for Cancer : A Review of Literature. Biomedical Journal of Scientific & Technical Research, 3(September). https://doi.org/10.26717/BJSTR.2021.38.006164

Sousa, S. F., Simões, M., & Filipović, N. (2020). Editorial: The Chemistry of Biofilms and Their Inhibitors. Frontiers in Chemistry, 8(August), 1–2. https://doi.org/10.3389/fchem.2020.00746

Tanzil, A. H., Sultana, S. T., Saunders, S. R., Shi, L., Marsili, E., & Beyenal, H. (2016). Biological synthesis of nanoparticles in biofilms. Enzyme and Microbial Technology, 95, 4–12. https://doi.org/10.1016/j.enzmictec.2016.07.015

Taraszkiewicz, A., Fila, G., Grinholc, M., & Nakonieczna, J. (2013). Innovative Strategies to Overcome Biofilm Resistance. BioMed Research International, 2013, 1–13. http://dx.doi.org/10.1155/2013/150653%0AReview

Toole, G. O., Kaplan, H. B., & Kolter, R. (2000). BIOFILM FORMATION AS MICROBIAL DEVELOPMENT.

Van Houdt, R., & Michiels, C. W. (2010). Biofilm formation and the food industry, a focus on the bacterial outer surface. Journal of Applied Microbiology, 109(4), 1117–1131. https://doi.org/10.1111/j.1365-2672.2010.04756.x

Veerubhotla, R., Varanasi, J. L., & Das, D. (2018). Biofilm formation within microbial fuel cells. In Progress and Recent Trends in Microbial Fuel Cells. Elsevier B.V. https://doi.org/10.1016/B978-0-444-64017-8.00012-9

Verderosa, A. D., Totsika, M., & Fairfull-Smith, K. E. (2019). Bacterial Biofilm Eradication Agents: A Current Review. Frontiers in Chemistry, 7(November), 1–17. https://doi.org/10.3389/fchem.2019.00824

Yu, Z., Rao, G., Wei, Y., Yu, J., Wu, S., & Fang, Y. (2019). Preparation, characterization, and antibacterial properties of biofilms comprising chitosan and ε-polylysine. International Journal of Biological Macromolecules, 141, 545–552. https://doi.org/10.1016/j.ijbiomac.2019.09.035

Zhang, K., Li, X., Yu, C., & Wang, Y. (2020). Promising Therapeutic Strategies Against Microbial Biofilm Challenges. Frontiers in Cellular and Infection Microbiology, 10(July), 1–16. https://doi.org/10.3389/fcimb.2020.00359

Zhao, X., Zhao, F., Wang, J., & Zhong, N. (2017). Biofilm formation and control strategies of foodborne pathogens: Food safety perspectives. RSC Advances, 7(58), 36670–36683. https://doi.org/10.1039/c7ra02497e

10

Biological Response to Biomaterials

P. W. Jayakumar Choudhury,[1] *Navya V. B.,*[6] *Saktishree Jena,*[2] *Prathap Somu,*[3] *S. Rajkumar,*[4] *Manoja Das*[5] and *Ravindra Kumar*[6,*]

1. Introduction

Biomaterial is a substance made up of organic, inorganic, or composite material that is engineered to interact with biological systems. These substances are introduced into the body tissues so that they can mimic bodily function for treatment or tissue-repairing purposes. Nowadays, biomaterials play an important role in medical instrumentation design and production. The science of biomaterials is new to society, and it is considered to be only 50–60 years old. The main criteria for selecting biomaterials include biocompatibility, cost, and chemical properties. As synthetic biomaterials show lower biocompatibility and adverse immune response, natural polymer-based biomaterials are used for effective response. The biocompatibility assessment is considered to measure the ability of the material to perform any host function without any adverse reactions like tumour formation, hypersensitivity reactions, anaphylaxis, etc. A foreign body reaction (FBR) is the host response towards the foreign invaders. When a biomaterial is implanted, body and body fluids

[1] P. G. Department of Zoology, S. C. S. Autonomous College, Puri – 752001, Odisha, India.
[2] Department of Biotechnology, Karpaga Vinayaga College of Engineering and Technology, Affiliated to Anna University, Madhuranhagam – 603308, Tamil Nadu, India.
[3] Department of Biotechnology, Saveetha School of Engineering, Tamil Nadu – 602107.
[4] Department of Chemistry, Karpaga Vinayaga College of Engineering and Technology, Affiliated to Anna University, Madhuranhagam – 603308, Tamil Nadu, India.
[5] Department of Biotechnology, Giet University, Gunupur, India.
[6] Computational Biology & Bioinformatics Lab, School of Biotechnology, National Institute of Technology, Calicut, Kozhikode – 673601, Kerala, India.
* Corresponding author: ravinsit06@gmail.com; ravindra@nitc.ac.in

interact with the implanted material; thus, initial reactions occur with blood clot formation followed by a coagulation process, thrombosis, complement activation, inflammatory mediators, invasion, ECM modification, vascularisation, etc. The host blood protein adsorbs onto the implant surface, and a fibrin matrix and a protective, stable fibrin capsule form at a later stage. Similarly, antigen-presenting and cytokine-releasing cells, the macrophages, and monocytes together form foreign giant cells during contact with bioimplant. Thus, the biocompatibility of the material was preferred as the primary objective during biomaterial selection to avoid adverse effects or its function failure (Bhat and Kumar 2013).

As biomaterials function a novel role in the host body, their selection and design were taken more care of. The physicochemical properties of the material, including toxicity, allergenicity, carcinogenic, etc., were compared for its application in host function (Saini 2015). Apart from these properties, its interaction and compatibility with blood proteins were also taken into consideration (M Sumita 1997). The first biomaterial used in the human body and became popular is tooth filling. Later on, the use of biomaterials in various body parts, such as joint replacement, hip implantation, and cardiovascular repairs, is becoming more familiar in clinical fields. These materials are usually inert, with minimal tissue responses (Namur et al. 2015). The success of this implant depends on the selection of material (Walker 2015).

No doubt, the use of biomaterial or devices improves the quality of life. The demand is also increasing day by day, not only for older people but also for younger ones, particularly those who are suffering from heart problems and inherited diseases (Hench et al. 1971; J Best 2007). Day by day, the continuous improvement of biomaterials is going on with respect to the bio-inertness of the materials and positive response towards the biological response (Danewalia and Singh 2016; Ferguson et al. 2014; Khan et al. 2015). The bioactive approach and interface between body cells and the implant are important with respect to the biocompatibility study. In this chapter, various biomaterials and their composition are elaborately discussed. The emphasis is also given to the clinical and immunogenic response of biopolymers with critical analysis of anti-rejection therapy.

2. What are Biomaterials

Despite the fact that some biomaterials have biological origins, the prefix 'bio' in the word 'biomaterial' refers to a condition in which a material becomes a part of a biological system in order to fulfil a specific purpose rather than being derived from one. The biological system here refers primarily to the human body. The term 'biomaterial' has been defined variously to rationally emphasise its role since the establishment of the interdisciplinary field of biomaterials more than half a century ago. These definitions became more precise over time.

2.1 Definitions

Biomaterial

According to the European Society for Biomaterials, in 1987, it was defined as 'a non-viable material used in a medical device, intended to interact with biological system' (Anonymous n.d.). After some deletions and additions, the definition was modified to 'a material intended to interface with biological systems to evaluate, treat, augment or replace any tissue, organ or function of the body' in 1999 (Williams 2009). According to a leading opinion paper by David F. Williams in 2009, the conclusive definition was 'A biomaterial is a substance that has been engineered to take a form which, alone or as part of a complex system, is used to direct, by control of interactions with components of living systems, the course of any therapeutic or diagnostic procedure, in human or veterinary medicine' (Williams 2009). As per the 2018 conference chaired by Professor David F. Williams, the following definition achieved consensus, 'A material designed to take a form that can direct, through interactions with living systems, the course of any therapeutic or diagnostic procedure' (Ghasemi-Mobarakeh et al. 2019).

Biocompatibility

Biocompatibility explains the tolerance of the biomaterials to function in the host body through interaction with blood and other body fluid components, cells, and organs without affecting the homeostasis and immune mechanism. If the implant is non-biocompatible, it may lead to adverse immune reactions and failure.

2.2 Materials That Can Be Used as Biomaterials

The substances with which the biomaterials are made up are broadly classified as metals, ceramics, synthetic, and biopolymers. Biomaterials can be nature-derived or human-engineered. They can be either biodegradable or non-biodegradable. After the advent of advanced technologies like nanotechnology and self-assembly process technologies, the boundaries between material classes have blurred nowadays.

Metallic Biomaterials

Metallic biomaterials are used largely in dental implantation, joint replacements, orthopaedic fixations, thus providing support to human tissues (Prasad et al. 2017). The greater availability of diverse metallic biomaterials helps in choosing a material to fulfil the specific objective of a particular treatment. Metallic biomaterials, which are chemically inert can be placed in the body permanently to replace a lost function. At the same time, biodegradable biometals are used as a temporary supporting framework to replace a lost function or for the regeneration of a lost tissue (Bhat and Kumar 2013; Henkel et al. 2013). The metallic biomaterials are divided into permanent and biodegradable biomaterials.

The metals and alloys used in permanent metallic implants, due to their high degree of corrosion resistance and wear resistance, exhibit long-term stability *in vivo* along with superior mechanical properties (Saini 2015). They are:

Stainless Steel: Stainless steel is a mixture of an iron alloy with chromium (Cr), nickel (Ni), and other metals along with carbon as a non-metallic additive. Chromium imparts corrosion resistance to stainless steel. Examples of kinds of stainless steel are Ni-free stainless steel and surgical stainless steel (316L), which contain 0.03% of carbon (Sumita 1997).

Titanium and Its Alloys: Titanium alloys exhibit a 50% more strength-to-weight ratio when compared to stainless steel. Titanium alloys contain metals like zirconium (Zr), tantalum (Ta), aluminium (Al), niobium (Nb), and vanadium (V), besides titanium, in different proportions and combinations.

Cobalt-Based Metallic Biomaterials: Cobalt-based metallic biomaterials possess a high degree of wear resistance when compared to titanium alloys.

Tantalum-Based Metallic Biomaterials: Tantalum (Ta) is considered the best biomaterial for certain applications due to its resistance to corrosion properties under highly acidic conditions (M Sumita 1997; Walker 2015). The Ti-Ta alloy is relatively better compared to pure tantalum when certain important properties are considered.

Even after being highly biocompatible due to the possession of certain attributes like high corrosion resistance and resistance to wear and friction, metallic ions leach out into the surrounding biological microenvironment in the long run from the metal surface due to the highly reactive nature of the surrounding biological system. This can lead to local tissue damage as well as systemic damage due to inflammation and metal hypersensitivity (Williams 2009).

In certain instances, using biodegradable biometals instead of permanent metallic implants produces desirable results. The most commonly used biodegradable biometals are magnesium (Mg), iron (Fe), and zinc (Zn) alloys that find cardiovascular and orthopaedic applications. A good amount of research is currently being done on these biometals (Walker 2015). In the case of this kind of bioresorbable biometal, biodegradation kinetics have to be regulated properly for best results.

Ceramic Biomaterials

Ceramic biomaterials or bioceramics are ceramic materials that exhibit physical, chemical, and surface properties that match the properties of various parts of the human skeletal system. Hence, bioceramics find application mostly in relation to the repair of joints, bones, and teeth, as well as to augment both soft and hard tissues (J Best 2007). Even bioceramics are used in soft tissue repair and regeneration. Bioceramics exhibit excellent biocompatibility, corrosion resistance, and high melting point. Its mechanical properties, poor plasticity in comparison to metallic biomaterials, and low electric and thermal conductivity make it a better alternative to metallic implants. Based on their interactions with biological tissues, these were categorised into the category of safe biomaterial for implants which does not elicit unfavourable host response (J Best 2007).

Bioinert Ceramics: Alumina and Zirconia are the most popular examples of bioinert ceramics. As the name suggests, the bioinert ceramics have lesser interaction with surrounding host tissues when placed and thus lack any bonding with the tissues of the host; hence, these lack bone-bonding ability. Due to this bioinert nature, these are used as good alternatives to metal alloys in the case of tooth implants and hip prostheses (J Best 2007).

Bioactive Ceramics: These are also called surface reactive ceramics. Calcium phosphate-based ceramics, calcium sulphate-based ceramics, bioactive glasses/

bioglasses, and bioactive glass ceramics are several major groups of bioactive ceramics.

Non-Bioresorbable Bioactive Ceramics: These ceramics exhibit 'Osteoconductivity' which is the bone-bonding ability due to their positive surface reactive nature. This category includes bioglasses and glass ceramics.

Hench and his colleagues established the concept of bioactive glass or bioglass® (BG) in the 1970s for applications related to bone replacement. BG is silicate (SiO_2) based, consisting of Na_2O-CaO-SiO_2 glass with P_2O_5 added. 45SiO_2-24.5CaO-24.5Na_2O–6P_2O_5 the composition (in wt%) of bioglass (45S5). A biologically active hydroxy-carbon apatite (HCA) layer is formed on the surface of these materials, which is structurally and chemically equivalent to bone's mineral phase and helps bond with the bone.

Glass ceramics are glassy crystals in amorphous and crystalline states. Glass ceramics exhibit relatively better mechanical properties, biocompatibility, and antibacterial properties (along with the addition of certain chemicals) than BG. Some common examples of bioactive glass ceramics are apatite-wollastonite glass ceramic (A-W GC), ceravital, wollastonite diopside (WD), etc.

Bioresorbable Bioactive Ceramics: These are biodegradable and bioresorbable. Bioresorbable bioactive ceramics are used for various medical applications as an alternative to polymeric and metallic biomaterials as they do not elicit inflammatory responses like those of metallic and polymeric biomaterials *in vivo* (Khan et al. 2015). This biodegradable implant material is resorbed and replaced by advancing natural tissue (generally bone). Thus, it helps in tissue regeneration. The most commonly used bioresorbable ceramics are calcium sulphate and calcium phosphates (CaP).

Various hydrate forms of calcium sulphate are used in bone defects as implants due to their higher abundance and resorption characteristics without chronic immune response (Ferguson et al. 2014). Different crystalline phases of calcium phosphates (CaP), including hydroxyapatite (HAp), tricalcium phosphate (TCP), amorphous hydroxyapatite (ACp), octacalcium phosphate, fluorapatite (FAp), etc., were also used in bone repair which shows higher performance. Bone grafts made of CaP degrade relatively slower than bone grafts made of calcium sulphate (Ferguson et al. 2017).

Polymeric Biomaterials

Polymeric biomaterials constitute a large class of biomaterials classified as natural and synthetic. Naturally-derived or biologically-derived polymeric biomaterials are synthesised by the living cells. Hence, their sources are animals, plants, algae, fungi, and microorganisms, and they are biodegradable and renewable. Synthetic polymeric biomaterials are either synthesised from bio-based or petroleum-based monomers. Synthetic polymeric biomaterials can be made either biodegradable or non-biodegradable, as well as renewable or non-renewable.

Synthetic polymeric biomaterials are preferred over naturally derived ones because of the tunability of their properties for the intended application. Based on environmental impact considerations, renewable sources are preferred over non-renewable sources.

Some common examples of natural polymeric biomaterials (biopolymers) are starch, chitin, chitosan, silks, elastin, collagen, etc. (Ivanova et al. 2014); Biopolymers are polyglycolic acid (PGA), polycaprolactone (PCL), polylactic acid (PLA), etc. (Lopes et al. 2012); petroleum-derived synthetic polymers are polyethylene (PE), polypropylene (PP), etc. (Brewis and Briggs 1981).

Carbon-Based Biomaterials

Carbon is the most abundant element in the human body after hydrogen and is regarded as the element of life.

Carbon-based nanomaterials are used for various biomedical applications as biomaterials. Carbon nanomaterials are the newest class of biomaterials that have been found application in several medical applications, such as tissue engineering, biomedical imaging, and cancer therapy, and even used in biosensors and therapeutic delivery systems (Klosterhoff et al. 2017). Despite the fact that carbon is the only element present, the diverse structure and chemistry exhibited by these are due to the variation in the arrangement of carbon atoms (Dash and Cudworth 1998), arising from different bonding patterns and hybridisation states of the same. The exceptional physical properties exhibited by these materials, like electric, thermal, mechanical, and optical properties, make them the most suitable biomaterials for medical/clinical applications.

The most commonly used carbon nanomaterials are graphene, graphene oxide, graphene quantum dots, and carbon nanotubes, which are used for drug delivery, cancer therapy, disease diagnosis, and to make biosensors. Carbon nanomaterials are also used and are further being investigated as coatings on orthopaedic implants (Hubbell 1995).

3. Biomedical Applications of Biomaterials

Biomaterials help to build implantable medical devices, prostheses, therapeutic drug delivery systems, diagnostic devices, tissue-engineered devices, extracorporeal devices, and scaffolds for tissue engineering. A tissue-engineered device is a combination of biomaterials, which has a tissue section combined with biomaterial.

According to the Food and Drug Administration (FDA), these biomaterial implants are those that get incorporated into the surface or inside the host body. There are two broad categories of medical implants: structural and functional. Structural implants include various kinds of artificial implants, such as in joints that structurally integrate with surrounding tissue at a later stage. Functional implants are medical devices such as cardiac pacemakers used to improve or replace a reduced or lost physiological function (Anderson 2001). Implantable drug delivery systems steadily release a therapeutic drug at a specific site in the body in order to maintain a steady concentration level. This site-specific-targeted drug delivery minimises side effects by reducing the required drug dose and, at the same time, increases the efficacy of the drug (Weber et al. 2011). Biomaterials in tissue engineering provide architectural framework comparable to the original extracellular matrix, ultimately encouraging cell growth and tissue regeneration (Martin and Leibovich 2005).

Following are some common biomedical applications of biomaterials:

- Cardiovascular Applications
  - Defibrillators (ICDs)
  - Pacemakers (Broughton et al. 2006)
  - Cardiac loop recorders
  - Prosthetic (artificial) heart valves
  - Coronary stents
  - Vascular stents
  - Cardiovascular Assist Devices (CAD)
  - Automated External Defibrillators (AED)
  - Blood vessel prostheses/prosthetic vascular grafts
  - Guide wires
  - Cardiac sensors
- Orthopaedic Applications
  - Joint/hip/shoulder/knee replacement
  - Spine-related
    - Minimally invasive fusion surgeries
    - Spinal fusion surgeries
    - Motion preservation and dynamic stabilisation surgeries
      - Interspinous spacers
      - Artificial discs
      - Pedicle-based rod systems
  - Fracture Fixation Devices
    - Pins
    - Screws
    - Wires
    - Rods
    - Bone plates
    - Intramedullary devices (IM nails)
  - Bioresorbable Tissue Fixation
    - Interference screws
    - Suture anchors
    - Meniscal repair tracks
    - Meshes

  - Viscosupplementation
  - Synthetic Bone Grafts
  - Artificial Ligaments and Tendons
  - Bone Cement
- Dental Applications
  - Dental implants
  - Dental bone grafts
  - Dental braces
  - Dental filling
  - Dental membranes/capping
- Ophthalmological Applications
  - Intraocular Lens
  - Synthetic Corneas
  - Contact Lens
  - Ocular Tissue Replacement
- Neurological Applications
  - Neural prosthesis
  - Drug tracking in CNS
  - CNS repair
  - Nerve conduits
- Cochlear Prosthetics
- Wound Healing
  - Surgical haemostats
  - Internal tissue sealants
  - Hernia meshes
  - Adhesion protein barriers
  - Skin substitutes
  - Wound closure devices
    - Staples
    - Sutures
- Tissue Engineering Applications
  - Tissue engineering scaffolds
- Plastic Surgery Applications
  - Wound healing
    - Engineered skin
    - Dressings
  - Soft tissue fillers or dermal fillers

- Craniofacial surgery (Brodbeck et al. 2003)
    - Osteosynthesis implants
    - Bone substitutes
  - Breast implants
  - Nose implants
  - Ear scaffolds and nose scaffolds
  - Prosthetic ears
- Gastrointestinal Applications (Wilson et al. 2005)
  - Biliary and pancreatic stents
  - Oesophageal stents
  - Gastric and oesophageal variceal obturation
    - Endoloops
    - Balloon tamponades
    - Self-expandable oesophageal stents
  - Resection devices for polypectomy
    - Endoloops
    - Hemoclips or endoclips
  - Bariatric surgery
    - Intragastric balloons (IGBs)
    - Endobarrier
- Urinary Applications
  - Urethral stents
  - Urethral catheters
- Drug Delivery
  - Enteric capsules
  - Drug-eluting stents/balloons
  - Transdermal adhesive patches
    - Reservoir
    - Polymer matrix
  - Transdermal microneedle patches
  - Drug delivery pumps
  - Nebulisers
  - Inhalers
  - Colloidal drug carriers
    - Liposomes
    - Nanoparticles
    - Micelles

        - Dendrimer
        - Hydrogel
    - Micro/nano drug delivery robots
    - Electronic drug delivery systems
    - Drug-eluting contact lenses
    - Ocular inserts
    - Vaginal and uterine rings
- Membranes in Extracorporeal Organ Support (ECOS) Devices
- Bioelectrodes
    - Recording electrodes
    - Stimulator electrodes
- Biomedical Sensors
    - Physical sensors
    - Chemical sensors
    - Bioanalytical sensors
    - Biopotential sensors
- Biomedical Adhesives and Sealants
- Biocompatible Coatings
- Cranioplasty
    - Cranial implants

4. Biological Response to Biomaterials

Biomaterials are commonly used nowadays as implantable prostheses, medical devices, tissue-engineered devices, and extracorporeal devices for various biomedical applications. Whatever may be the intended purpose, a biomaterial always comes into contact with the elements of a biological system and interacts with them through the interface or its surface.

The performance of any biomaterial is dependent on the proper regulation of these interactions, and this regulation enhances its biocompatibility and optimises its performance resulting in favourable host/tissue responses. When the biomaterial is not biocompatible/not made biocompatible, the interactions lead to unfavourable responses resulting in the failure of the prostheses or devices, which either harm or lead to the death of the recipient/patient. Hence, most of the materials cannot be used as biomaterials as such. They have to be made biocompatible to ensure they serve their intended purpose. Here, the emphasis will be on how the biological system responds to a biomaterial *in vivo*.

The process of implantation of biomaterial as well as its mere presence itself, triggers biological responses in the form of host tissue responses, which are inflammatory responses (innate immunity) followed by wound healing responses (Anderson and Cramer 2015).

4.1 Injury and Subsequent Events

Injury

Biomaterial, either a prosthesis or a medical device, causes injury when implanted surgically into a tissue or organ. This leads to disturbance in the homoeostatic mechanisms, triggering a cascade of host responses in a series that constitute the 'tissue response continuum'. The sequence of host cellular and tissue responses towards a biomaterial, constituting a 'tissue response continuum' following implantation, are blood components implant interaction, coagulation, thrombosis, complement activation, fibrin matrix formation, inflammatory mediators invasion, fibroblast recruitment, fibrin capsule around the implant (Anderson and Cramer 2015). Any material, when implanted into the human body, elicits the above host responses.

Acute wounds on the body's surface heal in a sequential and systematic manner and constitute four overlapping phases, namely haemostasis, inflammation, proliferation, and tissue remodelling (Gorbet and Sefton 2004).

These four successive phases coincide with the above-mentioned sequential steps of 'tissue response continuum' against the implantation. During the blood matrix interaction, haemostasis occurs, which involves the prevention of haemorrhage/bleeding, thus blocking the vascular breach through coagulation. The injury causes damage to the tissue cells through necrosis. This triggers inflammation, which is a crucial step for the subsequent wound-healing process (Henson and Johnston 1987). The haemostasis coincides with blood-material interactions and innate immune response; proliferation coincides with granular tissue formation and tissue remodelling coincides with foreign body reaction and fibrosis or normal healing (Anderson 2001; Anderson and Cramer 2015).

The extent of granulation, foreign body reaction, and fibrosis depend on the extent of injury, degree of cellular necrosis, basement membrane structure, blood-biomaterial interactions formation of a provisional matrix with some immune cells and inflammatory responses (Cotran and Pober 1990), and ultimately on the biocompatibility of the biomaterial.

Persistence of chronic inflammation for relatively longer periods of time, as well as continuous stimulation of chronic inflammation by various aspects of non-biocompatible biomaterial, leads to fibrosis, which is regarded as a negative outcome that finally leads to the failure of implantation which limits the longevity and functionality of the implanted biomedical device or prostheses (Anderson and Cramer 2015). Fibrosis is an example of excessive healing resulting from too much deposition of connective tissue, leading to highly modified architecture and loss of function (Hynes 1992). Another undesirable outcome is the appearance of non-healing chronic ulcers, which result due to insufficient deposition of connective tissue matrix, leading to the weakening of tissue architecture that can easily fall apart (Gorbet and Sefton 2004), which in the context of biomaterial implantation also regarded as a failure and is undesirable. Alternatively, in the case of biomaterials that have a high degree of biocompatibility, chronic inflammation is transient and leads to wound healing through tissue regeneration, which, to a greater extent, restores the original tissue architecture. This is called normal healing. Complete restoration

Table 1: Comparison of inflammation followed by wound healing responses between acute injury on the body surface and injury caused by implantation of a biomaterial.

<table>
<tr><th>Biomaterial Implantation</th><th colspan="2">Acute Injury on Body Surface/to Skin</th></tr>
<tr><td>Host Blood-Implant Material Interaction</td><td colspan="2" rowspan="2">Haemostasis</td></tr>
<tr><td>Provisional Matrix Formation</td></tr>
<tr><td>Acute and Chronic Inflammation</td><td colspan="2">Inflammation</td></tr>
<tr><td>Granulation Tissue Formation</td><td rowspan="2">Proliferation</td><td rowspan="3">Wound Healing</td></tr>
<tr><td>Foreign Body Reaction (FBR)</td></tr>
<tr><td>Fibrosis or Normal Healing</td><td>Tissue Remodelling</td></tr>
</table>

through true regeneration of original tissue architecture is not possible in the case of humans.

Blood-Biomaterial Interactions

Blood-biomaterial interactions start occurring immediately after biomaterial implantation. The damage caused by the injury to the tissue and its vasculature makes various molecular and cellular components of the injured tissue and the blood come into contact with the biomaterial surface. This interaction results in the adsorption of non-specific host proteins-proteins of extracellular matrix (ECM) and plasma/blood onto the biomaterial surface prior to its interaction with host cells. Besides proteins, lipids, sugars, and ions from the plasma are also adsorbed onto the biomaterial surface (Butcher 1991; Henson 1980). This takes place between seconds to minutes of implantation. These adsorbed proteins interact among themselves as well as with various receptors present on the inflammatory cells and initiate the host response cascades associated with biomaterial implantation. These proteins include albumins, globulins, complement proteins, fibrinogen, other coagulation factors, fibronectin, vitronectin, etc. (Klopfleisch and Jung 2017). Various aspects of adsorbed host proteins, such as their composition, concentrations, and conformations, determine the host response/reaction to the biomaterial (Butcher 1991), including activation complement system and coagulation cascades that initiate the inflammatory responses, survival of foreign body giant cells (FBGCs). The initial implant stage activates the acute inflammatory responses (Franz et al. 2011; Henson 1980), whereas various chemical mediators from injured/damaged tissue-mediated immune cells (cytokines) (colony stimulating factor-1 receptor-mediated) can even follow the chronic stage too (Table 1) (Doloff et al. 2017).

Provisional Matrix Formation

Blood coagulation and complement activation follow the blood protein (thrombus, fibrin, or blood clot) deposition onto the biomaterial surface. Many researchers have reported that biomaterial-associated blood coagulation is initiated by the intrinsic pathway of the blood coagulation cascade. This clot, which forms the provisional matrix, consists of inflammatory products released by platelets, endothelial cells, inflammatory cells, and via complement activation (Johnston 1988).

The provisional matrix, being a reservoir for cytokines, growth factors, and chemo-attractants, provides signals for the initiation of wound healing and ultimately acts as a platform for immune cell migration and adhesion (Gorbet and Sefton 2004), thus mediating the path for wound healing through foreign body reaction and fibrosis (alternatively wound healing through regeneration and restoration of original tissue architecture up to a large extent resulting in normal repair (Table 1) (Gorbet and Sefton 2004), in case of biocompatible biomaterial).

4.2 Inflammation

Inflammation is the body's natural and highly necessary response triggered by either pathogenic infection or tissue damage (Anderson and Jones 2007). In the case of implantation of a biomaterial, the inflammation is triggered due to tissue damage resulting from injury caused by surgery and trauma as well as by the biomaterial itself, which can lead to rejection through a hypersensitive reaction.

Damage-associated molecular patterns (DAMPs) from the damaged tissue cells induce sterile inflammation (Wahl et al. 1989). Sterile inflammation here refers to inflammation that does not involve pathogens. DAMPs, in association with pattern recognition receptors on immune cells, can activate and secrete chemical mediators like cytokines, chemokines, growth factors, oxygen free radicals, proteases, etc. (Browder et al. 2000).

The intensity of the inflammatory and wound healing responses depends on various aspects like the severity of the injury, the type of tissue or organ into which implantation is done, the physical dimensions of the prosthesis, or the chemical properties of the biomaterial used in the implant (Anderson 1988). Hence, the biocompatibility of the biomaterial also plays a vital role here.

Acute Inflammation

The presence of neutrophils and mast cells at the implant site is the initial response to acute inflammation which lasts between minutes and days. The duration depends upon the extent of the injury. Acute inflammation is characterised by oedema, which results from the exudation of fluid along with plasma proteins into the injury/implantation site. This is followed by the evasion of leucocytes from blood vessels to the injury site through ECM degradation (Brodbeck and Anderson 2009). Among the leucocytes that emigrated, neutrophils and monocytes are prominent, and neutrophils are predominant. The emigrated leucocytes exhibit adhesion through adhesion molecules present on their surface. These adhesion molecules are stimulated by complement component C5a, Leukotriene B_4 (LTB4), and Tumour Necrosis Factor-∞ (TNF-∞), whereas the complementary adhesion molecules are stimulated by Interleukin-I (IL-I) and TNF-∞. This enhances the interactions between monocytes and endothelial surfaces (Raghavendra et al. 2015). Even intracellular and cell-to-extracellular matrix interactions take place and are facilitated by a family of transmembrane glycoproteins called integrins (ABH 2012). CD11/CD18 family is a group of integrins that are present on the leucocyte surfaces. E-Selectins and ICAMs are also involved in the above interactions (Isa and Hobkirk 2000).

The localised leucocytes at the implant (injury) site exhibit the process of phagocytosis. In general, a phagocytic cell, after recognising the infectious agent, engulfs and degrades it. However, when they encounter the biomaterial, they cannot engulf it due to size disparity, which may cause 'frustrated phagocytosis' (Anselme 2000). Complement-activated fragment, C3b, and immunoglobulin G (IgG) are the two major opsonins that coat the biomaterial surface and aid in the frustrated phagocytosis by the phagocytic leucocytes. Neutrophils adhered to the opsonin-coated surfaces of the biomaterial may release enzymes extracellularly through exocytosis to degrade it (Anselme 2000). Mast cells release histamine, which promotes the recruitment of macrophages to the implant (injury) site (Mehdizadeh and Yang 2013). The recruited macrophages during acute inflammation are classically activated M1 macrophages that exhibit phagocytosis (Anderson 2001).

Chronic Inflammation

Chronic inflammation is characterised by the presence of cytokines monocytes, lymphocytes, antibodies, and increased blood flow to the injured site. The duration of chronic inflammation is one to a few weeks. In the case of biocompatible biomaterials, the duration of chronic inflammation is relatively short. It is also marked by the appearance of proliferating blood vessels and connective tissue (Doloff et al. 2017). Macrophages that are differentiated from infiltrated monocytes exhibit different phenotypes and polarisation states (Niinomi 2002) and mediate the wound healing process through antigen presentation phagocytic action, activating various immune cells via chemical mediators, thus regulating immune response (Anderson 2001). The lymphocytes involved in the chronic inflammation process mediate immune responses against the biomaterial (foreign body). Macrophages are regarded as the most important cells during chronic inflammation as they produce a large number of important classes of biologically active products. These include reactive oxygen species (ROS), chemotactic factors, cytokines, growth-promoting factors, neutral proteases, arachidonic acid metabolites, etc. The important growth factors, PDGF, TGF-β, TGF-alpha (Epithelial growth factor), and IL-1 and fibroblast growth factor are involved in blood vessel growth, fibroblast differentiation, and cellular proliferation (Anderson 2001; Cotran and Pober 1990; Franz et al. 2011; Henson and Johnston 1987).

4.3 Wound Healing

Granulation Tissue Formation

The name granulation tissue is derived from the soft, pink, granular appearance of healing wounds. Granulation tissue formations are marked by the presence of fibroblasts and macrophages and the presence of neovascularisation, thus initiating the wound healing process. Neovascularisation or angiogenesis is a process during which new blood vessels are formed from pre-existing damaged blood vessels. The fibroblast actively synthesises collagen and proteoglycans, and fibrous capsule formation also occurs. Apart from fibroblasts, myofibroblasts and fibrocytes are also involved in capsule formation, which helps in mediating implant isolation and its cellular interactions in host cells (Anderson 1988; Anderson and Cramer 2015).

Foreign Body Reaction

Foreign body response involves the host response to the implant, which involves the Foreign body giant cells (FBGCs) and granulation tissue component. The biomaterial composition and its surface-to-volume ratio determine the granulation tissue components (macrophages, fibroblasts, etc.) in the implant site during the FBR. The macrophages produced when adhered to the implant surface are derived from tissue invasion of monocytes, followed by FBGC formation by both fusions (Anderson and Cramer 2015).

Fibrosis

This marks the final stage of wound healing, in which the fibrous encapsulation surrounds the implant or biomaterial from all sides along with the foreign body reaction on its surface and separates from adjacent tissues in the host (Gorbet and Sefton 2004).

5. Designing of Biomaterials

As mentioned above, host responses during biomaterial implantation, adverse events during improper implants, and the compatible nature of the designing biomaterials play a major role. Biomaterials are designed using different types of metals, ceramics, polymers, composite materials, etc. After clinically successful implantation, the biomaterials were designed as either a single or a combination of suitable materials for a more effective procedure. Moreover, these kinds of biomaterials have various atomic arrangements in which different organ replacement applications in the human body can be expanded (Silver and Christiansen 1999). The various kinds of biomaterials and their design are demonstrated in the following segments.

(i) Biopolymers

Nowadays, biopolymers are the potential biomaterials used in various cardiovascular implants as an alternative to natural and regenerative medicine. Polymers are used in various biomedical fields, such as cardiac valves, bio-contact lenses, surgical and coating materials, tissue adhesives, breast prosthesis, and vascular grafts. The chemical composition, chain structure, and association of fundamental biomolecules specify the characteristics of biopolymers (Khan et al. 2015; Saini 2015). Moreover, to perform various biological applications, various biopolymer compositions with different physiochemical properties and biocompatibility are needed to meet the needs. The advantages of bio-polymeric materials over others include (i) quick fabrication, (ii) secondary processability, (iii) convenient mechanical strength and physicochemical properties, and (iv) low cost of production. Biopolymers are divided into two categories based on their origin: natural and synthetic (Bhat and Kumar 2013; Ivanova et al. 2014). The design of a synthetic biopolymer system includes polymeric units of siloxanes, ethylene, esters, urethanes, acrylics, amides, etc.

Though the designing and fabrication are easy, the major drawback is its low biodegradability and biocompatibility in almost all cases, and hence, their biomedical usage delivers an inflammatory side effect (Bhat and Kumar 2013). These limitations are overwhelmed by the utilisation of natural polymers. For instance, natural biopolymers such as guar gum, cellulose, alginate, and chitosan have already been used in regenerative medicine, drug delivery systems, and tissue engineering (Bhat and Kumar 2013).

(ii) Metals/Metal Oxides Materials

The metallic implant gets potential attention in clinical practice in the medical field for a prolonged duration. Several types of metals and metal alloys were used for various biomedical functions, such as cobalt-chromium, aluminium, zirconium-niobium, stainless steel, titanium, and tungsten alloys. The hasty progress in the medical/clinical field has generated focus to improve several medical bio-products prepared of metal, such as catheters, pacemakers, valves, dental materials and implants, medical and bone fascination devices, craniofacial plates for various applications (Niinomi 2002). The major norms in the selection of metal/metal oxide biomaterials are due to their corrosion resistance properties, biocompatibility, and low-cost productivity (Saini 2015). The function of the implant and the host tissue surrounding the implant are a few important points noted during the selection of metal for biomaterial construction. Stainless steel, like 316L, is the frequently utilised alloy in the bio-implants extending from cardiovascular to otorhinolaryngology. The metal's chemical properties have a potential significance in designing weight-holding orthopaedic and dental implants. CoCrMo combined metals-based alloys are used in orthopaedic artificial bone-joint implantation with higher wear resistance properties. The tensile strength and fatigue restriction of the metal/metal oxides compared with composites of polymeric and ceramics biomaterials make it an excellent choice as a biomaterial. However, its drawbacks include stress-strain failure, high cost, and corrosion in some cases. In addition, metals have lower strengths and elasticity than ceramics (M Sumita 1997). In biological fluids, after the metal implants, the surface of biomaterials can alter and biodegrade to discharge a few cell growth factors due to which implant surface and tissues interact (Williams 2009). Thus, the current studies focus on the future objective of getting more attention to understanding metal biomaterials and their structural features, thus improving their applications and drawbacks.

(iii) Composite Materials

Composite materials are designed as unique combinations that consist of more than one chemical/physical distinct constituent material regularly with numerous physical characteristics that make it more effective than others. Composites have three phases: bulk (matrix), reinforcement, and interphase (Iftekhar 2004). Due to their unique properties, they have been applied in various other applications where tissue growth is necessary. Composite scaffolds designed from the combination of bioglass materials and biopolymers with excellent mechanical properties close to bone strength help in some hard tissue (skull/hip/joint/bone, etc.) applications. Thus, composite materials can overcome the drawbacks of individual materials like metal ceramics in their biocompatibility, corrosion, brittleness, and fragile nature. During the designing of composite materials, the function 'linear expansion constant' should be taken into consideration. The constituents with similar linear expansion constant values are designed for its constructs. If it is different, the composite material construct will have deformities with void formation on the contact surface. Therefore, the design of composite material should be careful for biological applications (Bhat and Kumar 2013; Dorozhkin 2011).

(iv) Ceramics Materials

Ceramics are other sources for designing and developing biomaterials for biomedical applications. The usage of ceramics was heavily inspired by their inertness, good compressive strength, and wearable properties. Ceramics-based biomaterials developed for various biomedical applications such as cardiac valves, dental, orthopaedic, ear, and artificial knee implants, and surface modifications using ceramics can enhance the biocompatibility of metal-based implants. However, ceramics are applied to design biomaterials, but they are still not preferred compared to metals and polymers. Commonly, usage of ceramics is heavily limited due to the more brittleness and lack of tensile strength. However, phosphates-based ceramics are deeply applied to design unique biomaterials owing to their excellent biocompatibility and bone reformation; moreover, the biomaterials are commonly similar to the naturally occurring mineral composition of the bone materials (J Best 2007; Vallet-Regí 2001). The most commonly used ceramics are hydroxyapatites (HPs); these calcium phosphate-based biomaterials are used in various biomedical applications for the entire body system through wrapping of all sections of the skeleton. The primary applications of HPs include bone defects and fracture treatment, spinal therapies, and periodontal and joint replacement. On the other hand, HPs have been applied as bone fillers for skeleton imperfections and act as a soft and lightweight implant on sites of the nasal bone and middle ear (Hench et al. 1971; Jordan et al. 2002; Orlovskii, Komlev, and Barinov 2002). Moreover, HPs were utilised as a coating material for titanium and stainless steel-based implants and other metal implants to help in fixing it on the bone. In HPs, the basic metal outer layer and adjacent bone interact strongly; thus, extra care has been placed to reduce the delamination of coatings to overcome implant failure (Bermejo and Danzer 2010).

6. Conclusion

Tissue compatibility of biomaterials is the major concern over the utility and longevity of the device under consideration and to perform suitable functions in the host organ. Thus, the development of new biomaterials was evaluated with its tissue compatibility test as the main protocol, along with some other tests to detect the physiochemical and mechanical properties. Hemo-compatibility is also a major concern in developing blood-compatible materials to reduce the initial risk of failure. With the day-to-day rise in the need for bio-alternatives to mimic various biological functions, different biomaterials with unique characteristics are in search. The basic knowledge provided in this chapter about the implant, its host response, various biomaterials, and its design can help researchers in the future to discover novel biomaterials.

References

Anderson, James and Stephanie Cramer. 2015. Perspectives on the inflammatory, healing, and foreign body responses to biomaterials and medical devices, pp. 13–36 in Host Response to Biomaterials. Elsevier.

Anderson, James M. 1988. Inflammatory response to implants. ASAIO Journal 34(2): 101–7. doi: 10.1097/00002480-198804000-00005.

Anderson, James M. 2001. Biological responses to materials. Annual Review of Materials Research 31(1): 81–110. doi: 10.1146/annurev.matsci.31.1.81.

Anderson, James M. and Jacqueline A. Jones. 2007. Phenotypic dichotomies in the foreign body reaction. Biomaterials 28(34): 5114–20. doi: 10.1016/j.biomaterials.2007.07.010.

Anonymous. n.d. Williams D. Definitions in Biomaterials, Amsterdam. Oxford-New York-Tokio, Elsevier; 1987.

Anselme, K. 2000. Osteoblast adhesion on biomaterials. Biomaterials 21(7): 667–81. doi: 10.1016/S0142-9612(99)00242-2.

Bermejo, Raúl and Robert Danzer. 2010. High failure resistance layered ceramics using crack bifurcation and interface delamination as reinforcement mechanisms. Engineering Fracture Mechanics 77(11): 2126–35. doi: 10.1016/j.engfracmech.2010.02.020.

Bhat, Sumrita and Ashok Kumar. 2013. Biomaterials and bioengineering tomorrow's healthcare. Biomatter 3(3). doi: 10.4161/biom.24717.

Brewis, D. M. and D. Briggs. 1981. Adhesion to polyethylene and polypropylene. Polymer 22(1): 7–16. doi: 10.1016/0032-3861(81)90068-9.

Brodbeck, William G. and James M. Anderson. 2009. Giant cell formation and function. Current Opinion in Hematology 16(1): 53–57. doi: 10.1097/MOH.0b013e32831ac52e.

Brodbeck, William G., Erica Colton and James M. Anderson. 2003. Effects of adsorbed heat labile serum proteins and fibrinogen on adhesion and apoptosis of monocytes/macrophages on biomaterials. Journal of Materials Science: Materials in Medicine 14(8): 671–75. doi: 10.1023/A:1024951330265.

Broughton, George, Jeffrey E. Janis and Christopher E. Attinger. 2006. The basic science of wound healing. Plastic and Reconstructive Surgery 117(SUPPLEMENT): 12S–34S. doi: 10.1097/01.prs.0000225430.42531.c2.

Browder, Timothy, Judah Folkman and Steven Pirie-Shepherd. 2000. The hemostatic system as a regulator of angiogenesis. Journal of Biological Chemistry 275(3): 1521–24. doi: 10.1074/jbc.275.3.1521.

Butcher, Eugene C. 1991. Leukocyte-endothelial cell recognition: three (or More) steps to specificity and diversity. Cell 67(6): 1033–36. doi: 10.1016/0092-8674(91)90279-8.

Cotran, R. S. and J. S. Pober. 1990. Cytokine-endothelial interactions in inflammation, immunity, and vascular injury. Journal of the American Society of Nephrology 1(3): 225–35. doi: 10.1681/ASN.V13225.

Danewalia, Satwinder Singh and K. Singh. 2016. Magnetic and bioactive properties of MnO2/Fe2O3 modified Na2O-CaO-P2O5-SiO2 glasses and nanocrystalline glass-ceramics. Ceramics International 42(10): 11858–65. doi: 10.1016/j.ceramint.2016.04.108.

Dash, A. K. and G. C. Cudworth. 1998. Therapeutic applications of implantable drug delivery systems. Journal of Pharmacological and Toxicological Methods 40(1): 1–12. doi: 10.1016/S1056-8719(98)00027-6.

Doloff, Joshua C., Omid Veiseh, Arturo J. Vegas et al. 2017. Colony stimulating factor-1 receptor is a central component of the foreign body response to biomaterial implants in rodents and non-human primates. Nature Materials 16(6): 671–80. doi: 10.1038/nmat4866.

Dorozhkin, Sergey V. 2011. Biocomposites and hybrid biomaterials based on calcium orthophosphates. Biomatter 1(1): 3–56. doi: 10.4161/biom.1.1.16782.

Ferguson, J. Y., M. Dudareva, N. D. Riley et al. 2014. The use of a biodegradable antibiotic-loaded calcium sulphate carrier containing tobramycin for the treatment of chronic osteomyelitis. The Bone & Joint Journal 96-B(6): 829–36. doi: 10.1302/0301-620X.96B6.32756.

Ferguson, Jamie, Michael Diefenbeck and Martin McNally. 2017. Ceramic biocomposites as biodegradable antibiotic carriers in the treatment of bone infections. Journal of Bone and Joint Infection 2(1): 38–51. doi: 10.7150/jbji.17234.

Franz, Sandra, Stefan Rammelt, Dieter Scharnweber et al. 2011. Immune responses to implants – a review of the implications for the design of immunomodulatory biomaterials. Biomaterials 32(28): 6692–6709. doi: 10.1016/j.biomaterials.2011.05.078.

Ghasemi-Mobarakeh, Laleh, Davood Kolahreez, Seeram Ramakrishna et al. 2019. Key terminology in biomaterials and biocompatibility. Current Opinion in Biomedical Engineering 10: 45–50. doi: 10.1016/j.cobme.2019.02.004.

Gorbet, Maud B. and Michael V. Sefton. 2004. Biomaterial-associated thrombosis: roles of coagulation factors, complement, platelets and leukocytes. Biomaterials 25(26): 5681–5703. doi: 10.1016/j.biomaterials.2004.01.023.

Hench, L. L., R. J. Splinter, W. C. Allen et al. 1971. Bonding mechanisms at the interface of ceramic prosthetic materials. Journal of Biomedical Materials Research 5(6): 117–41. doi: 10.1002/jbm.820050611.

Henkel, Jan, Maria A. Woodruff, Devakara R. Epari et al. 2013. Bone regeneration based on tissue engineering conceptions — a 21st century perspective. Bone Research 1(3): 216–48. doi: 10.4248/BR201303002.

Henson, P. M. 1980. Mechanisms of exocytosis in phagocytic inflammatory cells. Parke-Davis Award Lecture. The American Journal of Pathology 101(3): 494–511.

Henson, P. M. and R. B. Johnston. 1987. Tissue injury in inflammation. oxidants, proteinases, and cationic proteins. Journal of Clinical Investigation 79(3): 669–74. doi: 10.1172/JCI112869.

Hubbell, Jeffrey A. 1995. Biomaterials in tissue engineering. Nature Biotechnology 13(6): 565–76. doi: 10.1038/nbt0695-565.

Huang, J. and S. M. Best. 2007. Ceramic Biomaterials. Tissue Engineering Using Ceramics and Polymers.

Hynes, Richard O. 1992. Integrins: Versatility, modulation, and signaling in cell adhesion. Cell 69(1): 11–25. doi: 10.1016/0092-8674(92)90115-S.

Iftekhar, Arif. 2004. Biomedical Composites. Standard Handbook of Biomedical Engineering and Design.

Isa, Z. M. and J. A. Hobkirk. 2000. Dental implants: Biomaterial, biomechanical and biological considerations. Annals of Dentistry University of Malaya 7(1): 27–35.

Ivanova, Elena P., Kateryna Bazaka and Russell J. Crawford. 2014. Natural polymer biomaterials: advanced applications, pp. 32–70 in New Functional Biomaterials for Medicine and Healthcare. Elsevier.

Johnston, Richard B. 1988. Monocytes and Macrophages. New England Journal of Medicine 318(12): 747–52. doi: 10.1056/NEJM198803243181205.

Jordan, David R., Seymour Brownstein, Steve Gilberg et al. 2002. Hydroxyapatite and calcium phophate coatings on aluminium oxide orbital implants. Canadian Journal of Ophthalmology 37(1): 7–13. doi: 10.1016/S0008-4182(02)80092-8.

Khan, Ferdous, Masaru Tanaka and Sheikh Rafi Ahmad. 2015. Fabrication of polymeric biomaterials: a strategy for tissue engineering and medical devices. Journal of Materials Chemistry B 3(42): 8224–49. doi: 10.1039/C5TB01370D.

Klopfleisch, R. and F. Jung. 2017. The pathology of the foreign body reaction against biomaterials. Journal of Biomedical Materials Research Part A 105(3): 927–40. doi: 10.1002/jbm.a.35958.

Klosterhoff, Brett S., Melissa Tsang, Didi She et al. 2017. Implantable sensors for regenerative medicine. Journal of Biomechanical Engineering 139(2). doi: 10.1115/1.4035436.

Lopes, M. Savioli, A. L. Jardini and R. Maciel Filho. 2012. Poly (Lactic Acid) production for tissue engineering applications. Procedia Engineering 42: 1402–13. doi: 10.1016/j.proeng.2012.07.534.

M, Sumita. 1997. Present status and future trend of metallic materials used in orthopedics. Orthop. Surg. (48): 927–34.

Martin, Paul and S. Joseph Leibovich. 2005. Inflammatory cells during wound repair: The good, the bad and the ugly. Trends in Cell Biology 15(11): 599–607. doi: 10.1016/j.tcb.2005.09.002.

Mehdizadeh, Mohammadreza and Jian Yang. 2013. Design strategies and applications of tissue bioadhesives. Macromolecular Bioscience 13(3): 271–88. doi: 10.1002/mabi.201200332.

Namur, Ricardo Sanson, Karla Miriam Reyes and Cláudia Eliana Bruno Marino. 2015. Growth and electrochemical stability of compact tantalum oxides obtained in different electrolytes for biomedical applications. Materials Research 18(suppl 2): 91–97. doi: 10.1590/1516-1439.348714.

Niinomi, Mitsuo. 2002. Recent metallic materials for biomedical applications. Metallurgical and Materials Transactions A 33(3): 477–86. doi: 10.1007/s11661-002-0109-2.

Orlovskii, V. P., V. S. Komlev and S. M. Barinov. 2002. Hydroxyapatite and hydroxyapatite-based ceramics. Inorganic Materials 38(10): 973–84. doi: 10.1023/A:1020585800572.

Prasad, Karthika, Olha Bazaka, Ming Chua et al. 2017. Metallic biomaterials: current challenges and opportunities. Materials 10(8): 884. doi: 10.3390/ma10080884.

Raghavendra, Gownolla Malegowd, Kokkarachedu Varaprasad and Tippabattini Jayaramudu. 2015. Biomaterials, pp. 21–44 in Nanotechnology Applications for Tissue Engineering. Elsevier.

Saini, Monika. 2015. Implant biomaterials: A comprehensive review. World Journal of Clinical Cases 3(1): 52. doi: 10.12998/wjcc.v3.i1.52.

Silver, Frederick H. and David L. Christiansen. 1999. Introduction to biomaterials science and biocompatibility, pp. 1–26 in Biomaterials Science and Biocompatibility. Springer New York.

Vallet-Regí, Maria. 2001. Ceramics for medical applications. Journal of the Chemical Society, Dalton Transactions (2): 97–108. doi: 10.1039/b007852m.

Wahl, S. M., H. Wong and N. McCartney-Francis. 1989. Role of growth factors in inflammation and repair. Journal of Cellular Biochemistry 40(2): 193–99. doi: 10.1002/jcb.240400208.

Walker, Michael Heiden Emily. 2015. Magnesium, iron and zinc alloys, the trifecta of bioresorbable orthopaedic and vascular implantation - A review. Journal of Biotechnology & Biomaterials 05(02). doi: 10.4172/2155-952X.1000178.

Weber, Benjamin, Leslie Saurer, Mirjam Schenk et al. 2011. CX3CR1 defines functionally distinct intestinal mononuclear phagocyte subsets which maintain their respective functions during homeostatic and inflammatory conditions. European Journal of Immunology 41(3): 773–79. doi: 10.1002/eji.201040965.

Williams, David F. 2009. On the nature of biomaterials. Biomaterials 30(30): 5897–5909. doi: 10.1016/j.biomaterials.2009.07.027.

Wilson, Cameron J., Richard E. Clegg, David I. Leavesley et al. 2005. Mediation of biomaterial–cell interactions by adsorbed proteins: A review. Tissue Engineering 11(1-2): 1–18. doi: 10.1089/ten.2005.11.1.

Yoruc, A. B. H. and B. C. Sener. 2012. A Roadmap of Biomedical Engineers and Milestones. Edited by S. Kara. InTech.

11

Host Response Immune Response to Polymeric Implants

Madhuparna Choudhury,[1] *Asmita Saha,*[1] *Divyashree Arun,*[1]
Madhusree Mitra[2,*] and *Shristi Ram*[1,*]

1. Introduction

Biomaterials have found profound interest in medical applications, such as therapeutic or other diagnostic purposes, irrespective of their natural or synthetic nature. These substances are generally acquired from natural resources or synthesised to be utilised in their natural form to facilitate healing or as a part of biomedical devices to assist, improve or mimic biological functions. Biomaterial science is, therefore, an important field of multidisciplinary research entailing material science, tissue engineering, medical science, biomedical and biomaterial engineering for scale-up purposes (Agrawal et al. 2023).

Biomaterials implementation plays an important role in various medical devices, such as catheters, cardiac pacemakers, glucose biosensors (Koschwanez and Reichert 2007), contact lenses, finger joint implants, knee replacement, hearing aids, etc. (Table 1).

There are various surgical methods such as abrasion, subchondral drilling, microfracture, osteochondral autograft transfer system (OATS) (also known as mosaicplasty), and autologous chondrocyte implantation (ACI) (Pestka et al. 2012; Albright and Daoud 2017; Solheim et al. 2018).

In the form of biological implants, these materials may cause foreign body reactions (FBRs) inside the host system. Since biomaterials are foreign substances

[1] Department of Microbiology, School of Basic and Applied Sciences, Dayananda Sagar University, City campus, Kudlu Gate, Bangaluru, 560068, Karnataka, India.

[2] Department of Forest Genetics and Plant Physiology, Swedish University of Agricultural Sciences (SLU), Umeå Plant Science Centre (UPSC), 901 83 Umeå, Sweden.

* Corresponding authors: Madhusree.mitra@slu.se; shristi-sbas@dsu.edu.in

Table 1: Composition of various medical devices and implants.

Medical Device	**Material**	**References**
Cerebrospinal fluid shunt	Silicone	Kalousdian et al. 1998
Cochlear implant	Silicon, platinum, titanium and ceramics. Iridium 90/10 is also used. Platinum is used as electrode contact.	Stöver and Lenarz 2009
Breast implant	Silicone gel	Abdul-Al et al. 2020
Dental implants	Cobalt-chromium-molybdenum-based alloys, titanium, iron-chromium-nickel-based alloys, and platinum, iridium, gold, palladium and alloys of these metals.	Amarnath et al. 2011
Hernia surgical mesh implant	PP, PTFE, ePTFE and polyester (POL)	Bilsel and Abci 2012
Essure® permanent contraceptive implant	Nickel-titanium, 316L stainless steel, platinum-iridium, and tin-silver solder) along Dacron (PET) fibres.	Aslan and Gilbert 2022
Hip implant	Titanium and its alloys, cobalt base alloys and stainless steels (bio-steels).	Choroszyński et al. 2017
Ophthalmology implant	Polyvinyl alcohol (PVA) gel, hydrogels, polymethyl methacrylate (PMMA), and hydroxy ethyl methacrylate (HEMA) hydrogel.	Ferraz 2022
Pelvic floor mesh	Synthetic polymers: Polypropylene (PP). Polyvinylidene fluoride (PVDF). Polycaprolactone (PCL). Resorbable methoxy polyethyleneglycol-polylactic-co-glycolic acid polymer (MPEG-PLGA). Polyamide. Polylactic acid (PLA). Natural polymers: Fibrinogen (Fg). Cellulose. Gelatin. Chitosan. Polymer blends.	Yuan et al. 2021
Cardiac pacemakers	Platinum group metal alloys or CoCr, polyurethanes and silicones.	Ozturk et al. 2020

to the host, the implantation of biomaterials may result in an immune response that is dependent on both the material and the host. This can also cause acute and chronic inflammations of the host tissues. Sometimes, disruption of host tissue may occur during biomaterial insertion through surgery or injection. Moreover, in the presence of injured host tissues, the implementation of biomaterial devices can trigger systemic inflammation. Therefore, another essential role of the biomaterial science discipline is to ensure desired material surface properties and other preventive measures to curtail the possibilities of FBRs.

The intensity of the host tissue reaction to implanted biomaterials varies based on various factors, including host, site of implantation, species, degree of trauma during implementation and the nature of the implanted material (Anderson et al. 2008). If the biomaterial is non-compatible to the host, the host's immune response will be elicited, trigereing the onset of phagocytosis, mediated by macrophages leading to localised or at times chronic inflammation. If an infection occurs due to the placement of the implant, there is a chance of the activation of polymorphonuclear leukocytes near the implant. If the implant is completely inactivated, with no appearance of macrophages, then only a thin collagenous layer encapsulates the implant.

Moreover, there is a possibility of tissue response, which can be minimal response or necrosis. In order to avoid the adverse effect of implantation, biomaterials with anti-inflammatory properties can be designed, and the desired outcome is achieved. There are two types of responses that can be obtained from biomaterials depending on their source, i.e., inner biomaterials such as metals, synthetic polymers, and ceramics, which are not part of the body and are known as foreign body reactions or responses. Macrophages are the key players associated with this type of reaction. Classically activated macrophages (M1) play a role in host defence by inducing inflammatory responses. M1 has the responsibility to remove foreign bodies, including bacteria, pathogens, and implanted biomaterials. Meanwhile, other types of macrophages, known as would healing or reparative macrophages (M2), are responsible for wound healing and remodelling. If the material is not biocompatible, it shows the formation of foreign body giant cells (monocytes and macrophages) in the biomaterial interface, leading to a chronic inflammatory state (Anderson et al. 2008).

2. Factors Affecting Host Response

2.1 Biomaterial-Host Response

An implant should be selected such that it remains *in situ* in the patient not only for a few months but for as long as it stays inside the host. This depends on the type of material that is being used. Over time, the implant can be destroyed, encapsulated or altered by the host immune system. During biomaterial implantation, the damage to the surrounding tissue activates the innate immune system. Neutrophils are the first immune cells to respond to damage and trigger other immune cells, such as macrophages and monocytes, leading to foreign body reactions (Badylak 2015).

After implantation, the blood and plasma proteins coat the surface of the biomaterial, which is known as the Vroman effect (Badylak 2015). The biomaterial does not come in direct contact with the host cells; instead, the proteins absorbed into the biomaterial form a protective matrix, which aids in attachment and clot formation while lowering the incidence of foreign body reactions.

There are three types of reactions that occur between the matrix and the host. Degradable implants activate a pro-remodelling M2 macrophage-associated response, which degrades the biomaterial and deposits required tissue in that area remodelling the site. Non-degradable synthetic implants cause M1 macrophages to initiate inflammation, foreign body reaction and scar tissue formation. For materials like metals, 'frustrated phagocytosis' occurs wherein the implant is eventually encapsulated (Badylak 2015). Small amounts of metals are expected to be present in metallic implants. However, regardless of their association with biological functions, an abundance of essential and/or non-essential metals can cause metal toxicity in the host system.

2.2 Host Factors

2.2.1 Age

Ageing is mainly characterised by changes in the immune system, which can alter the host's response to the biomaterial. The number of immune cells is not largely affected, but their efficiency, phagocytic capabilities, and response to foreign pathogens are reduced. This causes poor wound healing and an increased risk of infection around the implant. This can be termed as immunosenescence. Adult stem cells are responsible for tissue repair and wound healing. With ageing, the function of the stem cells decreases. B and T cells proliferate and mature from the stem cells, but with the onset of immunosenescence, their number decreases, leading to an increased risk of infections. Hence, the rate of implant rejection is much higher in older people (Badylak 2015).

2.2.2 Nutritional Status

Nutrition is of utmost importance when it comes to acceptance of biomaterial implants. The elderly population does not get enough micronutrients and macronutrients due to various reasons, including less nutritional requirements, slow metabolism, appetite loss, etc. Micronutrients like zinc, selenium and copper are important for homeostasis and inflammatory response (Badylak 2015). Malnutrition reduces T-cell production, proliferation of fibroblasts and tissue repair, leading to improper wound healing and implant rejection.

2.2.3 Anatomic Factors

Biomaterial implants can be placed almost anywhere in the host body, but anatomic placement plays a major factor in the immune response to the implant. This is because different areas of the body have different levels of vascularisation, and the microenvironment is also distinct. There will be contact with air if the implant is placed in the oesophagus; for internal organs, they will be in contact with blood, and for joint and other orthopaedic implants, the implant must be strong enough to support the weight and mechanical load of the individual. For example, the cornea being devoid of any blood or lymph vessels, does not evoke any harmful immune response, whereas biomaterials inserted near the heart of bones can cause high inflammatory response due to high vascularisation and complexity.

2.2.4 Comorbidities

Underlying diseases and comorbidities can profoundly impact the host cell's response towards biomaterials. They can weaken the immune system of the host and reject the implant. We will take the example of diabetes and obesity and their effect on host response.

2.2.4.1 Obesity

Obesity is characterised by the excessive accumulation of body fat, which can harm an individual's lifestyle. Obesity affects the host response to biomaterials in several ways, such as impairing wound healing, poor vascularisation, dehiscence of fascia and leakage of wounds after any surgical procedure. The macrophages and monocytes undergo changes (M1 to M2 macrophages), which cause chronic low-grade systemic pro-inflammatory response. The increase in adipose tissues increases the secretion of inflammatory mediators like interleukins and tumour necrosis factors, which can slow down wound healing (Boersema et al. 2016).

2.2.4.2 Diabetes

Diabetes is a metabolic disorder mainly characterised by increased blood sugar levels (hyperglycaemia). This can be due to various reasons, including failure of the Langerhans cells in the pancreas to produce insulin or the development of resistance against insulin in the host. The excess sugar in the blood reacts with the proteins and fat in the body and produces advanced glycation end products (AGEs). AGEs can interfere with the integration of the host cell and the biomaterial, preventing attachment of the implant. Diabetes also impairs wound healing and increases the risk of infection and bacteraemia (Lemarchand et al. 2023).

2.2.5 Chemotherapy and Radiation Therapy

Diseases that result from uncontrolled cell division, like cancer and tumours, require chemo and radiation therapies to prevent them from becoming life-threatening. These therapies mainly work by stopping cell proliferation in the affected area. However, since precisely determining the area of neoplasm formation is not possible, these therapies work on other areas, stopping the proliferation of normal body cells. Biomaterial implantation relies largely on the proliferation of immune cells, which can cause inflammatory responses and cause encapsulation. These therapies, when used in conjugation with biomaterial implantation, can render implantation inefficient (Badylak 2015).

3. Host Response to Biomaterials

The success or setbacks of a biomaterial and its efficiency eventually are mainly dependent on the host reaction to these biomaterials. Biocompatibility is a term used to define and characterise the host response of biomaterials, mainly medical devices. It simply means that these biomaterials will not cause any harm or produce any unfavourable effects on the tissues. If there are any sort of lethal effects associated with the biomaterial, it could lead to various *in vivo* adversities, which could be due to the material or host responses. Biocompatibility assessment is thus performed to determine any modifications in the host mechanisms that may determine their responses (Anderson and Cramer 2015).

3.1 Injury

For a biomaterial to be placed under *in vivo* conditions, insertions or any sort of implantations need to be done, all of which cause injury to the tissue. As soon as an injury is initiated, the host homeostatic mechanisms turn on. The degree of injury largely depends on the injury-causing procedures. The extent to which the host homeostatic mechanisms are activated is a potential estimate of the host response, which could be tissue, organ, or species dependent, and the biocompatibility of the biomaterial with the host. Following injury, several events take place, like acute inflammation, chronic inflammation, granulation tissue, provisional matrix formation and foreign body reaction (FBR) (Anderson and Cramer 2015).

3.2 Provisional Matrix Formation

When there is an injury or any sort of trauma in a tissue, an immediate inflammatory response is observed due to disturbance in the tissue. However, the extent and impact of this response depends on where the biomaterial is implanted or its biocompatibility (Gonzalez-Simon and Eniola-Adefeso 2012). In this situation, permeability and flow changes occur in the vascular system, leading to exudation, where blood cells and proteins flow out of the blood vessels to the site of injury. This series of events works to heal the site of injury immediately by replacing the injured tissues with new cells and scar tissues, as well as by thrombosis. As blood and its associated proteins accumulate and deposit on the surface of the biomaterial during this time, it is called provisional matrix formation. It is activated by thrombosis and complement systems, which structurally and compositionally support effective and fast healing. It contains different fibrins and mediators along with some associated proteins, forming a mesh-like network that helps with cell attachment and migration processes. Growth factors present help in the synthetic processes (Anderson and Cramer 2015).

3.3 Acute Inflammatory Response

Innate immunity initiates the acute inflammatory response at the site of injury, which is mainly the primary defence, recognition and clearing up of foreign materials by neutrophils and macrophages (Anderson and Cramer 2015). The start and severity of this response depends on the extracellular matrix (ECM) and blood proteins that adsorb on the surface of the biomaterial during an injury or implantation (Chung et al. 2017). This acute inflammatory response lasts for a truly short period. It involves oedema (exudation of blood fluid) and the unidirectional movement of main neutrophils (with the help of inflammatory agents and mediators), called chemotaxis, from the blood vessels into the site (Anderson and Cramer 2015). Mostly, phagocytosis by neutrophils is seen next if the biomaterial's shape, size and chemistry are favourable for endocytosis (Gonzalez-Simon and Eniola-Adefeso 2012). An indication of infection is the presence of acute inflammation at the site of injury or implantation (Anderson and Cramer 2015).

3.4 Chronic Inflammatory Response

With due course of time, when the number of neutrophils starts to decrease, the chemotactic factors begin to recruit monocytes at the injury site. This provides a secondary defence action and a short chronic inflammatory response by differentiating into dendritic cells or macrophages, which can ultimately lead to the non-acceptance of the biomaterial. The various reasons for the inflammation lasting for so long could be due to the nature of the biomaterial or persistent injury at the inflammation site (Gonzalez-Simon and Eniola-Adefeso 2012). Chronic inflammation can lead to hyperactivity of the fibroblast, leading to excessive deposition of ECM proteins. Thus, understanding the relationship between the immune cells and fibroblasts can help in designing an ideal biomaterial (Chung et al. 2017).

3.4.1 Macrophages (Mononuclear Phagocytes)

Macrophages are phagocytic cells derived from monocytes present throughout the tissues and have a considerably long life, which makes them available to be present and sustained during the chronic inflammatory response. They perform several important functions, including phagocytosis and antigen presentation, to activate adaptive immunity. Just like the neutrophils, the macrophages also phagocytise foreign particles or microorganisms and release pro-inflammatory cytokines and chemokines (Gonzalez-Simon and Eniola-Adefeso 2012), coagulation factors, reactive oxygen metabolites and other biologically active products (Anderson and Cramer 2015). They also clear up old cell aged cell debris.

3.5 Granulation Tissue

Granulation tissue colour is a strong indication of the healing response of inflammation at the site of injury. The pink and dark red coloration of the granulated tissue indicates healthy healing and wound infection, respectively. The monocytes and macrophages initiate the formation of this special tissue, which is derived from the fibroblast cells and vascular endothelial cells and depends largely on the type and extent of injury. This granulation tissue is formed from a process called angiogenesis (formation of new blood vessels) or neovascularisation, which includes the growth and differentiation of cells in the blood vessels. In the early stages, proteoglycans and later type III collagen prevalence are seen (Jones 2015; Huimin et al. 2023). Sometimes, even myofibroblasts are seen during the development of this tissue as well as during repair and healing mechanisms for injured tissues. There is also prominence in the presence of macrophages. The whole process of wound healing can be primary or secondary, wherein the primary process deals with injuries without any sort of contamination or infection done by clean surgical equipment, and in the secondary approach, larger granulation tissues and provisional matrices are formed due to a larger amount of loss of tissue. The wound-healing process by secondary intention is more prevalent (Anderson and Cramer 2015). Ultimately, these responses work to establish normalcy or homeostasis at the site of injury (Gonzalez-Simon and Eniola-Adefeso 2012).

3.6 Foreign Body Reaction

The foreign body giant cells or FBGCs (multinucleated structures that are formed by the fusion of many macrophages) (Jones 2015) and the granulation tissue together account for the foreign body reaction or FBR to biomaterials, and this is dependent on the nature of the biomaterial. Generally, fibrosis is seen around the biomaterial at the implant site, and macrophage activation and adherence are predominant during the initial stages of wound healing.

3.6.1 Fibrous Encapsulation

During the healing process after inflammation, various cells come together to form a fibrous capsule at the site of injury. Collagen fibres and fibroblasts form the strong base and the capsule structure with the help of proteoglycans and adhesion molecules present in these components. Eventually, certain glycoproteins like fibronectins, thrombospondins, etc., form the fibrous tissue, which encases the biomaterial and helps in the formation of new tissue layers (Whitaker et al. 2021). However, over-development of this capsule can interfere with the function of the biomaterial or rejection of the implant, and in such cases, surgical excision or removal of the biomaterial is desirable (Gonzalez-Simon and Eniola-Adefeso 2012).

3.6.2 Fibroblast

At the site of inflammation, leukocytes recruit macrophages and fibroblasts for proliferation and regenerative modelling of the fibrotic scar tissue (Figure 1). These fibroblasts perform the major function of synthesising the extracellular matrix (ECM), and the fibrous capsule and its functions are controlled by the different chemical signals and factors released in the matrix (Whitaker et al. 2021; Jones 2015).

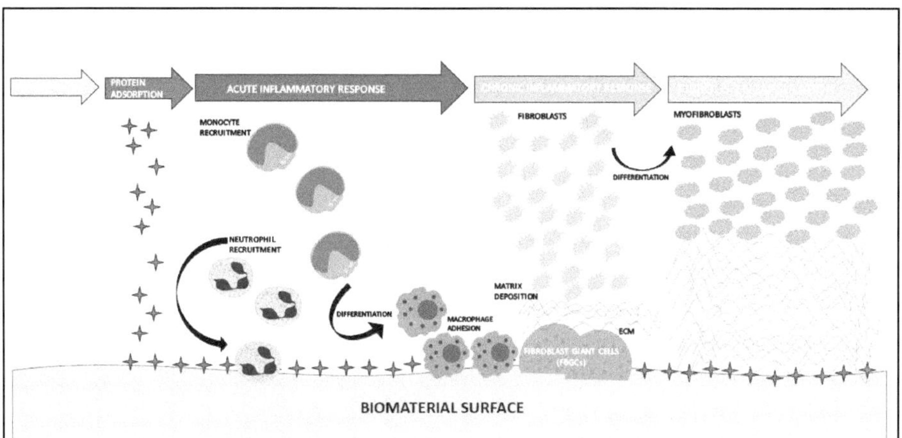

Figure 1: Mechanism of host response to biomaterials.

4. Immune Response

4.1 Naturally Occurring Biomaterials

Based on the nature of biomaterials, the degree of success of the implant depends on it. Naturally derived biomaterials are more advantageous than synthetic materials, i.e., biocompatibility, remodelling, and biodegradability. Damaged organs or tissues are replaced or treated with these biomaterials (Bose and Bandyopadhyay 2013).

4.1.1 Degradable Biomaterials

Degradable biomaterials are designed in such a way that these materials can degrade naturally inside the body after a certain period of implantation so that these materials can achieve a particular function. The use of degradable biomaterials shows resistance to infection by increased vascularisation. Polymeric materials are commonly used in degradable materials and have a wide range of implementation in various medical fields, including tissue engineering, drug delivery, etc. Magnesium and iron-based degradable materials are degraded inside the body with metabolism. For tissue engineering, orthopaedic and cardiovascular applications, magnesium alloy is mostly used (Bose and Bandyopadhyay 2013).

The natural matrix protein can be degraded by proteolysis, oxidation and hydrolysis. In the presence of tissue fluids and inflammatory cells, the polymer-based device, which is created as sensitive to water and protease, can contribute to degradation. Protease and oxidants have the ability to degrade implanted devices by degrading the polymer crosslink of the device. Leukocyte uses protease and reactive oxidants to degrade this implanted device (Mouthuy et al. 2016).

Materials that cause a severe immune response and the sensitivity of polymer crosslink degradation can be altered to design new devices with specific drug release profiles. Co-block polymer (poly lactic glycolic acid; PLGA) is the best example of this type of device, commonly using tissue fluids and innate immune response for the delivery of drugs (Mir et al. 2018). Early copolymer degradation and inflammation were observed in wound-closing sutures after the implantation of vascular stents (Gogolewvki 1992; Heller 1984). Other reasons for the degradation of implanted PLGA devices include the activation of giant cells, monocytes, neutrophils and hydrolysis (Van Der Giessen et al. 1996).

4.1.2 Cell-Grafted Biomaterial

Grafts are important for regenerating and repairing bone. The materials used for this are mainly based on the site of application. Various types of materials are used for the regeneration of bones. Transplanted cells are mainly autologous, allogeneic and xenogeneic (Rieder et al. 2016). Autologous cells are obtained from the same host who will get the transplant. Allogeneic cells are from the same species of host that will undergo transplantation. Xenogeneic cells are obtained from different species; they differ from the host (Kyyak et al. 2020). Grafted cells have therapeutic value. For example, it has the ability to produce islet cells that produce insulin and also help in tissue regeneration, growth and repair of injured cells. These cells can stimulate innate as well as adaptive immune responses.

Macrophages help to ingest the pathogen and digest it; the remaining digested fragments present on the cell surface pass a signal to stimulate other more effective immune cells. Major histocompatibility complex class II has a recognition site for T cells. Macrophages act on pathogens by presenting this MHC class II and costimulatory molecules, CD89 and CD86 (Thiele et al. 2003; Thiele et al. 2001), which activate T cells and cause the rejection of transplanted cells. Allogeneic and xenogeneic cells stimulate more immune responses than autologous cells because of the non-self features of these cells. This type of rejection can be prevented by implanting the cells within a polymer matrix. The pore size of the matrix should be such that only nutrients can pass through, preventing immune cells and inflammation.

4.2 Synthetic Biomaterials

Synthetic biomaterials are man-made and created in such a way that these materials should be compatible with the human body. These materials are easy to use and inexpensive compared to natural biomaterials; properties of biomaterials such as crystallinity, hydrophobicity and molecular weight are significant for biomaterials. Synthetic biomaterials have more predictable, degradable and physicochemical, mechanical properties. Synthetic biomaterials that are foreign to the body cells can cause severe inflammation sometimes. Important types of biomaterials include polycaprolactone, polyglycolide, polylactide, etc. (Aldana et al. 2023).

4.2.1 Orthopaedic Implants

Immune response to orthopaedic implants could be due to metal sensitivity or other components. The first successful joint replacement was done by Sir John Charnley, which was a hip replacement. Stainless steel and femoral components were the main components of these hip replacements. Mainly, cemented femoral components were used, and their biocompatibility was monitored by various methods, including clinical, radiological and histological analyses. Initially, acute inflammation was observed in that cemented prosthesis. The bone death that occurred during surgery was resolved within a week by fibrous tissue and fibrocartilage formation (Yao et al. 2015).

Another type of joint replacement, i.e., cementless replacement, mainly depends on initial ‘press fit’ to provide stability. This is added with splines, surface coating and screws so that these materials can attach to the host bone for a longer period. Based on the properties of the implant and host, as well as surgical methods used, it can determined which neighbouring bone will undergo remodelling. Sometimes, periprosthetic osteolysis occurs when the debris of polyethylene migrates along with the cementless terminal. Because of that, the rejection of the implant may occur (Gibon et al. 2017).

4.2.2 Pelvic Girdle Mesh

Most of the surgical meshes are synthetic or man-made. These meshes are commonly used for various medical conditions, including heart stent and hernia repair. Surgical meshes are used to support the weak or damaged tissue. Some of the meshes are made of animal tissue. Polypropylene and polyglycolic acid are used to treat pelvic floor disorder. After implementing these, it may cause pain, discomfort, bowel obstruction, nausea and also hernia recurrence. But this procedure is successful and can stay a lifetime in the body. After implantation of the meshes, immune cells migrate to the site of the implantation, resulting in the formation of a foreign body granuloma. As the meshes are foreign to the immune system, they stimulate innate immunity and activate macrophages that collect at the surface of the mesh. The stimulation of adaptive immune response activates both B lymphocytes and T lymphocytes, which are important but require more time to activate (Patnaik et al. 2016).

5. *In Vitro* and *In Vivo* Models for Testing Host Response

Foreign body reactions (FBR) are the major driving force for the successful implantation of biomaterials. However, due to their unpredictability, all biomaterials, regardless of being synthetic or natural, cause complications in the host (Saleh et al. 2017). Therefore, studying these responses in models is important to create an almost ideal biomaterial implant. There be devastating side effects and failures of these implants if their outcomes and compatibility are not known. *In vitro* and *in vivo* models can help determine the complications involved in transplantation and how to overcome them (Vasconcelos et al. 2019).

When an implant is introduced into the host system, non-specific protein gets adsorbed on its surface and inflammatory cells are recruited. Neutrophils are the first to arrive, releasing cytokines to recruit other inflammatory cells, especially long-lived macrophages, which drive the foreign body reactions (Saleh et al. 2017). These macrophages recognise the implant as foreign but cannot engulf them due to their size, so they fuse together and form multinucleated foreign body giant cells (FBGCs) (Sheikh et al. 2015). Soon, a fibrous capsule forms within three to four weeks and stops communication between the implant and the host. This macrophage-implant interaction can be used to study the FBR response to different biomaterials (Saleh et al. 2017; Wolf et al. 2015).

5.1 In Vitro Models

In vitro models focus on studying the foreign body reaction in a controlled environment. Models involving culturing of macrophage and fibroblasts are used since they are the major effector cells of FBR. Macrophages are primarily associated with inflammatory response, but fibroblasts are crucial for wound healing and fibrotic responses (Zhou et al. 2016). Macrophage interrogation of a biomaterial and co-culture of macrophages with fibroblast is shown in Figure 2 (Lock et al 2019).

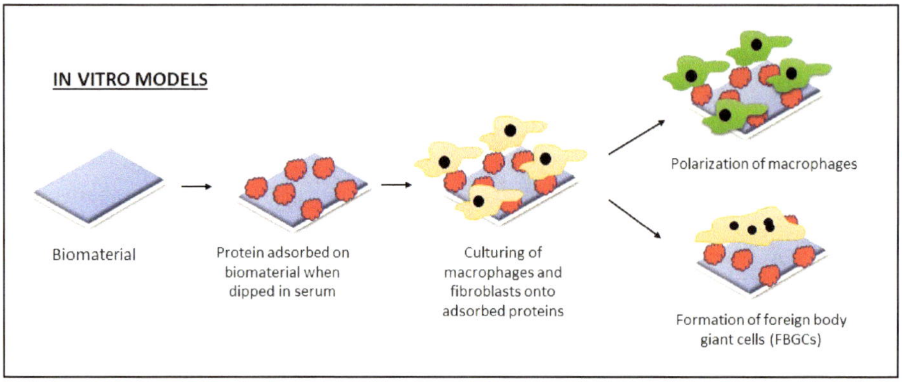

Figure 2: Mechanism of host response to biomaterials in *in vitro* models.

5.1.1 Macrophage Interrogation of a Biomaterial

The macrophages derived from blood or bone marrow monocytes are used for testing. Macrophage cell lines from rodents (murine) include cell lines RAW 264.7, J774A.1, and IC-21 (Chamberlain et al. 2015). Human monocyte-derived macrophages include cell lines of THP-1 and U-937. Several studies found that the cell lines RAW 264.7, J774A.1 and IC-21 responded to inflammatory stimulants by increased expression of pro-inflammatory cytokines but in low magnitudes, whereas THP-1 responded in greater magnitude to biomaterial extracts and an exogenous inflammatory stimulant. Hence, the choice of cell lines should be made while considering these factors as well (Saleh et al. 2017).

Macrophage polarisation occurs when macrophages can change their functions and gene expression in response to signals from their environment. Macrophages majorly have two polarised states, i.e., classically activated (M1) macrophages and alternatively activated (M2) macrophages. M1 macrophages are prompted by cytokines IFN-y and LPS and are involved in inflammatory functions, while M2 macrophages are induced by IL-4 and IL-3 and are responsible for wound healing (Chen et al. 2023). The polarisation of macrophages by biomaterials can determine the host response. The macrophages are cultured onto the biomaterial, and then they are checked for polarisation or other changes, and compatibility is determined. This shift from inflammatory to wound-healing macrophages is known as plasticity (Brown et al. 2012).

Before the macrophages are cultured onto the biomaterial surface, the biomaterials are dipped into a serum containing proteins, which get adsorbed onto the surface of the material, and onto these proteins, the macrophages initiate inflammatory responses. This helps us to understand the perfect surface of biomaterial for the adhesion of macrophages. Findings from this model include hydrophobic surface-mediated improved attachment efficiency of monocytes and macrophages when compared to neutral hydrophilic surfaces, although the attachment of inflammatory macrophages is higher on hydrophilic surfaces. An increase in surface roughness, decreased alignment, and micron-size surface features augment monocyte/macrophage attachment and polarisation (Saleh et al. 2017).

5.1.2 Co-Cultures of Macrophages and Fibroblasts

Although several research studies focus on pro-inflammatory responses observed in biomaterial-host model interaction studies, there is a scarcity of data about fibrotic reactions in such systems. In a research study performed by Klopfleisch in 2016, the effect of surface chemistry on fibrous capsule formation was studied using a mouse model. For that, they have implanted the co-cultures of macrophages and fibroblasts as self-assembling monolayers (SAMs) or micron-sized particles in the mice subcutaneous air pouches. However, the in-depth knowledge of fibrosis was still unclear (Hasan and Pandey 2018). An *in vitro* fibroblast/macrophage co-culture system was created using migration fence chambers, which include internal and external chambers, to overcome this problem. Macrophages and fibroblasts can be implanted separately into the chambers of the biomaterial, allowing their effects to be studied simultaneously. After 24 hours, the fence was removed, and cells were allowed to interact with each other. This allowed the study of fibroblast outgrowth, macrophage migration and interaction between the two cell types (Zhou and Groth 2018).

5.2 In Vivo Models

In vitro models alone cannot give a complete understanding of the host response to biomaterials as the actual host system is extremely complex, and the interaction at the *in vitro* level of macrophages cannot predict *in vivo* outcomes due to the inherent complexity, dynamics and donor variability of macrophage activation (Figure 3). Hence, *in vivo* models like rodents and larger animal models are used to mimic the environment of the implant. *In silico* analyses can link the data from *in vitro* and *in vivo* studies and give a complete understanding of the host response to a certain biomaterial implant. For FDA certification of a biomaterial, *in vivo* testing is compulsory. Three types of *in vivo* models, namely wild rat models, genetically modified rats, and large animal models, are discussed (Al-Maawi et al. 2017).

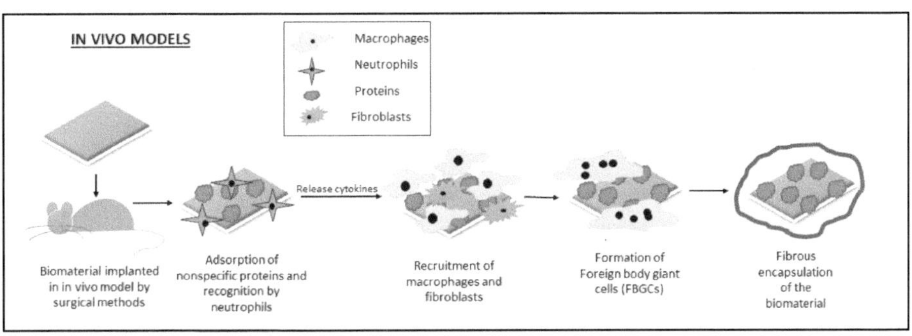

Figure 3: Host response to biomaterials in *in vivo* models.

5.2.1 Wild Rodent Model

These are the commonly used *in vivo* models. The subcutaneous and intramuscular sites in rats and mice provide the ideal environment for implantation testing since they are also highly vascularised. Devices can be easily implanted in these models by simple surgical methods. Several biomaterials can be evaluated on a single model for short-term degradation, biochemical studies, as well as long-term remodelling and functional outcome studies (Sicari et al. 2013).

The partial thickness rat abdominal wall defect model is more advantageous than subcutaneous and intramuscular models since it can be used for long-term studies (up to 2 years) and can anchor the implant more securely. Subcutaneous models are more suitable for short-term studies on non-degradable materials. A histomorphological analysis is the main advantage of the partial thickness rat abdominal wall defect model, which provides information about the rate of degradation and encapsulation of biomaterial, vascularity of the remodelled tissue, muscle growth, etc. (Sicari et al. 2013).

5.2.2 Genetically Modified Mouse Models

Mice can be genetically modified in several ways, such as making them deficient in T cells, natural killer cells or other immune cells. Introducing biomaterials into these modified mice can help determine the formation of FBGCs in these modified models. Mice lacking these immune cells did not limit the formation of FBGCs, indicating that they are not essential for FBR. Mice that were devoid of toll-like receptor-4, involved in detecting damaged tissues, resulted in fewer macrophages, but that still did not inhibit the formation of FBGCs. Mice lacking Rag2 and IL2rg lead to macrophage dysfunction, which prevented the formation of FBGCs and fibrous encapsulation of the biomaterial. These studies proved the importance of macrophages for proper FBR and implantation of biomaterial (Saleh and Bryant 2017; Klopfleisch 2016).

Genetically modified nondiabetic mice have helped in the detection of endothelial progenitor cells (EPCs). This helped to study the relationship between diabetes and EPCs, which is essential for understanding the progression of diabetes. The GFP (green fluorescent protein) mice express a green fluorescent protein in every cell except red blood cells. The bone marrow from GFP mice was inserted into C57BL/6 mice, forming a chimeric mouse model that was used to detect the fate of bone marrow-derived cells during wound healing (Caskey and Liechty 2013).

5.2.3 Large Animal Models

Rodents and humans do have many similar genes, but there is still a huge gap between the immune system of both. In order to bridge these gaps, larger animal models, like non-human primates (NHPs), are used. NHP models have immune systems that are more closely related to humans than rodents (Costa-Pinto et al. 2016). Table 2 summarises the advantages and disadvantages associated with different *in vitro* and *in vivo* models used to test the host response to biomaterials.

Table 2: Overview of the advantages and disadvantages associated with *in-vitro* and *in-vivo* models.

Types of Models		**Advantages**	**Disadvantages**
In vitro models	Macrophage interrogation of a biomaterial.	Macrophages are the major inflammatory cells involved in host biomaterial interactions, so studying the interrogation of biomaterial with macrophages will help in understanding the compatibility of the biomaterial.	Does not give a complete understanding of the host response as it does not include other immune cells involved in FBR.
	Co-cultures of macrophages and fibroblasts	Since both macrophages and fibroblasts are involved, it is a better indicator of tissue remodelling response to biomaterials.	It makes the protocol and determination of results more complex.
In vivo models	Wild rodent model	These models help bridge the gap between *in vitro* studies and their working on humans since rodents are more complex and have an immune system.	Wild rodents cannot help us understand how specific gene changes affect the host response. There are also ethical concerns and variability in results.
	Genetically modified mouse models	We can control the genetic factors that influence the host response, which makes these models specific and reliable.	They still do not fully represent the human model, which is more complex. There are a lot of ethical concerns and more time and money are required.
	Large animal models	They have high similarity to the human immune system, so they have high clinical relevance in biomaterial testing.	They raise ethical concerns and are very costly to maintain. Moreover, several technical challenges are faced due to their complexity and unpredictability.

6. Conclusion

Biomaterials can be natural or synthetic, and they can interact with biological systems. These materials are used for various purposes, such as medical and therapeutic applications. The human immune system is complex, and because biomaterials are foreign to the body, they may stimulate the immune system. The immune response can be acute or chronic, depending on the type of biomaterials, site of implant, surgical technique, molecular weight, degradation rate, etc. Sometimes, rejection of implants occurs as the implants are identified as foreign bodies to the host immune system. After implementing biomaterials, they first come in contact with extracellular matrix (ECM) protein and blood plasma. ECM protein attracts neutrophils and monocytes and causes inflammation through macrophage activation. The implant should be created in such a way that it minimises the immune response so that rejection of the implant can be avoided and can decrease the unnecessary risk to the host from implantation (Aamodt and Grainger 2016).

7. Future Prospects

An ideal biomaterial must not elicit the host immune response, should be long-lasting and have corrosion-free characteristics. In order to make this happen, there are studies and experiments actively going on with both *in vitro* and *in vivo* models. There is a strong need to understand the relationships between the host and the biomaterial due to their varied responses as well as between events like injury, inflammation, and fibrosis (Zhou and Groth 2018). A deep and clear understanding of the underlying processes responsible for the bioactivity and biocompatibility of biomaterials will facilitate future efforts to build effective and safe biomaterials. Also, the final outcome may be different in a multi-cell culture environment due to cellular crosstalk (Ellermann et al. 2023). To make these biomaterials more adaptable, they can be enriched with additional bioactive compounds so that they can regulate the immune responses and be a closer replica of the natural host environment. Hence, it can be concluded that to build efficiently integrative and biocompatible tissue constructs or biomaterials, various approaches have to be used, including immunoengineering (to facilitate incorporation into the host) and anti-inflammatory procedures so there are fewer chances of rejection of the implant biomaterials (Zhou and Groth 2018).

References

Aamodt, J. M. and D. W. Grainger. 2016. Extracellular matrix-based biomaterial scaffolds and the host response. Biomaterials 86: 68–82.

Abdul-Al, M., A. Zaernia and F. Sefat. 2020. Biomaterials for breast reconstruction: Promises, advances, and challenges. Journal of Tissue Engineering and Regenerative Medicine 14(11): 1549–1569.

Agrawal, R., S. Singh, K. K. Saxena et al. 2023. A role of biomaterials in tissue engineering and drug encapsulation. Proceedings of the Institution of Mechanical Engineers, Part E: Journal of Process Mechanical Engineering, 09544089221150740.

Albright, J. C. and A. K. Daoud. 2017. Microfracture and microfracture plus. Clinics in Sports Medicine 36(3): 501–507.

Aldana, A. A., J. Bauer and M. B. Baker. 2023. Synthetic biomaterials. In Tissue Engineering (pp. 173–212). Academic Press.

Al-Maawi, S., A. Orlowska, R. Sader et al. (2017, February). *In vivo* cellular reactions to different biomaterials—Physiological and pathological aspects and their consequences. In Seminars in Immunology. Academic Press 29: 49–61

Amarnath, G. S., B. C. Muddugangadhar, S. Tripathi et al. 2011. Biomaterials for dental implants: An overview. International Journal of Oral Implantology and Clinical Research 2(1): 13–24.

Anderson, J. M., A. Rodriguez and D. T. Chang. (2008, April). Foreign body reaction to biomaterials. In *Seminars in immunology*. Academic Press 20(2): 86–100

Anderson, J. and Cramer, S. 2015. Perspectives on the inflammatory, healing, and foreign body responses to biomaterials and medical devices. pp. 13–36. In Host Response to Biomaterials. Academic Press.

Aslan, C. and J. L. Gilbert. 2022. *In vitro* corrosion assessment of the Essure® medical device in saline, simulated inflammatory solution and neutral buffered formalin. Acta Biomaterialia, 147: 414–426

Badylak, S. F. (Ed.). (2015). Host response to biomaterials: the impact of host response on biomaterial selection. Academic Press.

Bayston, R. 2011. 6.629. Cerebrospinal Fluid Shunts.

Bilsel, Y. and I. Abci. 2012. The search for ideal hernia repair; mesh materials and types. International Journal of Surgery 10(6): 317–321.

Boersema, G. S., L. Utomo, Y. Bayon et al. 2016. Monocyte subsets in blood correlate with obesity related response of macrophages to biomaterials *in vitro*. Biomaterials 109: 32–39.

Bose, S. and A. Bandyopadhyay. 2013. Introduction to biomaterials. In Characterization of Biomaterials (pp. 1–9). Academic Press.

Brown, B. N., B. D. Ratner, S. B. Goodman et al. 2012. Macrophage polarization: an opportunity for improved outcomes in biomaterials and regenerative medicine. Biomaterials 33(15): 3792–3802.

Caskey, R. C. and K. W. Liechty. 2013. Novel animal models for tracking the fate and contributions of bone marrow derived cells in diabetic healing. Wound Regeneration and Repair: Methods and Protocols, 99–115.

Chamberlain, L. M., D. Holt-Casper, M. Gonzalez-Juarrero et al. 2015. Extended culture of macrophages from different sources and maturation results in a common M2 phenotype. Journal of Biomedical Materials Research Part A, 103(9): 2864–2874.

Chen, L., Z. Yao, S. Zhang et al. 2023. Biomaterial-induced macrophage polarization for bone regeneration. Chinese Chemical Letters 34(6): 107925.

Choroszyński, M., M. R. Choroszyński and S. J. Skrzypek. 2017. Biomaterials for hip implants–important considerations relating to the choice of materials. Bio-Algorithms and Med-Systems 13(3): 133–145.

Chung, L., D. R. Maestas Jr., F. Housseau et al. 2017. Key players in the immune response to biomaterial scaffolds for regenerative medicine. Advanced Drug Delivery Reviews 114: 184–192.

Costa-Pinto, A., T. C. Santos, N. M. Neves et al. 2016. Testing natural biomaterials in animal models. Biomaterials from Nature for Advanced Devices and Therapies, 562–579.

Ellermann, E., N. Meyer, R. E. Cameron et al. 2023. *In vitro* Angiogenesis in Response to Biomaterial Properties for Bone Tissue Engineering; a Review of the State of the Art. Regenerative Biomaterials, rbad027.

Ferraz, M. P. 2022. Biomaterials for ophthalmic applications. Applied Sciences 12(12): 5886.

Gibon, E., L. A. Córdova, L. Lu et al. 2017. The biological response to orthopedic implants for joint replacement. II: Polyethylene, ceramics, PMMA, and the foreign body reaction. Journal of Biomedical Materials Research Part B: Applied Biomaterials 105(6): 1685–1691.

Gogolewvki, S. 1992. Resorbable polymers for internal fixation. Clinical Materials 10(1-2): 13–20.

Gonzalez-Simon, A. L. and O. Eniola-Adefeso. 2012. Host response to biomaterials. Engineering Biomaterials for Regenerative Medicine: Novel Technologies for Clinical Applications, 143–159.

Hasan, A. and L. M. Pandey. 2018. Self-assembled monolayers in biomaterials. In Nanobiomaterials (pp. 137–178). Woodhead Publishing.

Heller, J. 1984. Biodegradable polymers in controlled drug delivery. Critical Reviews in Therapeutic Drug Carrier Systems 1(1): 39–90.

Huimin, X., C. Xin, L. Xuanzhe et al. 2023. Recent advances in decellularized biomaterials for wound healing. Materials Today Bio, 100589.

Jones, K. 2015. Fibrotic response to biomaterials and all associated sequence of fibrosis. In Host response to biomaterials (pp. 189–237). Academic Press.

Kalousdian, S., M. S. Karlan and M. A. Williams. 1998. Silicone elastomer cerebrospinal fluid shunt systems. Neurosurgery 42(4): 887–892.

Klopfleisch, R. 2016. Macrophage reaction against biomaterials in the mouse model–Phenotypes, functions and markers. Acta Biomaterialia 43: 3–13.

Koschwanez, H. E. and W. M. Reichert. 2007. *In vitro, in vivo* and post explantation testing of glucose-detecting biosensors: current methods and recommendations. Biomaterials 28(25): 3687–3703.

Kyyak, S., S. Blatt, A. Pabst et al. 2020. Combination of an allogenic and a xenogenic bone substitute material with injectable platelet-rich fibrin–A comparative *in vitro* study. Journal of Biomaterials Applications 35(1): 83–96.

Lemarchand, M., K. Thouin, T. De Serres-Bérard et al. 2023. *In vitro* glycation of a tissue-engineered wound healing model to mimic diabetic ulcers. Biotechnology and Bioengineering.

Lock, A., J. Cornish and D. S. Musson. 2019. The role of in vitro immune response assessment for biomaterials. Journal of Functional Biomaterials 10(3): 31.

Mir, M., M. N. Ali, A. Barakullah et al. 2018. Synthetic polymeric biomaterials for wound healing: a review. Progress in Biomaterials 7: 1–21.

Mouthuy, P. A., S. J. Snelling, S. G. Dakin et al. 2016. Biocompatibility of implantable materials: an oxidative stress viewpoint. Biomaterials 109: 55–68.

Ozturk, S., F. B. Ayanoğlu, M. Parmaksiz et al. 2020. Clinical and surgical aspects of medical materials' biocompatibility. In Handbook of Biomaterials Biocompatibility (pp. 219–250). Woodhead Publishing.

Patnaik, S. S., A. Borazjani, B. Brazile et al. 2016. Pelvic floor biomechanics from animal models. In Biomechanics of the Female Pelvic Floor (pp. 131–148). Academic Press.

Pestka, J. M., G. Bode, G. Salzmann et al. 2012. Clinical outcome of autologous chondrocyte implantation for failed microfracture treatment of full-thickness cartilage defects of the knee joint. The American Journal of Sports Medicine 40(2): 325–331.

Rieder, E., A. Steinacher-Nigisch and G. Weigel. 2016. Human immune-cell response towards diverse xenogeneic and allogeneic decellularized biomaterials. International Journal of Surgery 36: 347–351.

Saleh, L. S. and S. J. Bryant. 2017. *In vitro* and *in vivo* models for assessing the host response to biomaterials. Drug Discovery Today: Disease Models 24: 13–21.

Sheikh, Z., P. J. Brooks, O. Barzilay et al. 2015. Macrophages, foreign body giant cells and their response to implantable biomaterials. Materials 8(9): 5671–5701.

Sicari, B., N. Turner and S. F. Badylak. 2013. An *in vivo* model system for evaluation of the host response to biomaterials. Wound Regeneration and Repair: Methods and Protocols, 3–25.

Solheim, E., J. Hegna, T. Strand et al. 2018. Randomized study of long-term (15–17 years) outcome after microfracture versus mosaicplasty in knee articular cartilage defects. The American Journal of Sports Medicine 46(4): 826–831.

Stöver, T. and Lenarz, T. 2009. Biomaterials in cochlear implants. GMS Current Topics in Otorhinolaryngology, Head and Neck Surgery, 8.

Thiele, L., H. P. Merkle and E. Walter. 2003. Phagocytosis and phagosomal fate of surface-modified microparticles in dendritic cells and macrophages. Pharmaceutical Research 20: 221–228.

Thiele, L., B. Rothen-Rutishauser, S. Jilek et al. 2001. Evaluation of particle uptake in human blood monocyte-derived cells *in vitro*. Does phagocytosis activity of dendritic cells measure up with macrophages? Journal of Controlled Release 76(1-2): 59–71.

Van der Giessen, W. J., A. M. Lincoff, R. S. Schwartz et al. 1996. Marked inflammatory sequelae to implantation of biodegradable and nonbiodegradable polymers in porcine coronary arteries. Circulation 94(7): 1690–1697.

Vasconcelos, D. P., A. P. Águas, M. A. Barbosa et al. 2019. The inflammasome in host response to biomaterials: Bridging inflammation and tissue regeneration. Acta Biomaterialia 83: 1–12.

Whitaker, R., B. Hernaez-Estrada, R. M. Hernandez et al. 2021. Immunomodulatory biomaterials for tissue repair. Chemical Reviews 121(18): 11305–11335.

Wolf, M. T., Y. Vodovotz, S. Tottey et al. 2015. Predicting *in vivo* responses to biomaterials via combined *in vitro* and *in silico* analysis. Tissue Engineering Part C: Methods 21(2): 148–159.

Yao, Z., T. H. Lin, J. Pajarinen et al. 2015. Host response to orthopedic implants (metals and plastics). In Host response to biomaterials (pp. 315–373). Academic Press.

Yuan, M., M. Hu, F. Dai et al. 2021. Application of synthetic and natural polymers in surgical mesh for pelvic floor reconstruction. Materials and Design 209: 109984.

Zhou, G. and T. Groth. 2018. Host responses to biomaterials and anti-inflammatory design—a brief review. Macromolecular Bioscience 18(8): 1800112.

12

Biopolymers in Phytomedicines, Preparations, and Applications

Sudarshan Singh,[1,*] *Ozioma F. Nwabor,*[2] *Jayeoye J. Titilope*[3] and *Oladipupo O. Olatunde*[4]

1. Introduction

Phytomedicine is perhaps one of the oldest forms of medicine. Before the advent of orthodox and modern medicine, plants were used in the treatment of various kinds of illnesses. However, the lack of adequate understanding and empirical research evidence regarding the activities and mode of action of most plant bioactive compounds, coupled with the discovery of synthetic drugs, promoted skepticism toward the continued use of phytomedicine. This notwithstanding, up-to-date evidence associating synthetic and chemical-based drugs with emerging physiological disorders has prompted a redirection of medical drug discovery toward natural products, with plant phytochemicals as a major source of pharmacologically relevant compounds (Jamil et al. 2012; Farzaei et al. 2015; Rtibi et al. 2017; Huang et al. 2018). Plant-derived bioactive compounds have expressed vast biological activities, resulting in an awakening of phytomedicines. However, chemical and biological limitations such as insolubility, hydrophobicity, low bioavailability, and high toxicity are challenges confronting the application of phytomedicines (Patra et al. 2018; Krishnasis et al. 2016; Olotu et al. 2020). This has led to the search for an appropriate

[1] School of Allied Health Sciences, Walailak University, Nakhon Si Thammarat, 80160, Thailand.
[2] Division of Infectious Diseases, Department of Internal Medicine, Faculty of Medicine, Prince of Songkla University, Hat Yai, Songkhla 90112, Thailand.
[3] Department of Chemistry, Faculty of Physical Science, Alex-Ekwueme Federal University, Ndufu-Alike Ikwo, Ebonyi State, Nigeria.
[4] Department of Food and Human Nutritional Sciences, Faculty of Agricultural and Food Sciences, University of Manitoba, Winnipeg, Manitoba, R3T 2N2, Canada.
* Corresponding author: sudarshansingh83@hotmail.com

drug-delivery system that can efficiently transport phytomedicines and is capable of increasing the therapeutic index as well as the bioavailability of the phytomedicines. Biopolymers are important ingredients with various applications, especially in the pharmaceutical, medical, and food sectors, due to their biodegradability, stability, and biocompatibility. Advances in polymer science have fostered the use of natural biopolymers as carriers for active pharmaceutical ingredients, implants, wound covers, and food packaging materials. Natural biopolymers are produced from plants, animals, or microorganisms as polynucleotides, polypeptides, and polysaccharides. With recent progress in phytomedicine and the need for effective and healthy drug-delivery carriers, biopolymers have become increasingly relevant, with several studies prospecting for novel alternatives. Polymer carriers in phytomedicine are employed to safeguard and preserve the integrity of active phytocomponents (Pisoschi et al. 2018). In addition, the encapsulation of phytocomponents in polymer matrix masks unfavorable sensory attributes such as taste, color, and smell (Rassu et al. 2014; Aguiar et al. 2016; Arce and Ustunol 2018). These help increase consumer appeal and market acceptance of phytomedicines. Furthermore, polymer encapsulation of phytomedicines ensures that active components delivered via the oral route are shielded from acid and enzymatic degradations while transiting to the site of action.

Several biopolymers, including cellulose, chitin, chitosan, alginate, etc., are currently employed in phytomedicine formulation either as a drug-delivery channel or an active medicinal component. Chitosan, an abundant biopolymer derived from the deacetylation of chitin, has gained significance in drug delivery due to its solubility and mechanical properties (Nwabor et al. 2020; Singh et al. 2021). Encapsulation in a chitosan-based polyelectrolyte complex was employed as an appropriate delivery system to promote the bioavailability and aqueous stability of curcumin in an *in vitro* α-amylase inhibitory assay and *in vivo* anti-hyperglycemic assay in alloxan-induced rats (Akolade et al. 2017). Nano-encapsulation of madecassoside, a bioactive pentacyclic triterpenoid saponin isolated from *Centella asiatica* in the alginate-chitosan co-polymer matrix was employed as a therapeutic delivery system for the management of epilepsies and related stress-inflammatory disorder (Kunjumon et al. 2021). Microencapsulation of medicinal plants in a chitosan matrix enhanced the solubility of the extracts in simulated vaginal fluid and increased the bioavailability of bioactive compounds for the treatment of vulvovaginal candidiasis (Moreno et al. 2018). However, the choice of polymers for the encapsulation of phytomedicinal ingredients is affected by several factors, including the solubility of the polymer and phytomedicinal component, the gelling strength of the polymer, and preparation technique (Villemin et al. 2019; Cheow and Hadinoto 2011; Jyothi et al. 2010).

Delivery of phytomedicines through biopolymer matrix carriers has been achieved through various fabrication technologies. However, the choice of formulation technology depends on the purpose and intended delivery site of the activity. Microencapsulation technologies such as ionic gelation, spray drying, and freeze drying have been extensively employed as appropriate methods for delivering phytomedicines. However, with recent advances in nano-pharmaceuticals, the delivery of active components through nanosized dispersions and particles has increasingly become a vibrant research focus. The size and large surface area of nanoparticles promote the release of bioactive ingredients and increase the bioavailability and efficacy of the active component (Yang et al. 2020). Phytomedicinal nano-formulations have been employed to treat several disease conditions (Moradi et al. 2020; Paul et al. 2010; Zahrani et al. 2017). The use of biopolymers as a nano-drug delivery matrix overcomes the impending limitations of phytomedicine by enhancing its solubility, stability, and bioavailability. Loading ferulic acid into cyclodextrin nanosponges increased the solubility of ferulic acid as an anticancer phytomedicinal component (Rezaei et al. 2019). Phytomedicinal applications of ellagic acid, a phytochemical polyphenol with vast pharmacological benefits, are hampered by its poor water solubility and limited oral bioavailability. Several studies have demonstrated that nano-encapsulation of ellagic acid in biopolymers improves its solubility, bioavailability, and oral absorption and enhances its effects as an anticancer, antioxidant, and anti-hemorrhagic agent (Yağmur and Şahin 2020; Gopalakrishnan et al. 2014; Mady and Shaker 2017; Arulmozhi et al. 2013).

Considering the heightened acceptance of phytomedicine and the benefits of biopolymers in the effective delivery of phytomedicines to pharmacological sites of activities, this chapter elaborates on the role of biopolymers in phytomedicine, its preparation methods, and applications. The chapter provides an overview of progress in the applications of biopolymer as phytomedicine.

2. Sources and Application of Biopolymers in Drug Delivery

Biopolymers are polymers produced by living things, and they occur as natural or synthetic polymeric biomolecules produced by the covalent bonding of monomers. Biopolymers may occur naturally in plants, animals, bacteria, and fungi or synthesized chemically using biological materials such as amino acids, sugars, natural fats, or oils. Biopolymers can be categorized into three classes based on their origin:

(a) Agro-polymers from biomass obtained from plants, such as lignin, pectin cellulose, lignocellulose, and starch;

(b) Polymers from microbial or animal origin, such as collagen, microbial cellulose, xanthan gum, alginic acid, chitin, and chitosan;

(c) Chemically synthesized biodegradable polymers, such as poly(-hydroxy esters), including polycaprolactone, polyglycolide or poly(glycolic acid), polylactic acid, or polylactide, and poly(ethers), including polyethylene glycol and poly(vinyl alcohol).

In addition, biopolymers can be classified into three categories based on their monomeric units and structure:

(1) Polynucleotides: DNA (deoxyribonucleic acid) and RNA (ribonucleic acid)

(2) Polysaccharides: Examples include cellulose, chitosan, and chitin

(3) Polypeptides: Examples include collagen, gelatin, gluten, and whey.

Figure 1 presents the various sources of selected biopolymers used in pharmaceutical formulations to deliver phytomedicines. Commercially, biopolymers are produced from traditional isolation of biopolymers from plants, *in vitro*

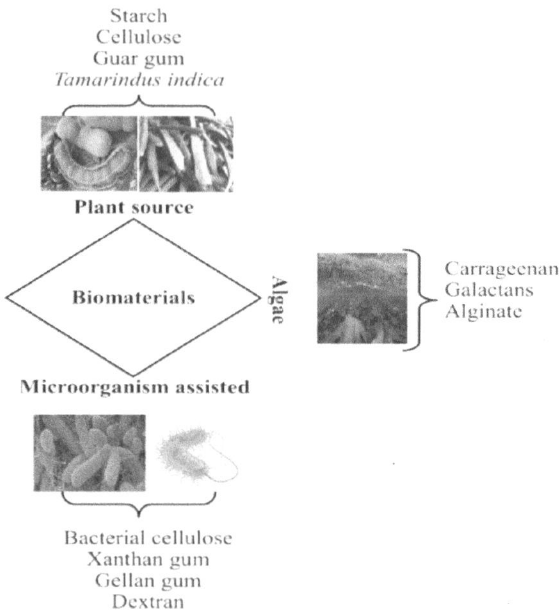

Figure 1: Source of biopolymers from animals, plant, and microorganism assisted.

synthesis of polynucleotides in cell-free systems, and large-scale fermentative or biotechnological production of biopolymers.

Developing biopolymers through economically viable methods for the formulation of active pharmaceuticals is an emerging alternative source with the potential to replace synthetic excipients. The wide availability and sustainability of natural resources is considered a possible bio-factory for developing prerequisite polysaccharides. Natural polysaccharides from different sources have long been studied and widely used in different areas, including the fabrication of pharmaceutical excipients, food, and the paper industry. Natural polymers are renewable, biocompatible, and biodegradable, with the ability to mimic the natural extracellular matrix environment. Amongst the plant-derived polysaccharides, cellulose is the most abundant and naturally available biopolymer, with a total annual biomass production of 1.5×10^{12} tons (Klemm et al. 2005). Plant polysaccharides consist of several α- or β-glycosidic bonded monosaccharide units that can be isolated using non-toxic organic solvents. Sources, compositional properties, and applications of naturally isolated biopolymers are presented in Table 1. Current industrial production of biopolymers employs engineered microbial cells as fermentation factories for the synthesis and large-scale manufacturing of biopolymers. Microbial production of biopolymers is an eco-efficient approach that does not harm the environment.

3. Fabrication Technologies for the Biopolymer in Drug-Delivery Systems

Nanotechnology and drug carrier approaches with innovative fabrication techniques have unlocked a new frontier in phytomedicine science using bio-based excipients.

Table 1: Polymers isolated from natural resources with their composition and pharmaceutical application

Biopolymer Source	Composition/Property	Application	References
Cassia uniflora	β-$_D$-glucose, Mw: 514 ± 110 KDa	Thickening agent	(Deore and Mahajan 2021)
Brazilian cacti	Xylose, galactose, and rhamnose	Emulsifier	(de Andrade Vieira et al. 2021)
Cydonia oblonga	L-arabinose, D-xylose, D-glactose, D-glucose, D-mannose, and Mw: 9.61 x 10^6 g/mol	Hydrogel	(Rezagholi et al. 2019; Guzelgulgen et al. 2021)
Corchorus olitorius	Rhamnogalacturonan-I, Mw: 51.2 ×10^3 KDa	Prebiotic	(Lee et al. 2021)
Opuntia ficus indica	- 8.0 mg of Indica xanthin per 100 gm of cactus pulp, 0.5 mg of betanin per 100 gm of cactus pulp, and 1.48 mg of betalains per 100 of pear	Colorant	(Carmona et al. 2021; Hernández-Carranza et al. 2019)
Plantago ovata	Arabinose 22.6 % and xylose 74.6%	Colorant, hydrogel	(Monge Neto et al. 2021; Yu et al. 2019)
Hibiscus rosasinensis L	Viscosity: 491.33 cPs and disintegration time 39.67 sec	Disintegrant and stabilizers	(Prabowo et al. 2021; Ash and Roy Biswas 2019)
Ornithogalum cuspidatum	Galactose, glucose, arabinose, galacturonic acid, and Mw: 1.7 × 10^6 kDa	Thickener	(Gheybi et al. 2021)
Hippeastrum vittatum	D-mannose, D-glucose, and β-1-4linked aldohexopyranose	Antidiabetic	(Attia et al. 2021)
Zea mays L.	Galactose, fucose, mannose, arabinose, xylose, and glucose	Excipients	(Nazari et al. 2020)
Dioscorea opposite	11.05% of glucose and 13.39 % of protein	Food product	(Ma et al. 2020)
Camelina sativa	Galactose, glucose, rhamnose, and xylose	Thickener	(Li et al. 2016)
Salvia hispanica L	Tetrasaccharides with branching of 4-O-methyl-α-D-glucuronic acid	Packaging	(Salazar et al. 2020)
Tamarindus indica and *Pithecellobium dulce*	β-arabinofuranosyl, β-glactose, β-glucose, α-xylose, and mannose	Prebiotic	(Sivakumar and Karuppaiyan 2020)
Manilkara zapota	Rhamnose and galactose	Excipients	(Sudarshan and Sunil B 2015)
Buchanania lanzan	Arabinose, rhamnose, fructose, and mannose	Excipients	(Singh and Bothara 2014a)

Table 1 contd. ...

Biopolymer Source	Composition/Property	Application	References
Diospyros melonoxylon Roxb	Rhamnose, fructose, and mannose	Excipients	(Singh and Bothara 2014b)
Guazuma ulmifolia Lam.	$\zeta - 51.1$ mV and reduced interfacial tension	Emulsifier	(Pereira et al. 2019)
Pereskia aculeate	Presence of minerals with endothermic melting point at 61°C	Emulsifier	(Silva et al. 2019)
Salvia hispanica L	Pseudo-plastic and shear-thinning behavior	Nutritional supplement	(Punia and Dhull 2019)
Ocimum basilicum L	Gel strength 69.39–238.01 g/cm^2 and enhanced tensile strength	Gel, composite	(Song and Kim 2019; Thessrimuang and Prachayawarakorn 2019; Mukherjee et al. 2019)
Aloe barbadensis Miller	Reduced degradation of gallic acid with EC_{50} of 545 mg	Microencapsulation	(Medina-Torres et al. 2019)
Brasenia schreberi	D – glactose (32%), D – glucuronic acid (19%), L – fucose (16%), D – mannose (14%)	Excipients	(Luo et al. 2019)
Pereskia aculeata Miller	Controlled water vapor permeability with excellent tensile strength	Film	(Oliveira et al. 2019)
Opuntia monacantha	Carbohydrates, including galactose, glucuronic acid, and arabinose	Stabilizer	(Dick et al. 2019)
Linum usitatissimum	Protein content of 12.5% with excellent viscosity and consistency	Excipients	(Castañeda-Cachay et al. 2019; Ding et al. 2018)
Portulaca oleracea	Presence of carbohydrate and saponins	Hydrogel	(Asnani et al. 2018)
Colocasia esculenta (L.)	Presence of carbohydrates with a semi-crystalline nature of mucilage	Excipients	(Mijinyawa et al. 2018)
Verbascum Thapsus	67.66% of radical scavenging efficacy	Excipients	(Babamoradi et al. 2018)
Cordia myxa	Carbohydrate 84.32% and protein 8.90%	Excipients	(Keshani-Dokht et al. 2018)
Abelmoschus esculentus	Nanocomposite showed low water vapor permeability and solubility	Film	(Araújo et al. 2018)
Colocasia esculenta	Glucose, galactose, arabinose, xylose, mannose, and glucuronic	Emulsifier	(Manhivi et al. 2018)
Cassia obtustifolia	Presence of carbohydrate and biocompatibility with mice	Excipients	(Deore and Mahajan 2018)

Cordia dichotoma G.	Presence of carbohydrates with 0.047 cPs of viscosity	Excipients	(Pawar and Jadhav 2015)
Hoheria populnea A.	Rhamnose, galactose, galacturonic acid, and glucuronic acid	Excipients	(Sims et al. 2018)
Lallemantia royleana	Galactose (36.28%), arabinose (35.96%), rhamnose (15.18%), xylose (7.38%) and glucose (5.20%), and Mw: 1.19 × 106 Da	Excipients	(Behbahani and Imani Fooladi 2018)
Dioscorea opposita Thunb	Mannose (62.52%) and glucose (23.45%)	Excipients	(Ma et al. 2018)
Hygrophila spinose	Presence of carbohydrate with excellent suspending efficacy	Stabilizer	(Krishna and Mohanan 2018)
Lallemantia iberica	Excellent water vapor and mechanical property	Composites	(Sadeghi-Varkani et al. 2018)
Hibiscus esculantus	Prolong gastric residence time	Excipients	(Mulchand et al. 2018)
Diospyros melonoxylon R	Prolong gastric residence time	Excipients	(Sudarshan et al. 2021)
Buchanania lanzan spreng	Prolong gastric residence time	Excipients	(Sudarshan et al. 2020)

Several studies reported that more than 70% of new drug entities fabricated in conventional form indicate low solubility, which limits the absorption of orally administered therapeutics (Rahman 2020). Phytomedicine and biopharmaceutical classified class II and IV drugs require attention during the fabrication process due to problems linked to solubility, which ultimately leads to failure in clinical trials. Enhancement in solubility, stability in colonic pH, and intestinal absorption, with improved therapeutic efficacy, can be achieved by utilizing novel drug delivery with carrier-based technology. Employment of innovative approaches, including nanoparticles and lipid-conjugated nanotechnology, is considered a potential and effective drug-delivery system for biomedicine. Beneficial features such as carrier size, protection of phytoconstituents from biological environments, stimuli-responsive release with improved drug distribution, targeted and intracellular delivery of active constituents, and ease of use as diagnostic and imaging agents enhance the utility of conjugated carriers in biomedicines. The recognition and approval of new pharmaceutical products are related to the fabrication technology and successful innovation, both commercially and remedially. The various types of nanocarriers and fabrication technologies involved with herbal drug delivery are depicted in Figure 2.

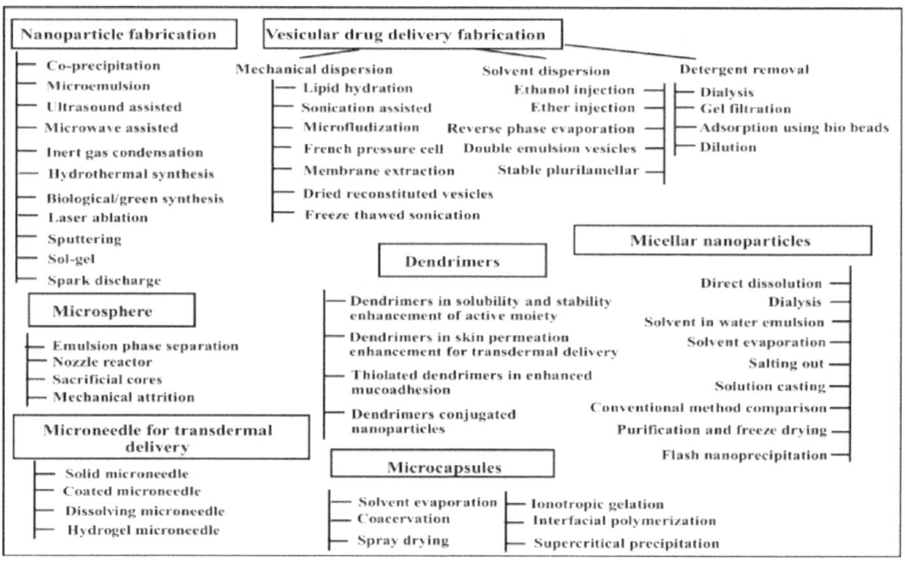

Figure 2: Formulation technology of various novel drug delivery systems incorporated with phytomedicine.

4. Potentials Phytomedicine With Biopolymers for Therapeutic Applications

Several bioactive materials have been identified in higher plants, with many originating from edible plants. These plant-derived bioactive compounds possess several active constituents, including polyphenols, flavonoids, alkaloids, saponins, and tannins reported with potential pharmacological efficacy (Figure 3). The phytoconstituent modulates several cellular pathways in the management of

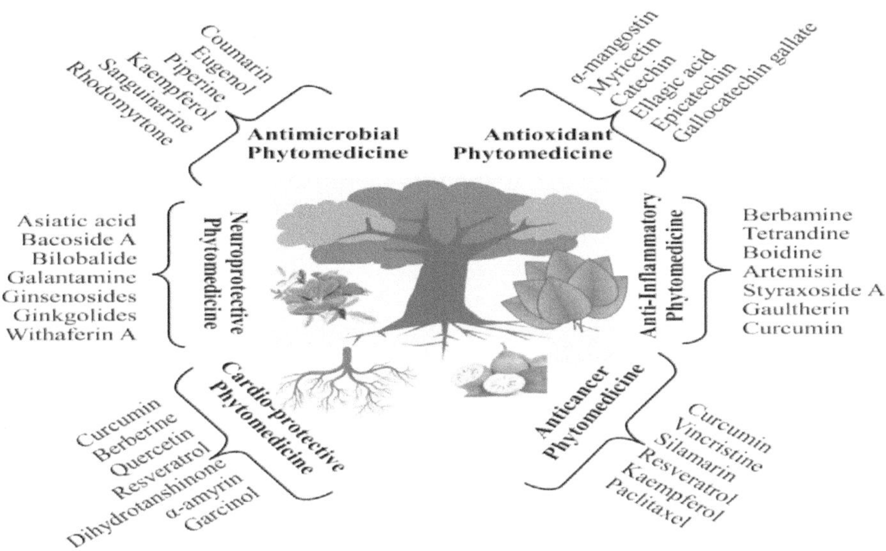

Figure 3: Potential phytomedicine with pharmacological efficacy.

various diseases, including cancer, Alzheimer's, topical inflammations, diabetes, dengue, etc. Despite the availability of synthetic active pharmaceuticals, herbs and herbal medicines are still preferred as major healthcare products in developing countries. In prospecting for new drug development, researchers have been drawn toward biologically active plant extract due to traditional claims of efficacy and pharmacological potential. Natural products are considered safe, and solvent-specific extracts exhibit concentration-dependent biocompatibility (Rahman 2020). However, phytomedicines face physiochemical obstacles similar to those of synthetic active pharmaceutical moiety, including low solubility and bioavailability. These problems can be overcome using nanotechnology with the use of cost-effective biopolymers. Research involved with biopolymers and natural product-incorporated nanotechnology has reported several advantages, including enhancement in solubility, pharmacological efficacy, improvement in tissue permeability, and stability (Figure 3). The rise in phytomedicine usage demands a hypothetical nanotechnology-based approach to develop a rational drug-delivery system for the appropriate use of natural products in several life-threatening diseases.

Pharmaceutical nanotechnology is the fastest-developing technique that enables the design of multifarious delivery systems, which significantly reduces the complications associated with phytomedicine. Incorporation of phytomedicine into a biopolymer-based nano drug delivery is an innovative approach to extend the physicochemical properties of phytomedicines with enhancement in bioavailability and therapeutic potential. For more than a generation, biopolymers have been used to deliver phytomedicine to biological tissue. Biopolymers may be natural or synthetic and are used in healthcare systems for upkeep, improvement, or supplant reparation of tissue or biological function. Biopolymer-based sutures originated from animal sinew and were the first historical use of implantable biopolymers, which later progressed with discoveries in tissue engineering, regenerative medicines, etc.

(Ige et al. 2012). This evolution was paralleled by a budding of professionalism and the establishment of biopolymer societies. The earliest biomaterials-related society was founded in 1954 as the American Society for Artificial Internal Organs to establish a platform for developing artificial kidneys and hearts (Ratner et al. 2004).

Biopolymer-incorporated phytomedicine offers multiple benefits in the treatment of both acute and chronic diseases through site-specific and target-oriented delivery of natural products. Several biopolymers incorporated nanotechnological products, including dermal scaffolds, lipid-based carrier systems, and aerosol nasal spray, have been reported for safe and effective tissue-targeted delivery of phytomedicine. Recently, several reports have highlighted outstanding research of biopolymer-incorporated phytomedicine in the treatment of cancer and Alzheimer's disease, showing enhanced permeability and bioavailability. Besides the various roles of biopolymers as excellent drug-delivery carriers, the biocompatibility, safety, and readily availability of biopolymers as a sustainable raw material are of economic benefits in drug formulation. In addition, several biopolymers have been reported to exhibit trace pharmacological activities, such as the antioxidant properties of collagen responsible for its cosmetic application as an anti-aging product and the antimicrobial effects of chitosan.

4.1 Bio-Based Polymeric Material With Phytomedicine in Topical Applications

Globally, billions of people are plagued by the burden of dermal disorders, resulting in severe consequences. Skin and dermal tissue-related complications are the most common and distressing forms of laceration, with thousands suffering when left untreated. Wound healing is a complex process comprising different mediators, including soluble mediators, blood cells, extracellular matrix, and parenchymal cells. Healing of wounds is broadly classified into three stages, which are the inflammation process, tissue formation, and remodeling of tissue (Chhabra et al. 2017). The inflammatory stage includes platelet accumulation, coagulation, and leukocyte migration, whereas re-epithelialization, angiogenesis, fibroplasia, and wound contraction are involved during tissue induction (Kirsner and Eaglstein 1993). Production of protein matrix and collagen lasts for one month in remodeling phase. The treatment of wounds with phytomedicine is gaining huge importance, either in semisolid gel form or incorporated in composites. Phytomedicine promotes the healing of wounds through different mechanisms, including modulation in wound healing, decrease in bacterial count, improvement in collagen deposition, and increase in fibroblast and fibrocytes (Thangapazham et al. 2016). However, such traditional treatments reflect disadvantages such as inappropriate healing of wounds due to exquisite shape and drying with exposure to air. Cross-linking of phytomedicine with biopolymers is a novel technique adopted by researchers in the fabrication of biodegradable copolymerized conjugate as a scaffold in tissue regeneration that mimics the extracellular tissue matrix. Scientific data indicated that phytomedicine-incorporated formulations have proven to be significantly useful in the treatment of various skin disorders, including coagulation, inflammations, fibroplasia, epithelization, collagenation, and wound contractions (Broughton et al. 2006).

Traditional wound dressings, such as bandages, cotton wools, and gauzes, have several disadvantages and are unable to provide adequate protection for dermal application. Thus, there is an urgent need to replace such traditional dressings with bio-conjugated phytomedicine as scaffolds to provide an optimum condition around the wound with the release of active constituents to facilitate wound healing. Phytomedicine-incorporated biopolymers have an astonishing advantage over traditional dressings due to negligible host response, high level of tunability, and biocompatibility, which are either reabsorbed or removed from biological systems. Recently reported phytomedicine with polymeric conjugate-based therapeutic paradigms for wound dressings and tissue regeneration are presented in Table 2.

4.2 Role of Polymeric Conjugates Toward Delivery of Anticancer Phytomedicine

Natural resources are vital sources of novel potential phytoconstituents with excellent therapeutic efficacy. The quest for the development of plant-based, safe, and effective drugs has led to an increased focus on exploring and harnessing nature, particularly for the discovery of active compounds in the treatment of life-threatening diseases, including diabetes, hypertension, and cancer. Cancer involves abnormal cellular growth with the potential to spread to other parts of the body and the ability to infiltrate and destroy normal body tissue. Globally, the rising prevalence of cancer has become a threat to public health and a great concern to the health sector, with overwhelming annual mortality and morbidity. Thus, research on anticancer drug formulation and development has become of paramount importance (Seyfried and Huysentruyt 2013). Approximately 70% of deaths due to cancer occur in low and middle-income countries. The annual number of new cases is projected to rise to 21.6 million in 2030. Moreover, the pecuniary influence of cancer is noteworthy. In the United States, the financial burden of cancer care by 2020 was estimated at $157.8 billion dollars (Yabroff et al. 2011). Nature is a potential source of anticancer drugs, with over 60% of currently used anticancer drugs derived from plants. Naturally occurring anticancer agents, including curcumin, vincristine, vinblastine, vindesine, vinorelbine, taxanes, podophyllotoxin, camptothecin, anthracyclines, trabectedin, epothilone, temsirolimus, and their derivatives are generally prescribed. The majority of clinically used anticancer phytomedicines available in conventional dosage form possess substantial limitations such as short biological half-life, low solubility with limited bioavailability, and several side effects. These drawbacks limit the efficacy of phytomedicine and constitute a driving force for the design of a biopolymer-based novel drug-delivery system to reduce the side effects with enhancement in the clinical efficacy of anticancer phytomedicine.

A surge in cancer diagnosis has been brought about by nanotechnology, with the competence to perceive even a distinct tumor cell. Individual cancer cells can be traced using biopolymer-based nanostructured materials such as nanoparticles, dendrimers, and nanosponges cross-linked with specific antibodies. Moreover, the incorporation of phytoconstituents in novel/nanotechnology drug-delivery systems fabricated using biopolymers is increasing daily in cancer treatment with enhanced clinical efficacy. In addition, biopolymer usage in the development of nanotechnology in the treatment of cancer provides the dual advantage of biodegradability as well as

Table 2: Pharmaceutical topical application of biopolymers incorporated phytomedicine.

Biopolymer Conjugate	Phytomedicine/Plant Extract	Application	References
Chitosan-poly(ethylene) oxide blend nano-fibrous mats	*Hypericum perforatum* L	Wound dressing	(Mouro et al. 2021)
Chitosan-poly(ethylene) oxide-cellulose nanocrystals nanofibers	*Acacia* plant-based extract	Wound dressing	(Ribeiro et al. 2021)
Chitosan-nanocellulose fiber	*Gleditsia triacanthos*	Wound dressing	(Marinas et al. 2021)
Chitosan-gelatin blend	*Sophora gibbosa*	Wound dressing	(Shalaby et al. 2021)
Chitosan-poly(ethylene) oxide blend	*Lawsonia inermis* and *Thymus vulgaris*	Implants	(Kharat et al. 2021)
Chitosan-poly(vinyl alcohol)	*Agrimonia eupatoria* L	Wound dressing	(Mouro et al. 2020)
Agarose/k-carrageenan	*Cryphaea heteromalla* (Hedw.) D. Mohr	Wound dressing	(Ditta et al. 2020)
Cellulose acetate/gelatin	*Zataria multiflora*	Wound dressing	(Farahani et al. 2020)
Sodium alginate/polyethylene glycol	*Satureja cuneifolia*	3D-printed scaffolds	(Ilhan et al. 2020)
Chitosan	*Cynara cardunculus*	Wound dressing	(Brás et al. 2020)
Gelatin	β-caryophyllene and lupeol from *Copaiba oleoresin*	3D-scaffolds	(Pascoal et al. 2020)
Chitosan-glucan complex with aloe vera	*Schizophyllum commune*	Wound dressing	(Abdel-Mohsen et al. 2020; Maru et al. 2013)
Modified bacterial nanocellulose	*Terminalia arjuna, Azadirachta indica, Withania somnifera, Tinospora cordifolia, Murraya koenigii*	Biomedical dressing	(Sharma and Bhardwaj 2020)
Pectin/starch	*Gunnera tinctoria* and *Ugni molinae*	Wound dressing	(Sabando et al. 2020)
Chitosan, poly-caprolactone, and polyethylene glycol	*Cydonia oblonga*	Scaffolds	(Izadyari Aghmiuni et al. 2020)

biocompatibility. Thus, nanotechnology could be used for superior cancer diagnosis with maximized delivery of phytomedicine to a molecular level for targeting tumor tissue/cells that ultimately improve the therapeutic management of cancer patients. The most recent report on anticancer phytomedicine with polymeric conjugate is presented in Table 3.

Table 3: Biopolymer conjugates with phytomedicine for delivery of anticancer therapeutics.

Polymer	Active Constituent/ Plant Extract	Application	References
Covalently linked Cyclodextrin	Curcumin	Anticancer activity against HeLa cell treated with encapsulated curcumin	(Möller et al. 2021)
Acrylic grafted guar gum	Curcumin	Hydrogel based carrier for treatment of cancer	(HaqAsif et al. 2021)
Chitosan/hyaluronic acid	Doxorubicin/cisplatin	Brest cancer chemotherapy	(Wang et al. 2019)
Chitosan/polylactic acid	Cisplastin/siRNA/ plasmid DNA	Resistant to human ovarian cancer cells	(Babu et al. 2014)
Chitosan	Demethoxycurcumin	Effective against A549 lung tumor	(Huang et al. 2015)
Chitosan	Bufalin	Brest cancer chemotherapy	(Tian et al. 2014)
Chitosan	Curcumin	Photodynamic therapy for cancer cell	(Tsai et al. 2018)
Lecithin-chitosan	Paclitaxel-cholesterol complex	Intra-tumoral injection for improving antitumor outcomes	(Chu et al. 2019)
Chitosan	Green tea polyphenol ephigallocatechin-3-gallate	Enhanced efficacy against prostate cancer cell	(Khan et al. 2014)
Carboxymethyl graft chitosan	Ricinoleic acid loaded rotenone	Enhanced bioavailability	(Feng et al. 2016)

4.3 Biopolymer as Carrier for Nasal to Brain Drug Delivery

Many drugs are effectively administered systemically, but the slow onset of non-intravenous administration with limited bioavailability limits their clinical efficacy. However, the delivery of systemic and local therapeutics via the nasal route has received significant attention due to its maximized efficacy. Moreover, during the last few decades, the delivery of potential drugs, such as protein, peptides, and phytomedicine, susceptible to enzymatic or acid degradation with hepatic metabolism, developed as a nasal drug-delivery system. These might be due to the low proteolytic activity, large surface area, and vascularity with porous endothelial membrane and the presence of a significant number of microvilli that help to enhance the permeability and avoid hepatic drug metabolism (Ozsoy et al. 2009).

Nose-to-brain drug delivery is one of the fastest emerging fields. However, the mechanism of action for efficient drug delivery to the brain is not yet completely known. Several studies have indicated the effective delivery of phytomedicine and its derivatives to the brain via the olfactory pathway, trigeminal pathway, and deposition in nasal mucosa. While the preferred dosage form for intranasal drug delivery is a solution, this approach has several disadvantages, making the selection of biopolymer in the fabrication critical. A combination of bio-adhesive materials with smart thermo-reversible polymers is used in the fabrication of intranasal drug delivery. These combinations of smart biopolymers include *Linum usitatissimum* L seed mucilage (Basu et al. 2009), drum-dried maze starch (Callens and Remon 2000), *Dellinia indica* L fruit mucilage (Kuotsu and Bandyopadhyay 2007; Dharmendra Kumar 2013), *Ficus carica* fruit gum (Basu and Bandyopadhyay 2010), gellan gum (Cao et al. 2009; Cao et al. 2007), chitosan (Kim et al. 2010), sodium alginate (Patil and Sawant 2009), hydroxypropyl methyl cellulose (Hansen et al. 2015), sodium carboxymethycellulose (Hansen et al. 2015), and poloxamer and derivatives (Adnet et al. 2020); these have been widely studied for delivery of therapeutics via intranasal. Guo and coworker reported that the root essential oil of *Radix bupleuri*, Chinese traditional medicine for the treatment of fever, pain, and inflammation associated with influenza, when incorporated into ion-activated gellan gum as *in situ* intranasal gel, demonstrated excellent antipyretic efficacy when administered *in vivo* compared to intranasal solution (Cao et al. 2007). Bonaccorso and coworkers compared the delivery and cellular uptake of curcumin alone and curcumin nanocrystals on olfactory unsheathing cells and reported an ameliorated permeability for curcumin nanocrystals (Bonaccorso et al. 2020). Rachana and coworker developed *Gingko biloba* extract loaded microemulsion for intranasal administration with biocompatibility on nasal epithelial cells and zero-order of controlled release (Singh et al. 2017)

4.4 Biopolymers With Phytomedicine in the Treatment of Alzheimer's

Alzheimer's is an irreversible, progressive neurodegenerative disorder and a frequent cause of dementia. However, this neural network degeneration appears long after the inception of the disease. The Centers for Disease Control and Prevention reported more than 5.8 million Americans in 2020 were affected by Alzheimer's disease, which accounts for 50–75% of globally (Matthews et al. 2019). In addition, a recent study projected that this figure will rise by triple folds to 14 million by 2060 (Matthews et al. 2019). The pathogenesis of Alzheimer's is not yet completely understood. However, genetic susceptibility and environmental-related issues are responsible for the late onset of intermittent Alzheimer's. Moreover, Alzheimer's condition is characterized by cognitive incompetence demonstrated through the accumulation of β-amyloid with the formation of hyperphosphorylated neurofibrillary tangles (tau protein) and dysfunction of the cholinergic system. In addition, Alzheimer's is also concomitant with gene mutation, including presenilins (PS1 and PS2) and apolipoprotein E. Several animal studies have been conducted to understand the causative mechanism of this disease for the development of drugs. Currently, the possible development of drugs is focused on the search for phytomedicines that can inhibit presenilins, β-amyloid, and apolipoproteins. Further approaches, including therapy with natural antioxidants (Grundman et al. 2002), anti-inflammatories (Moore and O'Banion 2002), and cholesterol-lowering mediators, have been reported in the treatment of Alzheimer's.

Several bioactive medicines have been tested due to the aftereffects of synthetic drug therapy. Phytomedicine, including capsaicin, curcumin, resveratrol, and hyperforin, have been reported to inhibit the effects of tau protein hyperphosphorylation with excellent pharmacological efficacy in animal studies. Moreover, the potential efficacy of active constituents originating from marine sources such as red algae (Syad and Devi 2015), sea cucumber (Li et al. 2019), and marine spoon worms have also been reported in the treatment of Alzheimer's. Other natural sources of phytomedicines with potential in the modulation of Alzheimer's are presented in Table 4. Nanotechnology-based interventions have opened an exciting area via theranostics of Alzheimer's disease in terms of tailored nanomedicine, which is able to target and deliver drugs across the blood-brain barrier. Polylactide-co-glycoside nanoparticles with surface modification by lactoferrin-conjugated N-trimethylated chitosan were employed as a mucoadhesive and targeted nano-medicinal for efficient intranasal delivery of Huperzine A to the brain for the treatment of Alzheimer's disease (Meng et al. 2018).

Table 4: Recent reports on the use of phytomedicine in treatment of neurodegenerative disorder.

Botanical Name	Phytoconstituents	Mode of Action	References
Withania somnifera	Withanolides, withanamides	Inhibit β-amyloid and acetylcholinesterase	(Mahrous et al. 2017)
Allium sativum	S-allylcysteine	Significant reduction in inflammatory response by reducing the activation of microglia and IL-1β	(Nillert et al. 2017)
Zingiber officinale	Shogaol, gingerol	Inhibition of GSK-3β	(Yerer et al. 2017)
Panax ginseng	Ginsenoside	Modulation of protein (CPLX2, SYN2, SNP25) expression	(Nie et al. 2017)
Coffea Arabica	Caffeine, chlorogenic acids	Modulation of nicotinamide, mononucleotide, and adenylyl transferase 2	(Ali et al. 2017)
Coriandrum sativum	Coumarin	Antioxidant, anti-inflammatory, and inhibition of extracellular regulated kinase signal	(Liu et al. 2016)
Ficus carica	γ-sitosterol	Dietary supplementation improved the cognitive and behavioral deficit	(Subash et al. 2016)
Nardostachys jatamansi	Valeranone nardosinone	Protective effect of stress-induced impairments	(Karkada et al. 2012)
Pleuroncodes planipes	Astaxanthin and docosahexaenoic acid	Protect PC 12 cells by up-regulation of heme oxygenase-I from β-amyloid induced cytotoxicity	(Reiserer et al. 2007)
Food oils and marine animals	Omega-3 fatty acid	Protective effect against Alzheimer's in rat brain	(Wang et al. 2018)

4.5 Role of Biopolymeric Lipid in Phytomedicine-Based Drug Carriers

Remediation with herbal medicine predates modern civilization. Phytomedicine refers to the practice of using plant parts and constituents for medicinal purposes based on traditional knowledge. The global annual market for herbal products is growing significantly and loomed to USD 60$ billion (Sendker 2017). Despite the advantages, pharmaceutical manufacturers are hesitant to invest in the development of phytomedicine as many natural compounds alone or in conventional delivery form could not pass the clinical phase trial study. This is largely due to problems associated with the poor solubility, limited absorption, and low bioavailability of most bioactive phytoconstituents. However, recent advances in drug-delivery carriers using biopolymeric compounds signal a significant improvement in the delivery of phytomedicines to pharmacological active sites. The use of biopolymers as drug excipients of carriers has been extensively investigated with promising outcomes. Screened and purified natural compounds in conjugation with lipids are gaining importance in healthcare management systems. Lipid-based carriers with bioactive components represent a rapidly developing technology used to deliver therapeutic agents in a regulated mode. Biopolymeric lipid carriers with phytomedicine drug-delivery systems are a proven, commercially viable approach for formulating pharmaceuticals, including topical, oral, parenteral, and pulmonary. Self-assembled lipid-based nanocarriers complexed with phytoconstituents exhibit several advantages, including enhanced solubility, stability, lipophilicity, biocompatibility, effective tissue targeting, cellular targeting, and molecular targeting. These carrier molecules shield the bioactive components from interaction with biological enzymes and systemic inactivation. Nano-carries can be broadly categorized based on carrier resources such as organic or inorganic carriers. Several novel drug deliveries such as liposomes, niosomes, pharmacosomes, transfersomes, ethosomes, archaeosomes, iscoms, vesosomes, virosomes, colloidosomes, herbosomes, phytosomes, cubosomes, nanospheres, solid lipid nanoparticles, nanostructure lipid carriers, lipid drug conjugate, microemulsion, nanoemulsion, and self-emulsifying drug-delivery system have been reported with the ability to deliver phytoconstituent. Table 5 shows the potential phytomedicine conjugate with biopolymer carrier reported for the treatment of life-threatening diseases.

5. Graft-Modified Polymeric Material With Phytomedicine in Biomedical Applications

Nature-derived polymeric materials and their derivatives are widely useful in the fabrication of various drug-delivery systems, including targeted and controlled release of phytomedicine with enhanced bioavailability. Pharmaceutical bio-based polymers possess unique properties such as non-toxic, biocompatibility, biodegradability, and ease of availability; however, the short shelf life and bacterial susceptibility limit the scope as pharmaceutical excipients. These deficits can be improved to achieve the desired properties by various modification techniques, including co-processing, spray drying, chemical modification, polyelectrolyte complex, and graft polymerization. Graft polymerization is a process in which monomers are covalently

Table 5: Biopolymeric-based carrier for delivery of phytomedicine.

Phytomedicine	Lipid-Based Drug-Delivery Carrier	Application	References
Berberine	Liposome	Inhibition of tumor growth in HepG2 tumor-bearing mice	(Lin et al. 2013)
Artemisinin	Chitosan, gelatin, and alginate nanocapsules	Anticancer	(Chen et al. 2009)
Curcumin	Poly(lactic-coglycolic acid) nanospheres	Prostate cancer	(Mukerjee and Vishwanatha 2009)
Berberine conjugated doxorubicin	Liposome	Anticancer	(Yahuafai et al. 2018)
Silybin	Phospholipid complexed nanoemulsion	Treatment of liver disorder	(Song et al. 2005)
Silymarin	Self-nanoemulsifying drug-delivery system	Enhanced bioavailability with hepatoprotective effect	(Tung et al. 2019)
Rutin	Floating alginate-chitosan microcapsule	Antioxidant and anti-inflammatory	(Xiao et al. 2008)
Camptothecin	Modified glycol chitosan nanoparticles	Anticancer	(Min et al. 2008)
Curcumin	Liposome with hydrogenated soya phosphatidylcholine	Enhanced antioxidant with hepatoprotective effect	(Maiti et al. 2007)
Gugulipid	Proniosome	Lipid-lowering	(Goyal et al. 2011)
Quercetin	Mircoemulsion	Anti-angiogenic	(Vicentini et al. 2008)

bonded and polymerized as side chains onto the backbone, with improvements in the physicochemical characteristics of the backbone (Figure 4). Copolymerization amends the symmetry of the polymeric chain and modulates both intramolecular forces and properties, such as glass transition temperature, crystallinity, solubility, elasticity, permeability, and chemical reactivity that results in the assimilation of favorable properties in a grafted polymer. Biopolymer grafting can be achieved through different initiation processes, including ceric ion initiation, persulfate initiation, irradiation, and hydrogen peroxide redox initiator. Moreover, the graft copolymerization of biopolymers is also obtained using monomers, including acrylic acid, acrylo-nitrate, acryl amide, and metha acrylate. The graft polymer with potential application in the delivery of phytomedicine is presented in Table 6.

Graft copolymer conjugates are widely used to deliver active constituents for the treatment of various disorders. Chitosan is the second most abundant naturally occurring pharmaceutical biopolymer derived as a deacetylated form of chitin (Lopes et al. 2020). The source of chitosan is widely distributed both in the animal and plant kingdom, including fungi, algae, echiruda, annelida, mollusca, cnidaria, achelminthes, entroprocta, bryozoa, phoronida, brachiopoda, arthropoda, and ponogophora

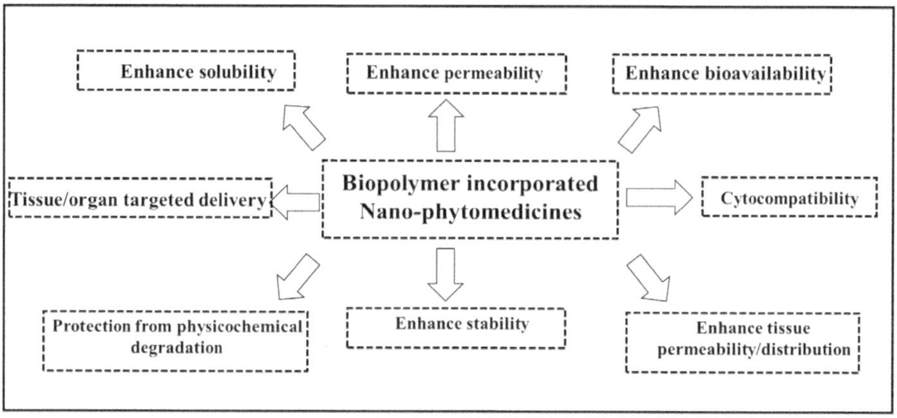

Figure 4: Improvement in various glitches encountered by phyto-pharmaceutical bio-based materials incorporated formulations.

(Ahmed 2015). In addition, the most common sources cited in the literature are shrimp, crab, lobster, crayfish, and oyster. Chitin occurs in three polymeric forms based on the differing degree of hydration, size of unit cell, and chitin chain per unit cell as α, β, and γ chitin. Chitosan has gained huge importance as a biopolymer for the fabrication of several medicated and consumer-end products. These graft biopolymers can be easily molded into the hydrogel, scaffold, and blend with other polymers, which offers numerous benefits with enhanced mechanical strength, biomimetic properties, and several other desired features for the development of phytomedicine incorporated delivery system (Figure 5). Recently, a polymerized precursor of chitosan grafted with *N*-isopropylacrylamide and oleic acid conjugated with folate micelles demonstrated a potential injectable application for targeting cancerous cells (Fathi et al. 2017). In addition, curcumin-loaded thermosensitive aqueous nanogel based on graft chitosan scaffold demonstrated improved bioavailability with biocompatibility (Luckanagul et al. 2018). Furthermore, chitosan-encapsulated curcumin and rutin nanoparticles demonstrated a non-fickian release profile with enhanced oral bioavailability in the treatment of tumor (Ramaswamy et al. 2017). Oh et al. (2021) reported grafting of chito-oligosaccharide using an ascorbic acid/hydrogen peroxide redox pair reaction with salicylic acid, functionalized with oxidized alginate and gelatin as wound healing hydrogel with improved antioxidant efficacy.

Alginate and its derivatives are naturally occurring biopolymers with linear unbranched polysaccharides produced from algae and bacteria consisting of D – mannuronic and L – guluronic acid units organized in an asymmetrical pattern. The presence of free hydroxyl and carboxyl groups along the backbone makes alginate an ideal candidate for the functionalization and improvement of desired properties. Among several ways of chemical modification, graft polymerization is one of the major employed techniques to develop the desired modified form of alginate (Figure 6). Free radical microwave-assisted graft polymerization of sodium alginate with vinyl sulfonic acid in the presence of the thiourea and backbone of polyacrylamide demonstrated excellent flocculation efficiency (Sen et al. 2010). Subhan et al. (2021) grafted a novel sodium alginate and bentonite doped poly (*N*-vinyl formamide-

Table 6: Graft biopolymer on the backbone of polysaccharides using various techniques with pharmaceutical applications.

Biopolymer Source	Grafting/Co-Processing Technique	Backbone Polymer	Application	References
Abelmoschus esculentus	Redox initiator	Acrylamide	Topical hydrogel	(Mishra et al. 2008)
Plantago psyllium	Redox initiator	Polyacrylamide	Flocculating agent	(Mishra et al. 2004)
Plantago psyllium	Redox initiator	Acrylonitrile	Sustained release delivery	(Mishra et al. 2003)
Trigonella foenum graecum	Redox initiator	Acrylamide	Tissue engineering	(Bal and Swain 2020)
Plantago psyllium	Microwave-assisted	Acrylic acid and acrylonitrile	Pharmaceutical beads	(Kumar et al. 2018)
Plantago psyllium	Redox initiator	Acrylic acid	Controlled release delivery	(Rao et al. 2017)
Tamarindus indica	Radical Polymerization	Methyl methacrylate	Drug-delivery system	(Mishra and Malhotra 2012)
Mimosa pudica	Redox initiator	Acrylamide	Drug delivery system	(Ahuja et al. 2011)
Abelmoschus esculentus	Redox initiator	Polyacrylonitrile	Excipients	(Mishra and Pal 2007)
Sterculia urens	Microwave-assisted	Acrylamide	Excipients	(Kaur and Gupta 2020)
Colocasia esculenta	Microwave-assisted	Acrylamide	Excipients	(Loveleenpreet Kaur 2020)
Ceratonia silique	Microwave-assisted	Acrylamide	Excipients	(Giri et al. 2015)
Manilkara zapota	Spray drying	Hypromellose, sodium carboxy methyl cellulose, and polyvinylpyrrolidone	Mucoadhesion	(Singh et al. 2020a)
Buchanania lanzan	Spray drying	Mannitol, lactose, and silicon dioxide	Excipients	(Singh et al. 2020b)
Chitosan	Anionic ring-opening polymerization	L-Alanine-N-carboxyanhydride	Tissue engineering	(Park et al. 2016)
Chitosan	Free radical polymerization	Acrylic acid and β-cyclodextrin	Controlled release of Curcumin	(Anirudhan et al. 2016)

co-acrylic acid) hydrogel with Fickian diffusion and excellent swelling capacity supported by Langmuir isotherm that could be useful as superabsorbent for removal of toxic dyes from industrial effluents. El-Ghoul and Alminderej (2021) reported sodium alginate cross-linked with *Carthamus tinctorius*, grafted with cellulosic dressing. The graft wound dressing demonstrated high porosity and biocompatibility

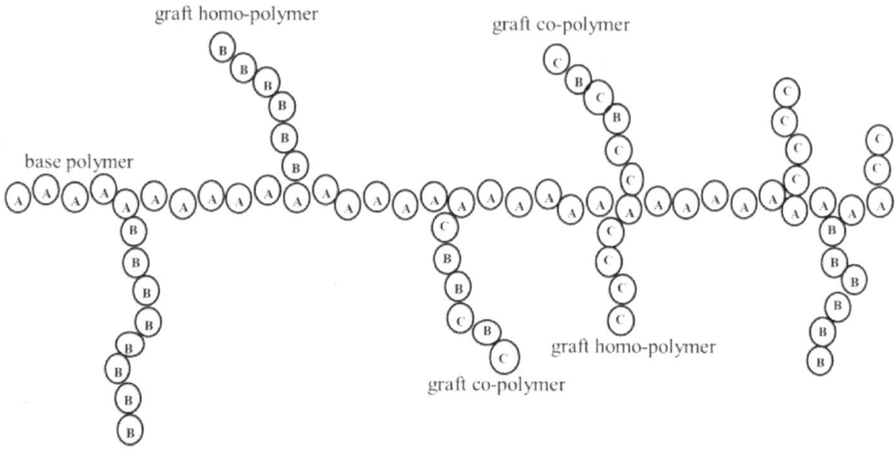

Figure 5: Graft polymerization of polysaccharides.

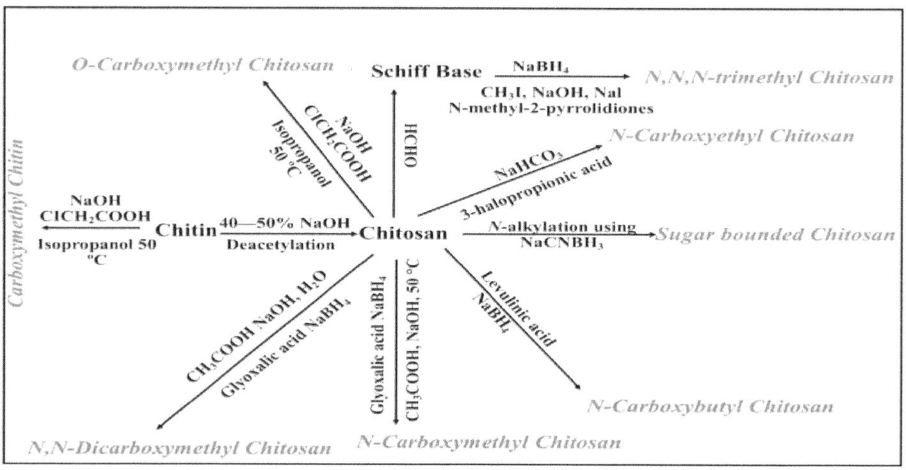

Figure 6: Graft chemically modified chitin and chitosan biopolymers.

on human cells with broad efficacy against potential wound pathogens. Han et al. (2019) fabricated electrospinning nanofibers using poly (vinyl alcohol) cross-linked sodium alginate and boric acid of plant lutein, a natural pigment that demonstrated enhanced hydrophilicity with controlled release of lutein for biomedical application.

6. Future Perspectives and Conclusions

Delivery of active components through biopolymer matrix and particulate systems can be achieved through a variety of techniques. The use of encapsulated biopolymer particles or composite matrices as drug-delivery systems for phytomedicines and other active ingredients is still an evolving research area. Current knowledge on the preparation of biopolymer particulate matter for the effective delivery of phytomedicine is still limited. The effective use of biopolymer as an appropriate

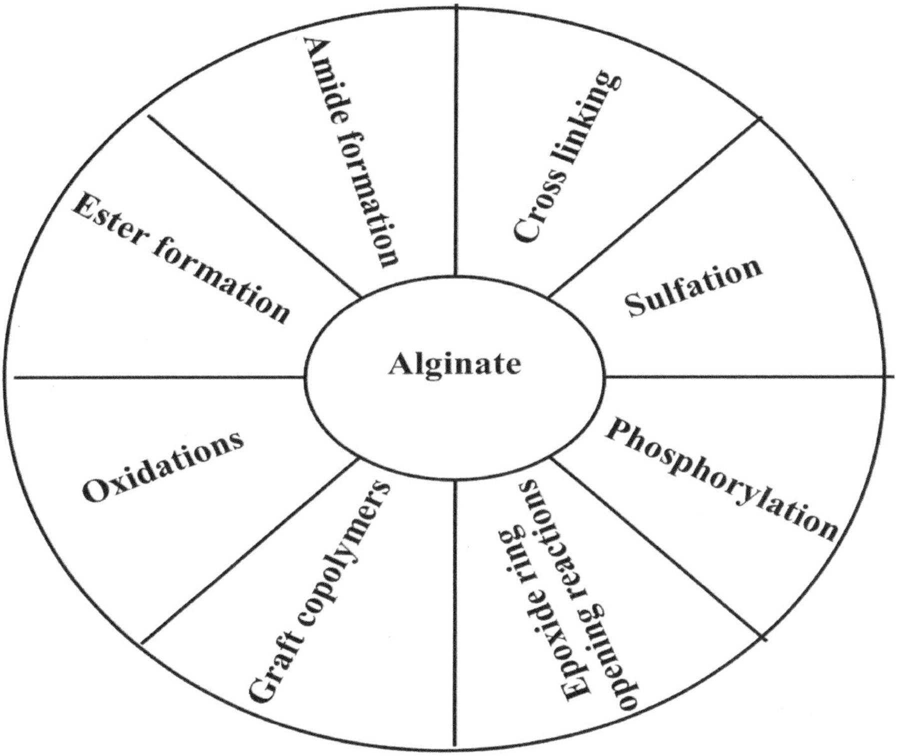

Figure 7: Chemical (–COOH and -OH) functionalization of alginate.

channel for the delivery of ingredients requires adequate knowledge of the polymer properties, its molecular and functional characteristics, as well as physicochemical mechanisms underlying particle formation and cross-linking. Excellent results have been reported through the use of biopolymers as drug carriers for phytomedicines. Future studies on particulate systems for phytomedicine delivery should focus on both novel techniques and the optimization of current methods. Although several methods have been reported in lab-scale production, adoption of these methods to industrial scale requires a thorough understanding of the various process parameters. In addition, material quality should be strictly maintained to ensure that only pharmaceutical-grade materials are used to enforce the safety of consumers and the quality of products. Research should be directed to cutting down on the high operational cost involved with the use of certain technologies and focus on facile and inexpensive methods. There is also a need to ensure improved encapsulation efficiency with increased stability of the matrix. Moreover, methods should be adapted based on the proposed site of delivery to promote the effectiveness of carriers and the efficacy of phytomedicinal products. This calls for the development of new techniques based on existing knowledge of biopolymer interactions and structures, as well as the inherent properties of the phytomedicine intended for delivery.

References

Abdel-Mohsen, A. M., J. Frankova, Rasha M. Abdel-Rahman et al. 2020. Chitosan-glucan complex hollow fibers reinforced collagen wound dressing embedded with aloe vera. II. Multifunctional properties to promote cutaneous wound healing. Int. J. Pharm. 582: 119349.

Adnet, Thomas, Anne-Claire Groo, Céline Picard et al. 2020. Pharmacotechnical development of a nasal drug delivery composite nanosystem intended for Alzheimer's disease treatment. Pharmaceutics 12(3): 251.

Aguiar, J., Berta Nogueiro Estevinho and L. Santos. 2016. Microencapsulation of natural antioxidants for food application–The specific case of coffee antioxidants–A review. Trends Food Sci. Tech. 58: 21–39.

Ahmed, S. and S. Ikram. 2015. Chitosan and Its derivatives: A review in recent innovations. Int. J. Pharm. Sci. Res. 6(1): 14–30.

Ahuja, Munish, Sumit Kumar and Monika Yadav. 2011. Microwave-assisted synthesis and characterization of polyacrylamide grafted co-polymers of Mimosa mucilage. Polymer Bulletin 66(9): 1163–1175.

Akolade, Jubril Olayinka, Hussein Oyelola B Oloyede and Paul Chidozie Onyenekwe. 2017. Encapsulation in chitosan-based polyelectrolyte complexes enhances antidiabetic activity of curcumin. J. Functional Foods 35: 584–594.

Ali, Yousuf O., Gillian Bradley and Hui-Chen Lu. 2017. Screening with an NMNAT2-MSD platform identifies small molecules that modulate NMNAT2 levels in cortical neurons. Scientific Reports 7(1): 43846.

Anirudhan, Thayyath S., Peethambaran L. Divya and Jayachandran Nima. 2016. Synthesis and characterization of novel drug delivery system using modified chitosan based hydrogel grafted with cyclodextrin. Chem. Eng. J. 284: 1259–1269.

Araújo, Antonio, Andrêssa Galvão, Carlos Silva Filho et al. 2018. Okra mucilage and corn starch bio-based film to be applied in food. Polymer Test 71: 352–361.

Arce, A. and Z. Ustunol. 2018. Effect of microencapsulated ferrous sulfate particle size on Cheddar cheese composition and quality. J. Dairy Sci. 101(8): 6814–6822.

Arulmozhi, V., K. Pandian and S. Mirunalini. 2013. Ellagic acid encapsulated chitosan nanoparticles for drug delivery system in human oral cancer cell line (KB). Colloids Surfaces B 110: 313–320.

Ash, Dipanjana and Gopa Roy Biswas. 2019. *Hibiscus* Leaf Mucilage as Stabiliser for Pharmaceutical Disperse Systems 11(2): 6–11.

Asnani, Geet P., Jagan Bahekar and Chandrakant R. Kokare. 2018. Development of novel pH–responsive dual crosslinked hydrogel beads based on Portulaca oleracea polysaccharide-alginate-borax for colon specific delivery of 5-fluorouracil. J. Drug Delivery Sci. Tech. 48: 200–208.

Attia, Eman Zekry, Marwa Fathy Khalifa, John Refaat Fahim and Mohamed Salah Kamel. 2021. Anti-diabetic potential of mucilage from *Hippeastrum vittatum* bulbs in streptozotocin-induced diabetic rats. South African J. Botany 136: 100–104.

Babamoradi, Nastaran, Shima Yousefi and Parisa Ziarati. 2018. Optimization of ultrasound-assisted extraction of functional polysaccharides from common mullein (*Verbascum thapsus* L.) flowers. J. Food Process Eng. 41(7): e12851.

Babu, A., Q. Wang, R. Muralidharan et al. 2014. Chitosan-coated polylactic acid nanoparticle-mediated combinatorial delivery of cisplatin and siRNA/Plasmid DNA chemosensitizes cisplatin-resistant human ovarian cancer cells. Mol. Pharm. 11(8): 2720–33.

Bal, Trishna and Sabyasachi Swain. 2020. Microwave assisted synthesis of polyacrylamide grafted polymeric blend of fenugreek seed mucilage-Polyvinyl alcohol (FSM-PVA-g-PAM) and its characterizations as tissue engineered scaffold and as a drug delivery device. DARU 28 (1): 33–44.

Basu, S. and A. K. Bandyopadhyay. 2010. Development and characterization of mucoadhesive in situ nasal gel of midazolam prepared with *Ficus carica* mucilage. AAPS Pharm. Sci. Tech. 11(3): 1223–31.

Basu, Shyamoshree, Subrata Chakrabatorty and Amal K. Bandyopadhyay. 2009. Development and evaluation of a mucoadhesive nasal gel of midazolam prepared with *Linum usitatissimum* L. seed mucilage. Scientia Pharm. 77(4): 899–910.

Behbahani, Behrooz Alizadeh and Abbas Ali Imani Fooladi. 2018. Shirazi balangu (*Lallemantia royleana*) seed mucilage: Chemical composition, molecular weight, biological activity and its evaluation as edible coating on beefs. Int. J. Biol. Macromol. 114: 882–889.

Bonaccorso, Angela, Maria Rosa Gigliobianco, Rosalia Pellitteri et al. 2020. Optimization of curcumin nanocrystals as promising strategy for nose-to-brain delivery application. Pharmaceutics 12(5): 476.

Brás, T., D. Rosa, A. C. Gonçalves et al. 2020. Development of bioactive films based on chitosan and *Cynara cardunculus* leaves extracts for wound dressings. Int. J. Biol. Macromol. 163: 1707–1718.

Broughton, George, II, Jeffrey E. Janis and Christopher E. Attinger. 2006. The basic science of wound healing. Plastic Reconst. Surg. 117(7S).

Callens, C. and J. P. Remon. 2000. Evaluation of starch-maltodextrin-Carbopol 974 P mixtures for the nasal delivery of insulin in rabbits. J. Control Release 66(2-3): 215–20.

Cao, S. L., E. Chen, Q. Z. Zhang et al. 2007. A novel nasal delivery system of a Chinese traditional medicine, Radix Bupleuri, based on the concept of ion-activated in situ gel. Arch. Pharm. Res. 30(8): 1014–9.

Cao, S. L., X. W. Ren, Q. Z. Zhang et al. 2009. In situ gel based on gellan gum as new carrier for nasal administration of mometasone furoate. Int. J. Pharm. 365 (1-2): 109–15.

Carmona, Juan Carlos, Paz Robert, Cristina Vergara and Carmen Sáenz. 2021. Microparticles of yellow-orange cactus pear pulp (*Opuntia ficus-indica*) with cladode mucilage and maltodextrin as a food coloring in yogurt. LWT 138: 110672.

Castañeda-Cachay, Ana, Nilthon Gutiérrez and Raúl Siche. 2019. Optimization of the extraction process of *Linum usitatissimum* mucilage using a sequential design. Scientia Agrop 10: 19–28.

Chen, Y., X. Lin, H. Park et al. 2009. Study of *artemisinin* nanocapsules as anticancer drug delivery systems. Nanomedicine 5(3): 316–22.

Cheow, Wean Sin and Kunn Hadinoto. 2011. Factors affecting drug encapsulation and stability of lipid–polymer hybrid nanoparticles. Colloids and surfaces B: Biointerfaces 85(2): 214–220.

Chhabra, Shruti, Naveen Chhabra, Avneet Kaur et al. 2017. Wound healing concepts in clinical practice of OMFS. J. Maxillofacial. Oral Sur. 16(4): 403–423.

Chu, Xiao-Yang, Wei Huang, Yu-Li Wang et al. 2019. Improving antitumor outcomes for palliative intratumoral injection therapy through lecithin- chitosan nanoparticles loading paclitaxel- cholesterol complex. Int. J. Nanomed. 14: 689–705.

de Andrade Vieira, Érica, Maristela Alves Alcântara, Nataly Albuquerque dos Santos et al. 2021. Mucilages of cacti from Brazilian biodiversity: Extraction, physicochemical and technological properties. Food Chem. 346: 128892.

Deore, U. V. and H. S. Mahajan. 2018. Isolation and characterization of natural polysaccharide from *Cassia Obtustifolia* seed mucilage as film forming material for drug delivery. Int. J. Biol. Macromol. 115: 1071–1078.

Deore, Ujwaldip V. and Hitendra S. Mahajan. 2021. Isolation and structural characterization of mucilaginous polysaccharides obtained from the seeds of *Cassia uniflora* for industrial application. Food Chemistry 351: 129262.

Dharmendra Kumar, Bhanupriya and Gupta SK. 2013. Development of domperidone nasal gel using natural mucoadhesive agnet obtained from the fruits of *Dellinia Indica* L. Pharmatutor: Article no. 1745.

Dick, Melina, Lucas Dal Magro, Rafael Costa Rodrigues et al. 2019. Valorization of *Opuntia monacantha* (Willd.) Haw. cladodes to obtain a mucilage with hydrocolloid features: Physicochemical and functional performance. Int. J. Biol. Macromol. 123: 900–909.

Ding, Huihuang H., Keying Qian, H. Douglas Goff et al. 2018. Structural and conformational characterization of arabinoxylans from flaxseed mucilage. Food Chemistry 254: 266–271.

Ditta, Lorena Anna, Estella Rao, Fiorenza Provenzano et al. 2020. Agarose/κ-carrageenan-based hydrogel film enriched with natural plant extracts for the treatment of cutaneous wounds. Int. J. Biol. Macromol. 164: 2818–2830.

El-Ghoul, Yassine and Fahad M. Alminderej. 2021. Bioactive and superabsorbent cellulosic dressing grafted alginate and *Carthamus tinctorius* polysaccharide extract for the treatment of chronic wounds. Textile Res. J. 91(3-4): 235–248.

Farahani, Hossein, Aboulfazl Barati, Mohammad Arjomandzadegan et al. 2020. Nanofibrous cellulose acetate/gelatin wound dressing endowed with antibacterial and healing efficacy using nanoemulsion of *Zataria multiflora*. Int. J. Biol. Macromol. 162: 762–773.

Farzaei, Mohammad Hosein, Roodabeh Bahramsoltani, Zahra Abbasabadi et al. 2015. A comprehensive review on phytochemical and pharmacological aspects of E *laeagnus angustifolia* L. J. Pharm. Pharmacol. 67(11): 1467–1480.

Fathi, Marziyeh, Parham Sahandi Zangabad, Ayuob Aghanejad et al. 2017. Folate-conjugated thermosensitive O-maleoyl modified chitosan micellar nanoparticles for targeted delivery of erlotinib. Carbohydrate Polym. 172: 130–141.

Feng, Bohua, Muhammad Aqeel Ashraf and Liufen Peng. 2016. Characterization of particle shape, zeta potential, loading efficiency and outdoor stability for chitosan-ricinoleic acid loaded with rotenone. Open Life Sciences 11(1): 380–386.

Gheybi, Nesa, Mir Khalil Pirouzifard and Hadi Almasi. 2021. *Ornithogalum cuspidatum* mucilage as a new source of plant-based polysaccharide: Physicochemical and rheological characterization. J. Food Meas Charact. 15: 2184–2201.

Giri, Tapan Kumar, Sanjeet Pure and Dulal Krishna Tripathi. 2015. Synthesis of graft copolymers of acrylamide for locust bean gum using microwave energy: swelling behavior, flocculation characteristics and acute toxicity study. Polímeros 25: 168–174.

Gopalakrishnan, Lalitha, Lakshmi Narashimhan Ramana, Swaminathan Sethuraman et al. 2014. Ellagic acid encapsulated chitosan nanoparticles as anti-hemorrhagic agent. Carbohydrate polymers 111: 215–221.

Goyal, C., M. Ahuja and S. K. Sharma. 2011. Preparation and evaluation of anti-inflammatory activity of gugulipid-loaded proniosomal gel. Acta Pol. Pharm. 68(1): 147–50.

Grundman, M., M. Grundman and P. Delaney. 2002. Antioxidant strategies for Alzheimer's disease. Proc. Nutr. Soc. 61(2): 191–202.

Guzelgulgen, Meltem, Dilce Ozkendir-Inanc, Umit Hakan Yildiz et al. 2021. Glucuronoxylan-based quince seed hydrogel: A promising scaffold for tissue engineering applications. International Journal of Biological Macromolecules 180: 729–738.

Han, Xinxu, Peipei Huo, Zhongfeng Ding et al. 2019. Preparation of lutein-loaded pva/sodium alginate nanofibers and investigation of its release behavior. Pharmaceutics 11(9): 449.

Hansen, K., G. Kim, K. G. Desai et al. 2015. Feasibility investigation of cellulose polymers for mucoadhesive nasal drug delivery applications. Mol. Pharm. 12(8): 2732–41.

HaqAsif, Afzal, Rahul R. Karnakar, Nagaraja Sreeharsha et al. 2021. pH and salt responsive hydrogel based on guar gum as a renewable material for delivery of curcumin: a natural anti-cancer drug. J. Polym. Environ. 29: 1978–1989.

Hernández-Carranza, Paola, Mariela Rivadeneyra-Mata, María E. Ramos-Cassellis et al. 2019. Characterization of red prickly pear peel (*Opuntia ficus-indica* L.) and its mucilage obtained by traditional and novel methodologies. J. Food Measurem. Charact. 13(2): 1111–1119.

Huang, W. T., M. Larsson, Y. J. Wang et al. 2015. Demethoxycurcumin-carrying chitosan-antibody core-shell nanoparticles with multitherapeutic efficacy toward malignant A549 lung tumor: from *in vitro* characterization to *in vivo* evaluation. Mol. Pharm. 12(4): 1242–9.

Huang, Weifeng, Junwei He, Muhammad Farrukh Nisar et al. 2018. Phytochemical and pharmacological properties of *Chaenomeles* speciosa: An edible medicinal chinese mugua. Evidence-Based Complem Alternative Med Article ID 9591845.

Ige, Oladeji O., Lasisi E. Umoru and Sunday Aribo. 2012. Natural Products: A Minefield of Biomaterials. ISRN Materials Science 2012: 983062.

Ilhan, E., S. Cesur, E. Guler et al. 2020. Development of *Satureja cuneifolia*-loaded sodium alginate/ polyethylene glycol scaffolds produced by 3D-printing technology as a diabetic wound dressing material. Int. J. Biol. Macromol. 161: 1040–1054.

Izadyari Aghmiuni, Azadeh, Saeed Heidari Keshel, Farshid Sefat et al. 2020. Quince seed mucilage-based scaffold as a smart biological substrate to mimic mechanobiological behavior of skin and promote fibroblasts proliferation and h-ASCs differentiation into keratinocytes. Int. J. Biol. Macromol. 142: 668–679.

Jamil, Maryam, Bushra Mirza, Abida Yasmeen et al. 2012. Pharmacological activities of selected plant species and their phytochemical analysis. J. Medicinal Plants Res. 6(37): 5013–5022.

Jyothi, N. Venkata Naga, P. Muthu Prasanna, Suhas Narayan Sakarkar et al. 2010. Microencapsulation techniques, factors influencing encapsulation efficiency. J. Microencapsulation 27(3): 187–197.

Karkada, G., K. B. Shenoy, H. Halahalli et al. 2012. Nardostachys jatamansi extract prevents chronic restraint stress-induced learning and memory deficits in a radial arm maze task. J. Nat. Sci. Biol. Med. 3(2): 125–32.

Kaur, Loveleenpreet and G. Gupta. 2020. Gum karaya-g-poly (acrylamide): microwave assisted synthesis, optimisation and characterisation. Int. J. Applied Pharmaceutics: 143–152.

Keshani-Dokht, Shaghayegh, Zahra Emam-Djomeh, Mohammad-Saeid Yarmand et al. 2018. Extraction, chemical composition, rheological behavior, antioxidant activity and functional properties of *Cordia myxa* mucilage. Int. J. Biol. Macromol. 118: 485–493.

Khan, N., D. J. Bharali, V. M. Adhami et al. 2014. Oral administration of naturally occurring chitosan-based nanoformulated green tea polyphenol EGCG effectively inhibits prostate cancer cell growth in a xenograft model. Carcinogenesis 35(2): 415–23.

Kharat, Zahra, Minoo Sadri and Mahboubeh Kabiri. 2021. Herbal extract loaded chitosan/peo nanocomposites as antibacterial coatings of orthopaedic implants. Fibers Polym. 22: 989–999.

Kim, S., S. K. Nishimoto, J. D. Bumgardner et al. 2010. A chitosan/beta-glycerophosphate thermo-sensitive gel for the delivery of ellagic acid for the treatment of brain cancer. Biomaterials 31(14): 4157–66.

Kirsner, R. S. and W. H. Eaglstein. 1993. The wound healing process. Dermatol. Clin. 11(4): 629–40.

Klemm, Dieter, Brigitte Heublein, Hans-Peter Fink et al. 2005. Cellulose: fascinating biopolymer and sustainable raw material. Angewandte Chemie. 44(22): 3358–3393.

Krishna, Anjali and Shan Mohanan. 2018. Formulation and evaluation of liquid oral suspension of paracetamol using newly isolated and characterized hygrophila spinosa seed mucilage as suspending agent. Asian J. Pharm. Clinical Res. 11: 437.

Krishnasis, C., S. Arun and R. Sundaram. 2016. Nano-technology in herbal medicines: A review. Int. J. Herb. Med. 4: 21–7.

Kumar, Deepak, Jyoti Pandey and Pramendra Kumar. 2018. Microwave assisted synthesis of binary grafted psyllium and its utility in anticancer formulation. Carbohydrate Polym. 179: 408–414.

Kunjumon, Renju, Gayathri Viswanathan, Devi Velayudhan Jayasree et al. 2021. Madecassoside encapsulated in alginate chitosan nanoparticles exerts anti-excitotoxicity effects in pilocarpine-induced seizure. Phytomedicine Plus 1(1): 100004.

Kuotsu, Ketousetuo and Amal Bandyopadhyay. 2007. Development of oxytocin nasal gel using natural mucoadhesive agent obtained from the fruits of *Dellinia indica*. L. Asia Life Sci. 33.

Lee, Hye-Bin, Seung-U. Son, Jang-Eun Lee et al. 2021. Characterization, prebiotic and immune-enhancing activities of rhamnogalacturonan-I-rich polysaccharide fraction from molokhia leaves. Int. J. Biol. Macromol. 175: 443–450.

Li, Ningbo, Guangyan Qi, Xiuzhi Susan Sun et al. 2016. Characterization of gum isolated from *Camelina seed*. Industrial Crops Products 83: 268–274.

Li, Qian, Hong-Xia Che, Cheng-Cheng Wang et al. 2019. Cerebrosides from sea cucumber improved Aβ1–42-induced cognitive deficiency in a rat model of alzheimer's disease. Mol. Nut. Food Res. 63(5): 1800707.

Lin, Y. C., J. Y. Kuo, C. C. Hsu et al. 2013. Optimizing manufacture of liposomal berberine with evaluation of its antihepatoma effects in a murine xenograft model. Int. J. Pharm. 441(1-2): 381–8.

Liu, Q. F., H. Jeong, J. H. Lee et al. 2016. *Coriandrum sativum* Suppresses Aβ42-induced ROS increases, glial cell proliferation, and ERK activation. Am. J. Chin. Med. 44(7): 1325–1347.

Lopes, Poliana Pollizello, Eduardo Hiromitsu Tanabe and Daniel Assumpção Bertuol. 2020. Chitosan as biomaterial in drug delivery and tissue engineering. pp. 408–422. *In*: Gopi, S., S. Thomas and A. Pius (Eds). Handbook of Chitin and Chitosan, Elsevier Inc, New York, USA.

Loveleenpreet Kaur and Gupta G D. 2020. Gum colocasia-g-polyacrylamide: Microwave-assisted synthesis and characterization. Int. J. Res. Pharm. Sci. 11(4): 6698–6706.

Luckanagul, Jittima Amie, Chutamart Pitakchatwong, Pahweenvaj Ratnatilaka Na Bhuket et al. 2018. Chitosan-based polymer hybrids for thermo-responsive nanogel delivery of curcumin. Carbohydrate Polym. 181: 1119–1127.

Luo, Qingying, Min Wu, Yanan Sun et al. 2019. Optimizing the extraction and encapsulation of mucilage from *Brasenia Schreberi*. Polymers 11(5): 822.

Ma, Fanyi, Dengke Wang, Yun Zhang et al. 2018. Characterisation of the mucilage polysaccharides from *Dioscorea opposita* Thunb. with enzymatic hydrolysis. Food Chem. 245: 13–21.

Ma, Fanyi, Ruijiao Wang, Xiaojing Li et al. 2020. Physical properties of mucilage polysaccharides from *Dioscorea opposita* Thunb. Food Chem. 311: 126039.

Mady, Fatma M. and Mohamed A. Shaker. 2017. Enhanced anticancer activity and oral bioavailability of ellagic acid through encapsulation in biodegradable polymeric nanoparticles. Int. J. Nanomed. 12: 7405.

Mahrous, Rahma, Doaa Ghareeb, Hoda Sherif et al. 2017. The protective effect of Egyptian *Withania somnifera* against alzeheimer's. Med Aromatic Plants 6: 285.

Maiti, Kuntal, Kakali Mukherjee, Arunava Gantait et al. 2007. Curcumin–phospholipid complex: Preparation, therapeutic evaluation and pharmacokinetic study in rats. Int. J. Pharm. 330(1): 155–163.

Manhivi, Vimbainashe E., Sonja Venter, Eric O. Amonsou et al. 2018. Composition, thermal and rheological properties of polysaccharides from amadumbe (*Colocasia esculenta*) and cactus (*Opuntia* spp.). Carbohydrate Polym. 195: 163–169.

Marinas, Ioana Cristina, Eliza Oprea, Elisabeta-Irina Geana et al. 2021. Valorization of *Gleditsia triacanthos* invasive plant cellulose microfibers and phenolic compounds for obtaining multi-functional wound dressings with antimicrobial and antioxidant properties. Int. J. Mol. Sci. 22(1): 33.

Maru, Sandip G. and S. Singh 2013. Physicochemical and mucoadhesive strength characterization of natural polymer obtained from leaves of *Aloe vera* 2(3): 303–308.

Matthews, Kevin A., Wei Xu, Anne H. Gaglioti et al. 2019. Racial and ethnic estimates of Alzheimer's disease and related dementias in the United States (2015–2060) in adults aged ≥ 65 years. Alzheimer's Dementia 15(1): 17–24.

Medina-Torres, L., D. M. Núñez-Ramírez, F. Calderas et al. 2019. Microencapsulation of gallic acid by spray drying with aloe vera mucilage (*Aloe barbadensis* miller) as wall material. Industrial Crops Products 138: 111461.

Meng, Qingqing, Aiping Wang, Hongchen Hua et al. 2018. Intranasal delivery of Huperzine A to the brain using lactoferrin-conjugated N-trimethylated chitosan surface-modified PLGA nanoparticles for treatment of Alzheimer's disease. Int. J. Nanomed. 13: 705.

Mijinyawa, Abubakar Hamisu, Geeta Durga and Anuradha Mishra. 2018. Isolation, characterization, and microwave assisted surface modification of *Colocasia esculenta* (L.) Schott mucilage by grafting polylactide. Int. J. Biol. Macromol. 119: 1090–1097.

Min, K. H., K. Park, Y. S. Kim et al. 2008. Hydrophobically modified glycol chitosan nanoparticles-encapsulated camptothecin enhance the drug stability and tumor targeting in cancer therapy. J. Control Release 127(3): 208–18.

Mishra, Anuradha, James H. Clark and Sunita Pal. 2008. Modification of Okra mucilage with acrylamide: Synthesis, characterization and swelling behavior. Carbohydrate Polym. 72(4): 608–615.

Mishra, Anuradha and Annu Vij Malhotra. 2012. Graft copolymers of xyloglucan and methyl methacrylate. Carbohydrate Polym. 87(3): 1899–1904.

Mishra, Anuradha and Sunita Pal. 2007. Polyacrylonitrile-grafted Okra mucilage: A renewable reservoir to polymeric materials. Carbohydrate Polym. 68(1): 95–100.

Mishra, Anuradha, Rajani Srinivasan, Malvika Bajpai et al. 2004. Use of polyacrylamide-grafted Plantago psyllium mucilage as a flocculant for treatment of textile wastewater. Colloid Polym. Sci. 282(7): 722–727.

Mishra, Anuradha, Rajani Srinivasan and Ramkumar Gupta. 2003. P. psyllium-g-polyacrylonitrile: synthesis and characterization. Colloid Polym. Sci. 281(2): 187–189.

Möller, Karin, Beth Macaulay and Thomas Bein. 2021. Curcumin encapsulated in crosslinked cyclodextrin nanoparticles enables immediate inhibition of cell growth and efficient killing of cancer cells. Nanomaterials 11(2): 489.

Monge Neto, André Álvares, Larissa Fonseca Tomazini, Amanda Gouveia Mizuta et al. 2021. Direct microencapsulation of an annatto extract by precipitation of *Psyllium* husk mucilage polysaccharides. Food Hydrocol. 112: 106333.

Moore, A. H. and M. K. O'Banion. 2002. Neuroinflammation and anti-inflammatory therapy for Alzheimer's disease. Adv. Drug Deliv. Rev. 54(12): 1627–56.

Moradi, Seyed Zachariah, Saeideh Momtaz, Zahra Bayrami et al. 2020. Nanoformulations of herbal extracts in treatment of neurodegenerative disorders. Frontiers Bioeng Biotech 8.

Moreno, María Alejandra, Laura Gómez Gómez-Mascaraque, Myriam Arias et al. 2018. Electrosprayed chitosan microcapsules as delivery vehicles for vaginal phytoformulations. Carbohydrate polym. 201: 425–437.

Mouro, C., C. P. Dunne and I. C. Gouveia. 2020. Designing new antibacterial wound dressings: development of a dual layer cotton material coated with poly(vinyl alcohol) chitosan nanofibers incorporating *Agrimonia eupatoria* L. extract. Molecules 26(1).

Mouro, Cláudia, Ana P. Gomes and Isabel C. Gouveia. 2021. Double-layer PLLA/PEO_Chitosan nanofibrous mats containing *Hypericum perforatum* L. as an effective approach for wound treatment. Polym. Advanced Tech. 32(4): 1493–1506.

Mukerjee, A. and J. K. Vishwanatha. 2009. Formulation, characterization and evaluation of curcumin-loaded PLGA nanospheres for cancer therapy. Anticancer Res. 29(10): 3867–75.

Mukherjee, Shubham, Dhruti Avlani, Sutapa Biswas Majee et al. 2019. Physicochemical and pharmaceutical characterisation of mucilage from sweet basil seed. Int. J. Applied Pharm. 11(special is): 12–17.

Mulchand, A. Shende and Rajendra P. Marathe. 2018. Pharmacokinetic studies of gastroretentive mucoadhesive matrices for diltiazem hydrochloride using some natural polysaccharides. Research J. Pharm. Tech. 11(4): 475–485.

Nazari, Meisam, Sophie Riebeling, Callum C. Banfield et al. 2020. Mucilage polysaccharide composition and exudation in maize from contrasting climatic regions. Frontiers Plant Sci. 11(1968).

Nie, L., J. Xia, H. Li et al. 2017. Ginsenoside Rg1 ameliorates behavioral abnormalities and modulates the hippocampal proteomic change in triple transgenic mice of alzheimer's disease. Oxid. Med. Cell Longev. 2017: 6473506.

Nillert, Nutchareeporn, Wanassanun Pannangrong, Jariya Umka Welbat et al. 2017. Neuroprotective effects of aged garlic extract on cognitive dysfunction and neuroinflammation induced by β-amyloid in rats. Nutrients 9(1): 24.

Nwabor, Ozioma Forstinus, Sudarshan Singh, Supakit Paosen et al. 2020. Enhancement of food shelf life with polyvinyl alcohol-chitosan nanocomposite films from bioactive Eucalyptus leaf extracts. Food Biosci. 36: 100609.

Oh, Gun-Woo, Se-Chang Kim, Tae-Hee Kim et al. 2021. Characterization of an oxidized alginate-gelatin hydrogel incorporating a COS-salicylic acid conjugate for wound healing. Carbohydrate Polym. 252: 117145.

Oliveira, Natália Leite, Adrise Aparecida Rodrigues, Isabelle Cristina Oliveira Neves et al. 2019. Development and characterization of biodegradable films based on *Pereskia aculeata* Miller mucilage. Industrial Crops Prod. 130: 499–510.

Olotu, Fisayo, Clement Agoni, Opeyemi Soremekun et al. 2020. An Update on the pharmacological usage of curcumin: has it failed in the drug discovery pipeline? Cell Biochem. Biophysics. 78: 267–289.

Ozsoy, Yildiz, Sevgi Gungor and Erdal Cevher. 2009. Nasal delivery of high molecular weight drugs. Molecules 14(9): 3754–3779.

Park, G. H., M. S. Kang, J. C. Knowles et al. 2016. Synthesis, characterization, and biocompatible properties of alanine-grafted chitosan copolymers. J. Biomater. Appl. 30(9): 1350–61.

Pascoal, Diego R. C., Eudes S. Velozo, Mara E. M. Braga et al. 2020. Bioactive compounds of *Copaifera* sp. impregnated into three-dimensional gelatin dressings. Drug Delivery Translational Res. 10(6): 1537–1551.

Patil, S. B. and K. K. Sawant. 2009. Development, optimization and *in vitro* evaluation of alginate mucoadhesive microspheres of carvedilol for nasal delivery. J. Microencapsul. 26(5): 432–43.

Patra, Jayanta Kumar, Gitishree Das, Leonardo Fernandes Fraceto et al. 2018. Nano based drug delivery systems: recent developments and future prospects. J. Nanobiotech. 16(1): 1–33.

Paul, Saili, Soumya Sundar Bhattacharyya, Naoual Boujedaini et al. 2010. Anticancer potentials of root extract of *Polygala senega* and its PLGA nanoparticles-encapsulated form. Evidence-Based Complementary and Alternative Medicine Article ID 517204.

Pawar, H. A. and P. Jadhav. 2015. Isolation, characterization and investigation of Cordia dichotoma fruit polysaccharide as a herbal excipient. Int. J. Biol. Macromol. 72: 1228–36.

Pereira, Gustavo Araujo, Eric Keven Silva, Nayara Macêdo Peixoto Araujo et al. 2019. Mutamba seed mucilage as a novel emulsifier: Stabilization mechanisms, kinetic stability and volatile compounds retention. Food Hydrocol. 97: 105190.

Pisoschi, Aurelia Magdalena, Aneta Pop, Carmen Cimpeanu et al. 2018. Nanoencapsulation techniques for compounds and products with antioxidant and antimicrobial activity-A critical view. Europ. J. Med. Chem. 157: 1326–1345.

Prabowo, Imam, Iskandarsyah and Rina Adriany. 2021. Characterization and concentration optimization of *Hibiscus rosa-sinensis* l. mucilage powder as superdisintegrant. Int. J. Applied Pharm. 13(2): 49–52.

Punia, Sneh and Sanju Bala Dhull. 2019. Chia seed (*Salvia hispanica* L.) mucilage (a heteropolysaccharide): Functional, thermal, rheological behaviour and its utilization. Int. J. Biol. Macromol. 140: 1084–1090.

Rahman, H. S., H. H. Othman, N. I. Hammadi et al. 2020. novel drug delivery systems for loading of natural plant extracts and their biomedical applications. Int. J. Nanomed. 15: 2439–2483.

Ramaswamy, S., L. P. Dwarampudi, M. Kadiyala et al. 2017. Formulation and characterization of chitosan encapsulated phytoconstituents of curcumin and rutin nanoparticles. Int. J. Biol. Macromol. 104(Pt B): 1807–1812.

Rao, Monica R. P., Laxmi Babrekar, Vaishali S. Kharpude et al. 2017. Synthesis and characterization of psyllium seed mucilage grafted with N,N-methylene bisacrylamide. Int. J. Biol. Macromol. 103: 338–346.

Rassu, Giovanna, Maria Nieddu, Paolo Bosi et al. 2014. Encapsulation and modified-release of thymol from oral microparticles as adjuvant or substitute to current medications. Phytomedicine 21(12): 1627–1632.

Hoffman, A. S. 2004. Application of smart polymer as biomaterial. pp. 107–115. *In*. Ratner, B. D., A. S. Hoffman, F. J. Schoen et al. (Eds.). Biomaterials Science: An introduction to materials in medicine. Amsterdam: Elsevier Acadmic Press.

Reiserer, R. S., F. E. Harrison, D. C. Syverud et al. 2007. Impaired spatial learning in the APPSwe + PSEN1DeltaE9 bigenic mouse model of Alzheimer's disease. Genes Brain Behav. 6(1): 54–65.

Rezaei, Atefe, Jaleh Varshosaz, Mehrafarin Fesharaki et al. 2019. Improving the solubility and *in vitro* cytotoxicity (anticancer activity) of ferulic acid by loading it into cyclodextrin nanosponges. Int. J. Nanomed. 14: 4589.

Rezagholi, Fatemeh, Seyed Mohammad Bagher Hashemi, Aliakbar Gholamhosseinpour et al. 2019. Characterizations and rheological study of the purified polysaccharide extracted from quince seeds. J. Sci. Food Agriculture 99(1): 143–151.

Ribeiro, Ana S., Sofia M. Costa, Diana P. Ferreira et al. 2021. Chitosan/nanocellulose electrospun fibers with enhanced antibacterial and antifungal activity for wound dressing applications. Reactive Funct. Polym. 159: 104808.

Rtibi, Kaïs, Imen Hammami, Slimen Selmi et al. 2017. Phytochemical properties and pharmacological effects of *Quercus ilex* L. aqueous extract on gastrointestinal physiological parameters *in vitro* and *in vivo*. Biomedicine Pharmacoth 94: 787–793.

Sabando, C., W. Ide, M. Rodríguez-Díaz et al. 2020. A novel hydrocolloid film based on pectin, starch and gunnera tinctoria and ugni molinae plant extracts for wound dressing applications. Curr. Top. Med. Chem. 20(4): 280–292.

Sadeghi-Varkani, Atina, Zahra Emam-Djomeh and Gholamreza Askari. 2018. Morphology and physicochemical properties of a novel *Lallemantia iberica* mucilage/titanium dioxide bionanocomposite. Polymer Testing 67: 12–21.

Salazar Vega, Ine Mayday, Patricia Quintana Owen and Maira Rubi Segura Campos. 2020. Physicochemical, thermal, mechanical, optical, and barrier characterization of chia (*Salvia hispanica* L.) mucilage-protein concentrate biodegradable films. J. Food Sci. 85(4): 892–902.

Sen, Gautam, Ram Prakash Singh and Sagar Pal. 2010. Microwave-initiated synthesis of polyacrylamide grafted sodium alginate: Synthesis and characterization. Journal of Applied Polymer Sci .115(1): 63–71.

Sendker, J. and H. Sheridan. 2017. History and current status of herbal medicines. pp. 11–23. *In:* D. P. Pelkonen O., P. Vuorela and H. Vuorela (Eds.). Toxicology of herbal products. Switzerland: Springer, Cham.

Seyfried, Thomas N. and Leanne C. Huysentruyt. 2013. On the origin of cancer metastasis. Critical Reviews Oncog. 18(1-2): 43–73.

Shalaby, K., E. M. Mostafa, A. Musa et al. 2021. Enhanced full-thickness wound healing via *Sophora gibbosa* extract delivery based on a chitosan/gelatin dressing incorporating microemulsion. Drug Dev. Ind. Pharm. 47(2): 215–224.

Sharma, Chhavi and Nishi K. Bhardwaj. 2020. Fabrication of natural-origin antibacterial nanocellulose films using bio-extracts for potential use in biomedical industry. International Journal of Biological Macromolecules 145: 914–925.

Silva, Sérgio Henrique, Isabelle Cristina Oliveira Neves, Natália Leite Oliveira et al. 2019. Extraction processes and characterization of the mucilage obtained from green fruits of *Pereskia aculeata* Miller. Industrial Crops Prod. 140: 111716.

Sims, Ian M., Alan M. Smith, Gordon A. Morris et al. 2018. Structural and rheological studies of a polysaccharide mucilage from lacebark leaves (*Hoheria populnea* A. Cunn.). Int. J. Biol. Macromol. 111: 839–847.

Singh, Manisha, Surya Singh and Rachana R. 2017. Development, characterization and cytotoxicity evaluation of *Gingko biloba* extract (EGB761) loaded microemulsion for intra-nasal application. J. Applied Pharm. Sci. 7: 024–034.

Singh, Sudarshan and Sunil Bothara. 2014a. Morphological, physico-chemical and structural characterization of mucilage isolated from the seeds of *Buchanania lanzan* Spreng. Int. J. Health Allied Sci. 3: 33.

Singh, Sudarshan and Sunil B. Bothara. 2014b. Physico-chemical and structural characterization of mucilage isolated from seeds of *Diospyros melonoxylon* Roxb. Brazilian J. Pharm. Sci. 50: 713–725.

Singh, Sudarshan, Ozioma F. Nwabor, Julalak C. Ontong et al. 2020. Characterization of a novel, co-processed bio-based polymer, and its effect on mucoadhesive strength. Int. J. Biol. Macromol. 145: 865–875.

Singh, Sudarshan, Ozioma F. Nwabor, Julalak C. Ontong and Supayang P. Voravuthikunchai. 2020. Characterization and assessment of compression and compactibility of novel spray-dried, co-processed bio-based polymer. J. Drug Delivery Sci. Tech. 56: 101526.

Singh, Sudarshan, Ozioma Forstinus Nwabor, Dwi Marlina Syukri et al. 2021. Chitosan-poly(vinyl alcohol) intelligent films fortified with anthocyanins isolated from *Clitoria ternatea* and *Carissa carandas* for monitoring beverage freshness. Int. J. Biol. Macromol. 182: 1015–1025.

Sivakumar, Nandhineeswari and Kavitha Karuppaiyan. 2020. Extraction and characterization of water-soluble polysaccharides from Tamarindus indica and Pithecellobium dulce seeds. J. Food Process Eng. 43(3): e13251.

Song, Ka-Young and Young-Soon Kim. 2019. Effect of mucilage extracted from Basil (*Ocimum basilicum* L.) seeds on physicochemical and rheological properties in low-fat milk protein gel. J. Food Processing Pres 43(11): e14191.

Song, Y. M., Q. N. Ping and Z. H. Wu. 2005. Preparation of silybin nanoemulsion and its pharmacokinetics in rabbits. J. China Pharm. Univ. 36: 427–431.

Subash, S., M. M. Essa, N. Braidy et al. 2016. Consumption of fig fruits grown in Oman can improve memory, anxiety, and learning skills in a transgenic mice model of Alzheimer's disease. Nutr. Neurosci. 19(10): 475–483.

Subhan, Hanif, Sultan Alam, Luqman Ali Shah et al. 2021. Sodium alginate grafted poly(N-vinyl formamide-co-acrylic acid)-bentonite clay hybrid hydrogel for sorptive removal of methylene green from wastewater. Colloids and Surfaces 611: 125853.

Sudarshan, Singh and Bothara Sunil B. 2015. *In vivo* mucoadhesive strength appraisal of gum *Manilkara zapota*. Brazilian Journal of Pharmaceutical Sciences 51: 689–698.

Sudarshan, Singh, R. Dodiya Tanvi, Dodiya Rajesh et al. 2021. *In vivo, ex vivo* and *in vitro* mucoadhesive strength assessment of potential pharmaceutical bio-resource polymer from *Diospyros melonoxylon* Roxb seeds. Int. J. Pharm. Sci. Nanotech. 14(1).

Sudarshan, Singh, R. Dodiya Tanvi, Singh Sangeeta et al. 2020. Bio-based polymer isolated from Seeds of *Buchanania lanzan* Spreng with potential use as pharmaceutical mucoadhesive excipient. Int. J. Pharm. Sci. Nanotech. 13(4).

Syad, A. N. and K. P. Devi. 2015. Assessment of anti-amyloidogenic activity of marine red alga *G. acerosa* against Alzheimer's beta-amyloid peptide 25–35. Neurol. Res. 37(1): 14–22.

Thangapazham, Rajesh L., Shashwat Sharad and Radha K. Maheshwari. 2016. Phytochemicals in wound healing. Advances in Wound Care 5(5): 230–241.

Thessrimuang, Naruenart and Jutarat Prachayawarakorn. 2019. Development, modification and characterization of new biodegradable film from basil seed (*Ocimum basilicum* L.) mucilage. J. Sci. Food Agriculture 99(12): 5508–5515.

Tian, X., H. Yin, S. Zhang et al. 2014. Bufalin loaded biotinylated chitosan nanoparticles: an efficient drug delivery system for targeted chemotherapy against breast carcinoma. Eur. J. Pharm. Biopharm. 87(3): 445–53.

Tsai, Wen-Hsuan, Kun-Hua Yu, Yi-Cheng Huang et al. 2018. EGFR-targeted photodynamic therapy by curcumin-encapsulated chitosan/TPP nanoparticles. Int. J. Nanomed. 13: 903–916.

Tung, N. T., C. S. Tran, H. A. Nguyen et al. 2019. Formulation and biopharmaceutical evaluation of supersaturatable self-nanoemulsifying drug delivery systems containing silymarin. Int. J. Pharm. 555: 63–76.

Vicentini, F. T., T. R. Simi, J. O. Del Ciampo et al. 2008. Quercetin in w/o microemulsion: in vitro and in vivo skin penetration and efficacy against UVB-induced skin damages evaluated *in vivo*. Eur. J. Pharm. Biopharm. 69(3): 948–57.

Villemin, Elise, Yih Ching Ong, Christophe M. Thomas et al. 2019. Polymer encapsulation of ruthenium complexes for biological and medicinal applications. Nature Reviews Chem. 3(4): 261–282.

Wang, Lijun, Hongguang Fan, Jingchun He et al. 2018. Protective effects of omega-3 fatty acids against Alzheimer's disease in rat brain endothelial cells. Brain Behavior. 8(11): e01037.

Wang, Yaping, Junmin Qian, Ming Yang et al. 2019. Doxorubicin/cisplatin co-loaded hyaluronic acid/chitosan-based nanoparticles for *in vitro* synergistic combination chemotherapy of breast cancer. Carbohydrate Polym. 225: 115206.

Xiao, L., Y. H. Zhang, J. C. Xu et al. 2008. Preparation of floating rutin-alginate-chitosan microcapsule. Chinese Traditional Herbal Drugs 39: 209–212.

Yabroff, K. Robin, Jennifer Lund, Deanna Kepka et al. 2011. Economic burden of cancer in the United States: estimates, projections, and future research. Cancer Epidemiology, Biomarkers Prevention 20(10): 2006–2014.

Yağmur, Nuray and Saliha Şahin. 2020. Encapsulation of ellagic acid from pomegranate peels in microalgae optimized by response surface methodology and an investigation of its controlled released under simulated gastrointestinal studies. J. Food Science 85(4): 998–1006.

Yahuafai, J., T. Asai, N. Oku et al. 2018. Anticancer Efficacy of the combination of berberine and pegylated liposomal doxorubicin in meth a sarcoma-bearing mice. Biol. Pharm. Bull. 41(7): 1103–1106.

Yang, Bingyan, Yixin Dong, Fei Wang et al. 2020. Nanoformulations to enhance the bioavailability and physiological functions of polyphenols. Molecules 25(20): 4613.

Yerer, Mukerrem Betul, Mehmet Kaan Tiryaki and Eren Demirpolat. 2017. GSK-3beta inhibitory effects of 6-gingerol and 6-shogaol help to the recovery of SHSY-5Y cells after amyloid beta1–42 oligomer or aggregate toxicity. J. Cellular Biotech. 2: 145–157.

Yu, Long, Gleb E. Yakubov, Elliot P. Gilbert et al. 2019. Multi-scale assembly of hydrogels formed by highly branched arabinoxylans from Plantago ovata seed mucilage studied by USANS/SANS and rheology. Carbohydrate Polym. 207: 333–342.

Zahrani, K., F. Imansari, Tania Surya Utami et al. 2017. Release profile of andrographis paniculata leaf extract nanocapsule as α-glucosidase inhibitors. IOP Conference Series: Materials Science and Engineering 214; 012020.

13

In Vitro Toxicity Assessment Techniques for Biomaterials

Thrupthi, Swetha Sadhan, Varsha Kaushik, Anwesha Patel and *Shristi Ram**

1. Introduction

A biomaterial is created specifically to interact with biological systems for therapeutic or diagnostic purposes in medicine. The interdisciplinary field of biomaterials, which has recently been affected by tissue engineering and material science, includes medicine, biology, physics, and chemistry. Biomaterials have been around for about 50 years. In the development of biomaterials, it is imperative to ensure their safety for human usage. To access the biological processes of biomaterials when they come into touch with the body, various *in vitro* tests are performed (Inayat-hussain et al. 2009). The use of medical devices has been employed extensively in the past many decades, but their use has been increasingly observed for the past 40 years. In this regard, the ever-growing population and demands to enhance a quality of life using various technologies have also increased, and the use of these biomedical applications has been observed in various fields of science and technology, including cardiovascular surgery, general surgery, wound healing, etc. (Eloy 2012).

Toxicology is crucial in the development of biomaterials to guarantee the security of the materials chosen for usage in devices (Inayat-hussain et al. 2009). *In vitro* tests are carried out to encourage biological responses to substances when they come into contact with the body. It is widely divided into three categories: cytotoxicity, method of cell death, and genotoxicity assessment of biomaterials, all of which provide

Department of Microbiology, School of Basic and Applied Sciences, Dayananda Sagar University, City campus, Kudlu Gate, Bengaluru, 560068, Karnataka, India.

* Corresponding author: shristi-sbas@dsu.edu.in

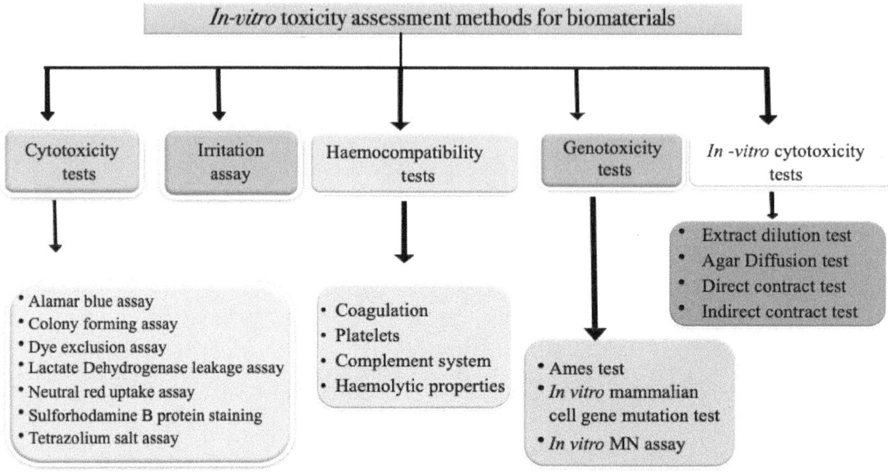

Figure 1: Summary of various methods employed for *in vitro* evaluation of biomaterials.

significant information on *in vitro* toxicity for cell death assessment. The most often utilized indicator of the *in vitro* toxicity of biomaterials is the suppression of cell proliferation (Wieslander et al. 1993). An overview of the *in vitro* tests discussed in this chapter is represented in Figure 1.

2. Pre-Testing Biomaterial Characterization

The first phase of an intricate procedure to authenticate a material or device designed for clinical purposes includes performing standardized regulatory assessments in laboratory settings and live organisms. These assessments adhere to established programs that evaluate biocompatibility and safety, aiming to acquire the essential approval for human applications.

3. *In vitro* Cytotoxicity Assay (ISO 10993-5:2009)

3.1 *Extract Dilution Test*

The extraction dilution exposure method detects toxins leached from medical devices (Srivastava et al. 2018). The test assesses both qualitative and quantitative cytotoxicity. Extract dilution test uses various methods such as tetrazolium salt-based assay, 3-(4, 5-dimethylthiazolyl-2)-2, 5-diphenyltetrazolium bromide) (MTT), (3-(4,5-dimethylthiazol-2-yl)-5-(3-carboxymethoxyphenyl)-2-(4-sulfophenyl)-2H-tetrazolium) (MTS), water-soluble tetrazolium salt (WTS) assay, and neutral red (NR) assay. MTT and NR assays are similar in chemical cytotoxicity evaluation. It detects toxins leached from exposed surfaces using a test material extraction solution and evaluation of extracts based on morphological changes of cells (Shafiee et al. 2021). The test detects mild to strong cytotoxicity (Srivastava et al. 2018).

3.2 Direct Contact Test

In this method, the solid biomaterial is used in the direct contact approach with *in vitro*-produced mammalian cells (Figure 2). The morphological alterations and variations in cell quantity are seen during the cytotoxic test, which may accurately depict the effects of testing medical devices on cells (Li et al. 2015). The vast majority of materials may be evaluated using direct contact testing.

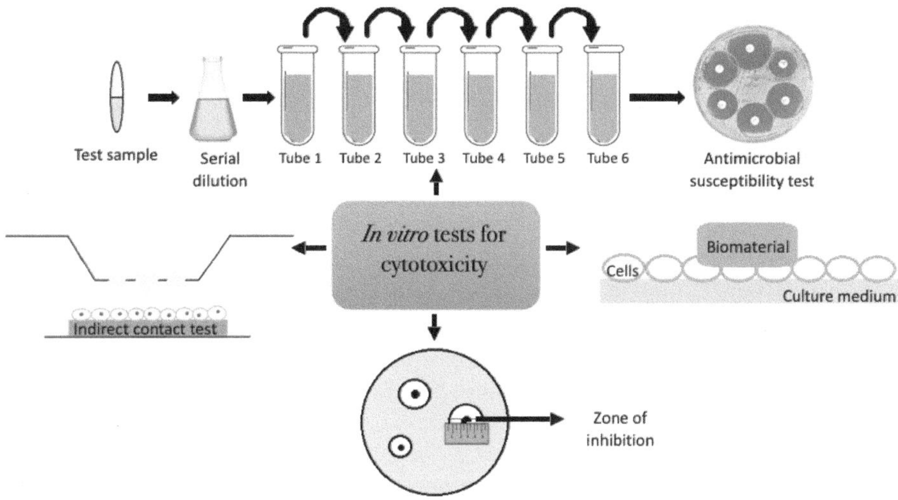

Figure 2: The discs infused with antimicrobial agents causes the formation of clear zones around itself called the "zone of inhibition". This occurs on incubation of discs on agar plate with pregrown culture of test organism. Measuring this clear zone determines the "zone of inhibition".

3.3 Agar Diffusion Test

The agar diffusion test, also known as the disk diffusion method or Kirby-Bauer test, is an antibiotic susceptibility test. In 1956, W. M. M. Kirby and colleagues from the University of Washington School of Medicine and the King County Hospital introduced the single disk method as a proposed technique for testing antimicrobial susceptibility (Hudzicki 2009). It is used to determine the antibiotic properties of a compound. If bacterial growth is inhibited by an antibiotic *in vitro*, then it can be used for clinical treatment of diseases involving that bacterium. The purpose of this procedure is to ascertain the resistance or sensitivity of pathogenic aerobic and facultative anaerobic bacteria. The medium facilitates the radial movement of the antibiotic compound (whose concentration reduces as it moves away from the circumference of the disk) and creates "zones of inhibition" or "clear zones" of bacterial growth (Figure 2).

3.4 Indirect Contact Test

The interaction of leachable by-products with cell monolayer without direct contact is assessed using an indirect contact test. Materials placed in transwell microplate insert and cultured in culture media and incubation of the treatment group is done at 37°C and 5% carbon dioxide for 24 hours before the evaluation of cytotoxicity with methoxynitrosulfophenyl-tetrazolium carboxanilide (XTT) cell metabolic activity assay and fluorescence microscopy (Wang et al. 2013). Indirect contact tests include agar diffusion test or agarose overlay test and molecular filtration (Figure 2). Agar overlay test is cheap, simple, and can be used for a variety of medical devices/materials. It does not accurately represent biological barriers in the body. Molecular filtration can be used to investigate the cytotoxicity of medical devices (Li et al. 2015).

4. *In vitro* Genotoxicity Testing (ISO 10993-3:2014)

4.1 Ames Test

Ceramic-based biomaterials have been extensively employed in recent years for orthopedic and dental purposes (Baino et al. 2015). Each new formulation of a ceramic-based material or the advancement of current biomaterials must first demonstrate their safety and effectiveness before their clinical performance can be evaluated. Risk and reward are used to define safety and effectiveness, respectively. The behavior of bioceramics upon implantation in organisms allows for their categorization into distinct groups. These groups include bioinert bioceramics (like alumina and zirconia), resorbable and bioactive bioceramics (such as hydroxyapatite and biological glasses), and bioceramics capable of triggering targeted cellular responses at the molecular level. These versatile materials serve as scaffolds for tissue engineering applications (Navarro et al. 2008). All of these bioceramics are reportedly biocompatible, although varying levels of bioactivity have been discovered when examining the ability to interact with the biological environment to enhance biological integration with tissues. The first biomaterials with bioactivity were glass ceramics, which had stimulatory effects on bone cells and direct chemical interaction with host tissues (Han et al. 2007).

In addition to the chemical characteristics of ceramic composition, ion release, oxidation, and corrosion are key variables impacting biocompatibility. *In vitro* and *in vivo* testing are required for inquiry into the functional qualities of the biomaterials once their safety has been evaluated (Fini and Giardino 2003).

The ISO10993-5:2009 test techniques are intended to measure the reaction of cultivated mammalian cells in close proximity to the test material or with a solution-based assay utilizing physiological factors. Gene mutation in bacteria and genetic alterations in cultured mammalian cells are the two most straightforward and sensitive *in vitro* techniques. The use of *in vivo* testing is only permitted if it is supported by science and if the outcomes of the *in vitro* tests show possible genotoxicity (search of chromosomal aberration). Ames (Salmonella/microsome mutagenicity assay) and the micronucleus tests are the two *in vitro* procedures that are employed most frequently (Torricelli et al. 2017).

The Ames test employs several *Salmonella typhimurium* strains that display frameshift or point mutations within the genetic sequences needed to manufacture histidine (OECD test Guideline 471, 1997). When looking for probable genotoxic carcinogens, the Ames test is a sensitive screening method. Due to the fact that a

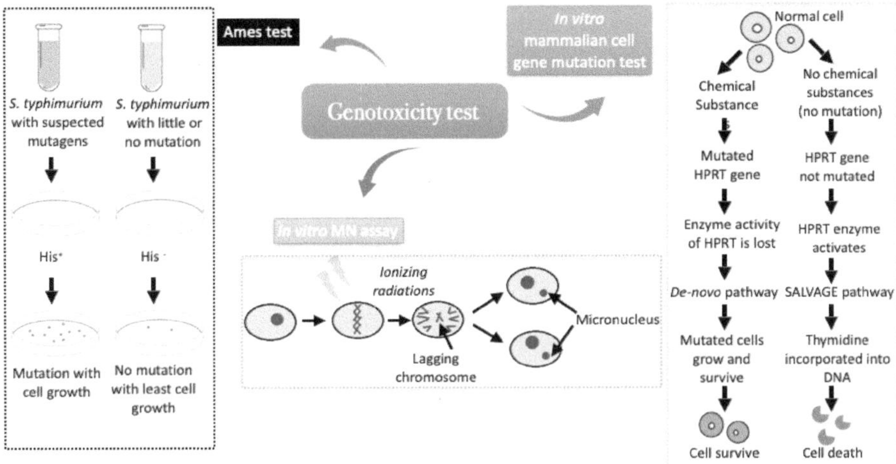

Figure 3: Ames test using Salmonella thypimurium, MN assay and gene mutation test are In vitro genotoxicity tests used for the potential toxic effects of the biomaterial on the genetic material of the cells. Abbrevation used: HGPRT (Hypoxanthine-guanine phosphoribosyltransferase); His (Histidine).

mutagen in the Ames test does not always pose a risk to people, a positive result—despite the strong correlation—can be challenging to interpret for the specific instance at hand. When a material is introduced to the culture media, it induces a back mutation that allows the bacteria to synthesize histidine once again (Figure 3). The mutagenicity of a drug is proportional to the number of colonies seen (Torricelli et al. 2017).

In order to ensure mutagenic capabilities, the Ames test is a straightforward, reliable procedure carried out employing several bacterial strains, such as *Salmonella typhimurium* and *Escherichia coli*. The basic goal of the Ames test is to induce a new mutation in a gene that is currently mutating, which will replace the existing mutation and restore the gene's function. This test is also known as the "Reversion Assay" because of the newly produced mutation that prevents colonies of histidine-producing cells from developing in the new mutant cells (Vijay et al. 2018; Vasdev et al. 2022). An important amino acid necessary for cell proliferation is histidine.

4.2 In vitro Mammalian Cell Gene Mutation Test (OECD 476)

The "mammalian cell gene mutation assay" is a method to detect the chemical substances that cause mutations in genes. It uses the HPRT (hypoxanthine-guanine phosphoribosyl transferase) gene as a marker. The HPRT gene is involved in the purine and pyrimidine base pair synthesis. Hypoxanthine phosphoribosyl transferase (HPRT, EC 2.4.2.8) catalyzes the synthesis of 5'-IMP and 5'-GMP involves the condensation of 5'-phosphoribosyl-l-pyrophosphate (PRPP) with the purine bases hypoxanthine and guanine, respectively (Stout and Caskey 1985). This marker gene, which is involved in a metabolic pathway, is exposed to the test chemicals. If the protein produced by the marker gene is altered, it changes the phenotype of the mammalian cell. This change in phenotype is indicative of a mutation. Hence, this assay method provides an effective way to study the mutagenicity of a chemical substance.

4.3 *In vitro* MN Assay

The MN assay stands for micronucleus assay. It is an *in vitro* genotoxicity test (toxicity test used to determine the substance's potential to cause genetic damage). This test has been in use for over 40 years and has successfully detected the genotoxicity of various chemical and physical agents. The *in vitro* micronucleus test is employed as a method for evaluating genotoxicity, specifically for the detection of micronuclei (MN) within the cytoplasm of interphase cells. Micronucleus are extra nuclei formed either due to chromosomes without a centromere or chromosomes that fail to reach the poles during the anaphase of cell division. The *in vitro* MN assay can be used to estimate the DNA damage by detecting micronucleus. The existence of micronuclei in a cell represents genetic damage. In this way, the MN assay can determine the potential of any substance to cause damage to chromosomes that are not favorable to the cell.

In the present day, we are exposed to various physical, chemical, and biological agents that might interfere with the normal functioning of our cells (cause diseases like neurodegenerative diseases, cardiovascular disease, chronic obstructive pulmonary disease, and cancer), some of which can be lethal. Some agents may be genotoxic and not lethal. These lead to damage in DNA, which sometimes gets carried on to the next generation. It is, therefore, essential to test the genotoxicity of any biomaterial to avoid health risks. MN assay is just one of the *in vitro* genotoxicity tests.

This test uses cells with a good metabolizing capability; if not, the cells are treated with sources for metabolic activation. After this, the cell is exposed to test chemicals and allowed to grow. Defects like chromosomal damage and altered cell division are observed after growth, which leads to the formation of micronuclei. After interphase, these cells are harvested, which is followed by staining and observation for micronuclei. Micronuclei should be identified in cells that have completed the process of mitosis after being exposed to the test chemical.

The ideal temperature for cell culture maintenance is 37°C and a humidified atmosphere of 5% carbon dioxide. These cultures should be checked for *Mycoplasma* contamination and stability of modal chromosome number. If such contamination or changes in chromosomes are observed, the cells cannot be used for the assay. The cells from the stock culture are transferred to the culture medium for exponential growth. Lymphocytes (whole blood treated with heparin or separated lymphocytes) can be cultured and used.

The MN assay is efficient and promising but also laborious and time-consuming. In order to combat this, the process can be automated. Automation is advantageous as it makes the assay more reliable. This is because manual errors in identifying the micronuclei can be eliminated. It also makes it possible to test thousands to ten thousand cells instead of one cell at a time.

5. *In vitro* Irritation Assay (ISO 10993-10:2021)

In spite of the fact that medical equipment is being developed at an ever-increasing rate, it is important to consider the safety, efficacy, and potential risks of every product utilized. The idea behind this is pretty clear; just like any other product meant for human use, such as vaccinations, new medications must pass specific testing and adhere to specific regulations before being made available on the market.

Similarly, the same idea also applies to medical devices. *In vitro* irritation assay (ISO10993-10:2021) is basically an international standard that mainly aims at tests for irritation and delayed-type hypersensitivity (Figure 4). In order to assure biocompatibility and safety, regulatory bodies from many nations have been established to assess products made for human use.

During the regulatory clearance process, manufacturers are obligated to conduct comprehensive safety testing of their completed devices in both preclinical and clinical phases. These testing procedures align with established regulations stipulated by esteemed organizations such as the US Food and Drug Administration (FDA), the International Organization for Standardization (ISO), or the Japanese Ministry of Health and Welfare (JMHW).

The ISO 10993 "Biological Assessment of Medical Devices" is a reference to the fundamental rules guiding biological testing and evaluation within a risk-management process. It is one of the International Standards (UNI EN ISO 10993-10: 2021). ISO tests should be used with interpretation and judgment by skilled and knowledgeable individuals, as well as numerous elements important for the assessment. The ISO 10993 assessment by qualified and experienced experts, as well

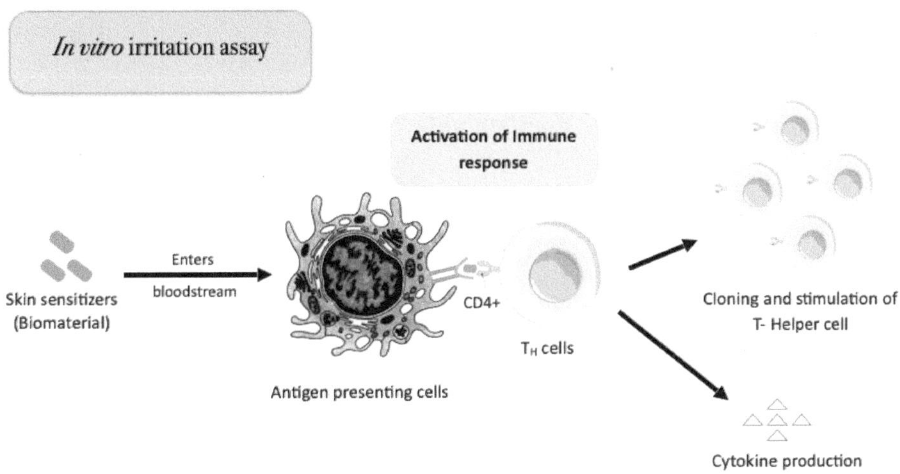

Figure 4: An overview of the immune response triggering after the skin sensitizers derived from a biomaterial enters the host bloodstream. Skin sensitizers are processed by Antigen presenting cell (macrophages) and stimulates TH (Helper) cells, which leads to proliferation of T cells (cloning) and cytokines production. In vitro skin irritation tests is utilized to assess the toxicity of biomaterial on host cells by testing its irritability and hypersensitivity.

as many other elements relating to the subject matter, the anticipated uses, the most recent information offered by scientific publications, and prior clinical experience, must be taken into account (Torricelli et al. 2017).

The *in vitro* irritation assay (ISO10993-10:2021) standard explains the materials used, the procedure for evaluating the medical devices, and the possible risk factors associated with skin sensitization. The use of *in vitro* and *in silico* methodologies for the first evaluation of skin irritation is covered by ISO 10993-10:2021.

From the pre-testing perspective, the identification of the substance to be examined must be established before beginning any examination of biological safety. In light of this, it is crucial to characterize the biomaterial accurately. Even before testing has been done, it is also essential to understand a material's chemical makeup in order to realize any significant threats connected to specific components that are still in the early stages of development (De Jong et al. 2012).

The existence and severity of delayed hypersensitivity are often determined at a certain time. Many of those *in vivo* and *in vitro* tests can be used to accomplish this. The strategy of inhibiting the spread of macrophages was deemed to be a suitable one; it is essentially based on the observation that the spread of peritoneal macrophages from animals displaying delayed hypersensitivity is noticeably suppressed in the presence of particular antigen (Fauve and Dekaris 1968; Dekaris et al. 1971). A few of this method's characteristics were especially helpful. Firstly, researchers take peritoneal cells from living animals without first irritating the peritoneal cavity, allowing for the ability to wash the peritoneum every day without hurting the animal. In addition, there's no serum in the incubation medium, fully eliminating variability caused by various serum components, and the test is accomplished; the immunological reactivity of washed cells may be tested in 90 minutes. Lastly, as peritoneal cells may be directly monitored using phase contrast microscopy, any noticeable cellular anomalies that may occur throughout the experiment can be easily seen. Animals were classified based on two antigens: one that induced the usual delayed-type hypersensitivity (tuberculin) and the other that elicited both delayed and rapid reactivity as well as the production of circulating antibodies (diphtheria toxoid) (Dekaris et al. 1972). Determining the time of course of each reaction and the formation of circulating antibodies for each antigen allowed researchers to compare the dynamics of the delayed hypersensitivity reaction to two antigens in each individual animal as well as the creation of antibodies against one of them (Veselić et al. 1973). Delayed allergic responses are distinguished by their sluggish progression and by their capacity to be passively transmitted by lymphoid cells rather than serum. These characteristics have been regarded as supporting Karush and Eisen's hypothesis of delayed hypersensitivity, which is mediated by antibodies attached to cells (Levine 1965).

Now, various biological experiments have been conducted on device extracts, including cell toxicity, irritative potential, allergic response, generalized acute/chronic toxicity, genetic toxicity, blood cell lysis, and chemical analysis or chemical profiling. The standards for biological testing that establish acceptance criteria or criteria for prioritizing the biological response include cell toxicity, irritative potential, allergic response, and blood cell lysis. Results in different levels for these tests are usually chosen based on years of device testing expertise. To determine whether a substance is acceptable for systemic toxicological evaluation, clinical, biochemical, hematologic, and histopathological measurements should not deviate biologically significantly from control groups. In the same manner, local tolerance is evaluated based on semi-quantitative ratings of the constituent parts of the histological reaction. Thus, blood compatibility complement activation is advised to be performed using both reference materials that have already been clinically used. There are no specified threshold acceptance standards, nevertheless. Similar to this, the appraisal of the genotoxicity of drugs and chemical substances is founded on the interpretation of results supported by 70 years of professional expertise. A minor but persistent rise in the expression of the gene known to cause genotoxicity when compared to control levels should be carefully examined in a number of tests for devices to determine if metabolic activation is required or not. This approach could be suitable for both eukaryotic and prokaryotic experiments (Eloy 2012). The foundation for creating standards is science. Standards for biocompatibility must be used to evaluate the performance and safety of medical equipment. Biological assessment should definitely include the risk analysis method, a thorough understanding of a device's chemical components, clinical conditions of usage, and clinical data review.

6. Cytotoxicity Tests

The methods used in the cytotoxicity evaluation of biomaterials are shown in Figure 5.

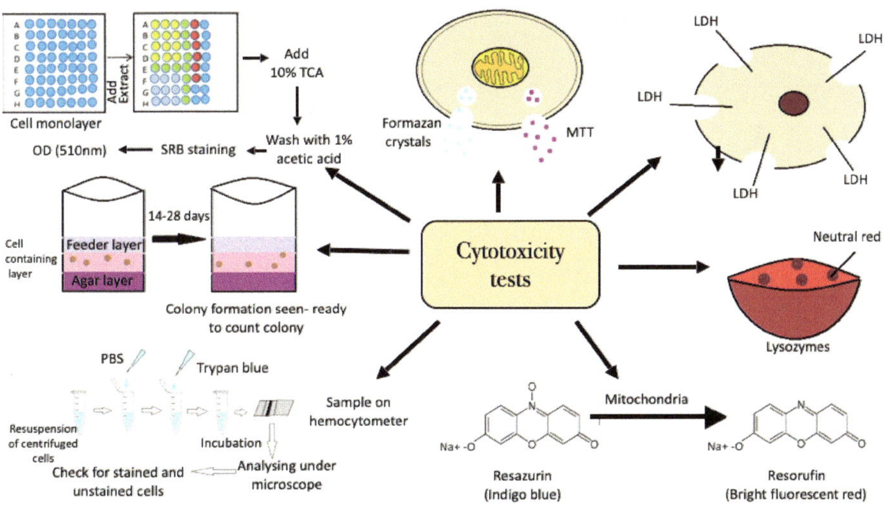

Figure 5: Methods employed in the cytotoxicity evaluation of biomaterials are tetrazolium salt assay, LDH (Lactate Dehydrogenase) leakage assay, neutral red uptake assay, alamar blue assay, colony forming assay, dye exclusion assay and sulforhodamine B (SRB) protein staining. Abbrevations used: MTT: 3-(4,5-dimethyl-thiazol-2-yl)-2,5-diphenyltetrazolium; LDH: Lactate dehydrogenase; TCA: Trichloroacetate.

6.1 Tetrazolium Salt Assay

Compounds of tetrazolium salts are frequently used to assess viability. Tetrazolium salts fall into different groups: MTT, XTT, and WST-1. In order to evaluate metabolic activity in active cells, the MTT/XTT/WST-1 test makes use of several tetrazolium salts. MTT 550 nm, XTT 492 nm, and WST-1 420–480 nm are the required wavelengths for each kit in order to detect metabolic activity (Figure 5).

6.1.1 MTT Assay

The tetrazolium-based MTT test has long been considered the gold standard of cytotoxicity assays because it is extremely sensitive and may be used for high-throughput screening (Tonder et al. 2015). In the MTT test, the tetrazolium salt 3-(4,5-dimethyl-thiazol-2-yl)-2,5-diphenyltetrazolium bromide (MTT) is avidly absorbed into cells and subsequently reduced in a mitochondrial-dependent procedure that generates a formazan product. MTT is a cationic salt that may enter living eukaryotic cells by electrostatically interacting with the anionic plasma membrane (Freeberg et al. 2019). Only *in vitro* live cells are analyzed by the tetrazolium-based calorimetric assay (MTT test), and the outcome is directly correlated with the quantity of viable cultivated cells (Figure 5). It has been employed in cancer research, immunological investigations, and, most recently, biocompatibility analysis. To adapt the MTT technique to an *in vitro* investigation of biomaterial-cell interactions, they made a few modest adjustments (Ciapetti et al. 1996). For instance, using the MTT test, the cytotoxicity of biomaterials, including hydroxyapatite, chitosan, and carbon nanotube scaffolds that interact with mesenchymal stem cells to aid in bone regeneration is evaluated (Paretsis et al. 2021).

6.1.2 XTT Assay

In order to assess cell viability, proliferation, and cytotoxicity, the XTT test is employed to quantify cellular metabolic activity. Cell proliferation, viability, and cytotoxicity may all be measured in a non-radioactive manner using the Cell Proliferation Kit II (XTT). Cells in suspension or adhering are cultivated in 96-well microplates as the sample material. The colorimetric, non-radioactive XTT (sodium 3'-[1-(phenylaminocarbonyl)-3,4-tetrazolium]-bis (4-methoxy6-nitro) benzene sulfonic acid hydrate) assay system is used to measure the amount of a substance in a sample (Paull et al. 1988). Since XTT (2,3-bis(2-methoxy-4-nitro-5-sulfophenyl)-5-carboxanilide-2H-tetrazolium) is soluble in the majority of aqueous media, it can be employed as an alternative to MTT (Figure 5). An advantageous feature of the XTT test is that it is also appropriate for cells in suspension. The test is based on the mitochondrial production of NADH via trans-plasma membrane electron transport and an electron mediator, which is then used to reduce XTT extracellularly (McGaw et al. 2014). In recent years, a great deal of *in vitro* and *in vivo* research has been done on magnesium based biomaterial for use in degradable implants. This study while adhering to the accepted methodology DIN EN ISO 10993-5:2009, enables an optimal and consistent *in vitro* evaluation of the cytocompatibility of magnesium based biomaterial. The preferred test cell line and medium, respectively, are mouse fibroblast line L-929 and MEM augmented with 10% FCS, penicillin/streptomycin, and 4 mM l-glutamine. To measure the proliferation and to determine the viability of cells, researchers performed the XTT test (Jung et al. 2015).

6.1.3 WST-1

In order to detect cell proliferation, cell viability, and cytotoxicity in mammalian cells, WST-1 (2-(4-iodophenyl)-3-(4-nitrophenyl)-5-(2,4-disulfophenyl)-2H tetrazolium monosodium salt) Assay Reagent offers a straightforward, precise, and ready-to-use test. The WST-1 test procedure is based on converting formazan, a product of cellular mitochondrial dehydrogenases, from the tetrazolium salt WST-1. The activity of the mitochondrial dehydrogenases increases with viable cell count, which also increases the quantity of formazan dye produced (Figure 5). The medium for cell culture receives the salt that is water soluble. The test assesses the metabolic activity of cells by causing a color change throughout the incubation time that is directly proportional to the quantity of mitochondrial dehydrogenase in the cell culture (Aslantürk 2018). The most used technique to determine the *in vitro* cytotoxicity of compounds is the WST-1 assay. Nevertheless, the interference of the examined compounds, such as carbon nanotubes or Mg particles, can have an impact on tetrazolium-based assays. Here, they describe a fresh Mn material interference with the WST-1 test. The WST-1 reagent has a number of benefits over the other tetrazolium salt-based cell proliferation reagents, MTT, and XTT, including water solubility, speed, increased stability, and sensitivity (Scarcello et al. 2020).

6.2 Lactate Dehydrogenase (LDH) Leakage Assay

LDH is a widely distributed protein that is ubiquitous and exhibits its peak activity in essential organs, including the heart, liver, muscles, kidneys, lungs, and blood cells (Figure 5). LDH is essential for glycolysis and gluconeogenesis, which are catalyzed by an oxidoreductase, and there is a reversible reaction occurring where lactate is converted to pyruvate and, at the same time, NADH and NAD+ undergo interconversion. LDH is composed of muscle and heart subunits. LDH is divided into five isomeric types with variable enzymatic activities based on the component of a subunit from LDH-1 to LDH-5 measured by *in vitro a*ssay (Wu et al. 2021). In terms of distribution, LDH1 is primarily located in the heart, and LDH2 is prevalent in the reticuloendothelial system, encompassing the liver, spleen, and lungs, as well as red blood cells. LDH3 is specifically found in the lungs, LDH4 in the kidneys, and LDH5 in the liver and skeletal muscle. The activities of serum LDH and their dynamic ranges are compared among various human diseases, being observed that out of 172,933 patients, it was found in 48 clinically defined diseases and 9,528 healthy individuals. LDH is an enzyme that exhibits stability and is present in all cells. During cellular damage, LDH is released into the supernatant of cell cultures. *In vitro* LDH activity serves as a sensitive, accurate, and cost-effective alternative to assays that rely on radioisotopes or dyes. LDH is responsible for producing lactic acid during exercise. LDH catalyzes lactate to pyruvate, and NADH reduces INT to form red formazan, which can be measured at 490 nm (Kumar et al. 2018). LDH plays a vital role in the anaerobic metabolic pathway by facilitating the bidirectional conversion of lactate to pyruvate (Farhana and Lappin 2023). Various conditions such as liver disease, anemia, heart attack, bone fractures, muscle trauma, cancers, and infections like encephalitis, meningitis, encephalitis, and HIV can lead to an increase in LDH in the blood. LDH is a cytosolic enzyme that is a reliable indicator of cellular toxicity. The cell culture medium is exposed to LDH when there is damage to the plasma membrane, leading to the release of the enzyme. LDH test is low risk but can cause bruising and soreness. However, commercially available LDH assay kits have limited reproducibility and high cost (Kaja et al. 2015). LDH helps in the quantitative determination of cell viability and late-stage apoptosis. The assessment of LDH leakage, which measures the activity of the widely present cytosolic enzyme, serves as a means to evaluate the integrity of cell membranes. This approach is employed to assess the extent of membrane damage induced by irritants (Kumar et al. 2018).

6.3 Neutral Red Uptake Assay

Through the neutral red uptake assay, which assesses both cell viability and proliferation in the aftermath of exposure to biomaterials, we can quantify the cytotoxicity of the employed biomaterials (Ciapetti et al. 1996). In this test, the ability of live cells to incorporate and bind a weak cationic dye called neutral red within their lysosomes is utilized (Ates et al. 2017). Due to the modified membrane characteristics of dying cells, neutral red (NR) cannot be absorbed by cells anymore. The cells are quickly rinsed with PBS after three hours of exposure to the dye. The cells' internalized dye is then released in an acidified ethanol solution. At OD 540 nm, released neutral red is evaluated in proportion to the number of viable cells. Unexposed cell viability, as measured by OD540 nm, is set to 100%. Viability curves may be created based on the absorption measurements to ascertain the test chemical concentration required to inhibit NRU by 50% (Repetto et al. 2008). A cell viability test called the neutral red absorption assay enables *in vitro* assessment of xenobiotic-induced cytotoxicity. For the assessment of viral cytopathogenicity, immuno-toxicity, and the identification of hazardous substances, *in vitro* tissue culture experiments utilizing the neutral red dye were created. As a chemosensitivity test for cell viability, the neutral red absorption cytotoxicity assay was created at Rockefeller University. The foundation of it is the supravital dye neutral red's capacity to attach to and be incorporated by healthy cells. The weakly cationic dye has the ability to permeate nonionic cell membranes and accumulate within lysosomes. Inside the lysosomes, it engages in electrostatic, hydrophobic interactions with anionic and/or phosphate groups present in the lysosomal matrix; subsequently, living cells are treated with an ethanol solution that has been acidified, resulting in the removal of the dye. A spectrophotometer is then used to measure the solubilized dye's absorbance (Repetto et al. 2008).

Based on the data, it is evident that the NRU, Alamar Blue assay (RES), and SRB assays display minimal variability within the linear range, whereas the MTT assay demonstrates the highest variation. This implies that the former assays are more adept at detecting slight alterations in cell number compared to the MTT assay.

6.4 *Alamar Blue Assay*

An integral part of testing synthesized biomaterials is determining their effect on cell viability, cell proliferation, and cytotoxicity detection. Alamar blue is an assay reagent for cell viability and cytotoxicity. This test has been in use for more than 50 years and is considered one of the best techniques because of its accuracy in predicting the adverse effects of test samples or biomaterials and detecting their toxic threshold.

The Alamar blue bioassay has been used to test various biological and environmental systems, as well as a diverse number of cells. Some of these cell types are bacterial cells, fungi, yeast, protozoa, and cultured mammalian and piscine cells. This uses a dye called resazurin (IUPAC name: 7-hydroxy-10-oxidophenoxazin-10-ium-3-one), which is cell-permeable, water-soluble, stable in culture medium and non-toxic and a weak blue fluorescent indicator. This enters the cell and is acted upon by the cellular redox enzymes, which convert it to resorufin. Through the acceptance of electrons from NADPH, FADH2, FMNH2, NADH, and cytochromes, this process triggers a prominent color transition from indigo blue to a vivid fluorescent red. The resulting red color diffuses out of the cells and into the culture medium, facilitating convenient measurement via colorimetry or fluorometry at a wavelength of 590 nm (Kumar et al. 2018). The ongoing transformation of resazurin into resorufin by viable cells provides a reliable and quantitative assessment of cell viability. Resazurin can be used to monitor cells continuously in a culture medium.

6.5 *Colony-Forming Assay*

A colony-forming assay, also known as "clonogenic assay of cells", is an *in vitro* technique to determine the proliferation and differentiation of cells. In simple terms, it is a method that is used to study the capability of each cell to grow into a colony. A group of cells can be considered a colony if it consists of a minimum of 50 cells. This method can be used to test each cell's ability to divide and grow in number (Figure 5). The colony-forming assay is one of the best standards for finding out a cell's reproductive death (a state where a cell loses its ability to divide due to accumulation damage in its DNA but can carry on some of its basic functions). This assay is useful for studying the characteristics of a cell under the influence of certain chemicals (e.g., cytotoxic agents), drugs, ionizing radiations, or genetic manipulations. The utilization of parametric cell survival curves is widely recognized as the gold standard for assessing cell reproductive death after exposure to ionizing radiation. These curves depict the correlation between radiation doses and the percentage of surviving colonies, facilitating the precise quantification of cell survival (Braselmann et al. 2015). Puck and Marcus, in 1956, described the colony-forming assay. They published a seminal paper using a cell culture technique that involved growing mammalian cells in a culture medium with necessary growth factors. These growth factors were provided as "feeder" cells. They can also be grown in a conditioned or complex culture media for a suitable environment that eliminates the need to use "feeder" cells. The clonogenic cell assay has been used for many decades to study the behavior of a variety of cells. The assays are capable of identifying cells that still possess the ability to generate a substantial number of offspring, even after exposure to treatments that may induce cell reproductive death (Franken et al. 2006). A typical colony-forming assay includes:

(a) Treating the monolayer of cells in tissue culture flasks;
(b) Preparing cell suspensions followed by plating them;
(c) Colonies need to be fixed and stained;
(d) Incubate for a suitable period.

6.5.1 Colony-Forming Assay Setup

There are two ways to perform the colony-forming assay:

6.5.1.1 Cells Plated After Treatment

Cells are treated in the parent culture plate, which is followed by re-plating them into varied dilutions to detect the colony-forming ability. The cells can be exposed to chemicals, radiation, drugs, etc. After treatment, the cells can be re-plated immediately (IP: Immediate Plating) or can be plated after a while (DP: Delayed Plating). An advantage of DP is that it gives time for repair to the treated cells, which yields more accurate results. In the field of radiobiological research, this method is specifically utilized to explore the mechanisms of repair for both potentially lethal and sublethal damage (Franken et al. 2006). The time given for DP is usually between 6–24 hours. The cells in the cell suspension are to be counted using a Coulter counter after the harvesting of treated cells. The culture should be incubated until sufficiently large colonies are developed in the control plate.

6.5.1.2 Cells Plated Prior to Treatment

This approach involves plating the cells from a stock culture into dishes of varied dilutions and letting them get attached to the dishes. This generally takes around 2 hours. The cells are treated after this. It is noteworthy that the cells will start replicating and form colonies soon after being plated. Therefore, the treatment has to be done before the cell proliferates. If treatment is done after the cells begin to divide, it will result in an enormous colony growth. The dishes are incubated (37°C) until six potential cell divisions. This is a good method for rapid screening of cells and determining the susceptibility to various treatments.

In order to combat this, a method called soft agar colony forming assay was developed in which the cell culture is suspended in a layer of soft agar (low agar concentration) that sits on a layer of solid agar (higher agar concentration). This allows three-dimensional movement of the transformed cells.

6.6 Dye Exclusion Assay

In order to assess the proportion of living and non-viable cells, dye exclusion assays are utilized. These tests are based on the idea that intact plasma membranes in living cells prevent the entry of certain dyes but allow in dead cells (Avelar-Freitas et al. 2014). Several cytotoxicity studies of chemicals and drug screening procedures involve *in vitro* cell viability and cellular toxicity experiments with grown cells. In recent times, there has been a rise in interest in the application of these tests. Tests involving human cells may hold greater applicability than certain animal models *in vivo* experiments, and *in vitro* cellular toxicity and/or cell viability assays offer several benefits, including speed, reduced price, and capacity for automation (Aslantürk 2018). They do not yet possess the technological sophistication to replace animal testing completely, which has significant drawbacks. At the end of the experiment, it is essential to understand how many cells are still viable and/or how many cells are dead. Current practices in the disciplines of toxicology and pharmacology utilize an extensive variety of cytotoxicity and cell viability experiments. Nowadays, a diverse array of studies examining cytotoxic effects and cell viability are employed within the domains of toxicology and pharmacology. Detailed inspection of the assay technique is required before deciding on the type of interaction (Aslantürk 2018).

There are various categories for cytotoxicity and cell viability tests; in this section, these assays are categorized by the sorts of endpoints that are assessed (color changes, fluorescence, luminescence, etc.). Broadly, these tests can be categorized into:

(a) Dye exclusion assays such as trypan blue, eosin, Congo red, and erythrosine B assays;

(b) Colorimetric assays;

(c) Fluorometric assays; and

(d) Luminometric assays.

Viable cells exclude dyes in the dye exclusion method, but dead cells do not. Although the staining process is rather straightforward, running experiments on a significant sample volume is challenging and time-intensive. By dye exclusion, membrane integrity can be measured.

These dyes include eosin, Congo red, erythrosine B, and trypan blue, among many others. Trypan blue has seen the highest use out of all the dyes listed (Aslantürk 2018).

Under the dye exclusion assay, we have two widely used approaches: (i) trypan blue dye exclusion assay and (ii) erythrosine B dye exclusion assay. Trypan blue dye exclusion assay is employed to count the number of alive and/or dead cells in a cell suspension. A substantial negatively ionized molecule is trypan blue. The trypan blue dye exclusion experiment is based on the idea that whereas dead cells lack entire cell membranes, living cells do. In this assay, test substances are serially diluted and cultured with adherent or non-adherent cells for a range of periods, whereas an important dye for counting live cells has already been developed: erythrosine B. This dye exclusion assay's fundamental workings are comparable to the trypan blue dye exclusion assay's. Although being a different bio-safe vital dye for cell counting, erythrosine B is not frequently employed to determine whether a cell is alive or dead (Aslantürk 2018).

6.7 Sulforhodamine B Protein Staining

One of the most widely used techniques for *in vitro* cytotoxicity screening is Sulforhodamine B protein staining, which was established in 1990. The assay is determined by the ability of SRB to bind to protein elements of cells that have been trichloroacetic acid-fixed to tissue culture plates (TCA). Under conditions that are only mildly acidic, SRB dissociates into a brilliant pink amino acid remnant. In accordance with the stoichiometric nature of SRB binding, The quantity of dye obtained from cells that have been stained correlates directly with the mass of the stained cells. The test can be run in a 96-well format due to the SRB staining's intense intensity. By achieving a signal-to-noise ratio of 4.83 at a cell density of 5,000 cells per well, researchers have demonstrated the assay's ability to detect densities as low as 1,000–2,000 cells per well. The assay exhibits a sensitivity level comparable to that of widely employed visible dye staining techniques in fluorescence analysis. The SRB technique has also been proven to be successful in analyzing the interactions between irradiation and chemotherapy, as well as for evaluating the *in vitro* susceptibility of cancer cells to radiation, with sensitivity similar to that of the conventional clonogenic assay (Vichai and Kirtikara 2006). The SRB assay's sensitivity is higher than that of the Lowry or Bradford assays and on par with a number of fluorescence assays. Colorimetric, nondestructive, and indefinitely stable are the characteristics of the SRB assay's endpoint. The SRB assay is a useful and sensitive method for determining the cytotoxicity caused by drugs as well as for determining how clonogenicity is measured. The SRB approach is additionally quick and affordable. As a result of these practical developments, the SRB test is now a suitable and sensitive assay for determining drug-induced cytotoxicity, even in large-scale applications (Voigt 2005). Table 1 discusses the advantages and disadvantages associated with various cytotoxicity assays.

7. Haemocompatibility Assay (ISO 10993-4:2017)

One of the most crucial requirements for a blood-contacting biomaterial's effective *in vivo* application is its hemocompatibility. Hence, prior to clinical applications, thorough *in vitro* tests in accordance with ISO 10993-4 are necessary. Blood-contacting biomaterials' clinical utility is constrained by a number of important factors, including haemocomptibility. These substances are in direct touch with blood, a complex "organ" made up of erythrocytes, leukocytes, and platelets (which make up 1% of the blood), which is composed of 55% plasma and 44% erythrocytes. To avoid the activation and destruction of blood components, it is essential to thoroughly investigate any harmful interactions between newly produced materials and blood (Weber et al. 2018). Cardiovascular implants are increasingly significant in regenerative medicine. New polymer-based biomaterials that prevent adverse responses, especially thrombotic events, are desperately needed to enhance the safety and functionality of blood-contacting implants (Jung et al. 2013). The thrombotic reaction that is created when a substance or device comes into contact with blood and causes complement activation, platelet response, and the beginning of the blood coagulation cascade is known as haemocomptibility (Figure 5; Mulinti et al. 2018). The events unfolding in haemocomptibility are:

Table 1: Overview of advantages and disadvantages related to various cytotoxicity assays.

Assays	Advantages	Disadvantages
MTT assay	• Simplicity, safety, excellent repeatability, and widespread usage in cytotoxicity and cell viability testing. • It is preferable to alternative cytotoxicity assays like the dye exclusion technique (Aslantürk 2018).	• Fluorescent and luminous approaches are more sensitive than the absorbance method of detection utilized by the MTT test and other tetrazolium reduction assays (i.e., MTS, XTT, WST-1, and WST-8). • The assay's ability to be miniaturized for HTS applications is greatly constrained by its lack of sensitivity. • Background absorbance levels and assay artifacts are exacerbated by test settings that interfere with the chemical or enzymatic degradation of MTT.
XTT assay	• It is quick, sensitive, user-friendly, and secure approach. • It is extremely sensitive and accurate (Aslantürk 2018).	• It is known that a wide range of chemical substances might hinder the MTT test. • The majority of them are reducing substances that cause the MTT to be reduced to formazan non-enzymatically (Aslantürk 2018).
WST-1	• It is frequently employed to carry out both cytotoxicity and cell viability studies since it is simple to use, safe, and highly reproducible (Aslantürk 2018).	• It requires a 2-hour incubation period as normal, and it is currently unknown if the influence of the testing agents at various time periods on the trend of relative cell viability can be captured by a single addition of WST-1 (Aslantürk 2018).
Neutral red uptake assay	• The NRR test is a reliable indicator of lysosomal damage. · The benefit of this test is its speed and ease of assessment (Aslantürk 2018).	• According to reports, environmental parameters, including temperature and salinity, either have little to no impact on the NRR assay or have no impact at all. • Contaminants have the most impact on NRU assay (Aslantürk 2018).
Colony forming assay	• It can be used to study stem cells, tumor cells, sensitivity of cells to radiation, and other physical and chemical agents.	• This only provides information about the colony growth of normal cells. • These cells cannot grow without proper anchorage. Transformed cells are cells that have lost this property and do not have any "adhering ability." • These cells can grow and proliferate regardless of their substrate. The colony-forming ability of such cells cannot be studied through colony-forming assay.
Sulforhodamine B protein staining	• The SRB assay has a number of advantages, including the fact that this type of interference is rarely a problem and that cellular staining is not influenced by cellular metabolic activity.	• It requires fewer steps to optimize assay conditions for particular cell lines than the MTT assay, but its use is currently restricted to manual or semiautomatic screening because of the many washing and drying steps, which are currently not automatable. • This approach offers a sensitive and effective screening tool, particularly for use in laboratories with less sophisticated equipment (Vichai and Kirtikara 2006).

1. Coagulation: The activation, adhesion, and aggregation of platelets, as well as the deposition and maturation of fibrin, are all components of the coagulation process. Upon damage to the endothelium lining a blood artery, coagulation starts nearly immediately (Figure 6). The innate coagulation pathway is activated by touch when plasma proteins interact with artificial surfaces. The plasma kallikrein-kinin system, also known as the contact-phase system, is made up of the non-enzymatic coagulation factor high molecular weight kininogen (HMWK), three serine proteinases – factors XII and XII and plasma prekallikrein (PK) – as well as three serine proteases, factors XII, XI, and XII. Blood is converted from Factor XII to the active enzyme Factor XIIa when it comes into contact with synthetic, negatively charged surfaces such as kaolin, glass, collagen, silica, or dextran sulfate. HMWK is changed into bradykinin by Factor XIIa, while PK is transformed into active kallikrein (Weber et al. 2018).
2. Platelets: Local prothrombotic substances like tissue factor and bound platelet secretion products both function as stimulants for platelet activation (Figure 5). Platelet activation can be triggered by several routes. Large numbers of platelets circulate in the bloodstream under normal physiological circumstances for 7–10 days in a dormant state. A healthy endothelium monolayer serves as a protective boundary separating the bloodstream from the surrounding tissues and inhibits platelet adhesion and activation by releasing substances, including nitric oxide (NO) and prostacyclin (PGI2). When biomaterial is exposed to the blood, platelets may get unintentionally activated, which might cause thrombotic problems. Haemocomptibility tests thus include a critical component that analyses platelet activity (Weber et al. 2018).
3. Complement System: The possibility of the complement system being activated by a medical device is assessed using the complement activation test. This humoral immune response may be brought on by the ingestion of a foreign substance and can have negative outcomes, including tissue damage and inflammation. Under the laws governing the testing of biomaterials designed for blood contact, the study of complement activation is a pertinent standard. The presence of C3a,

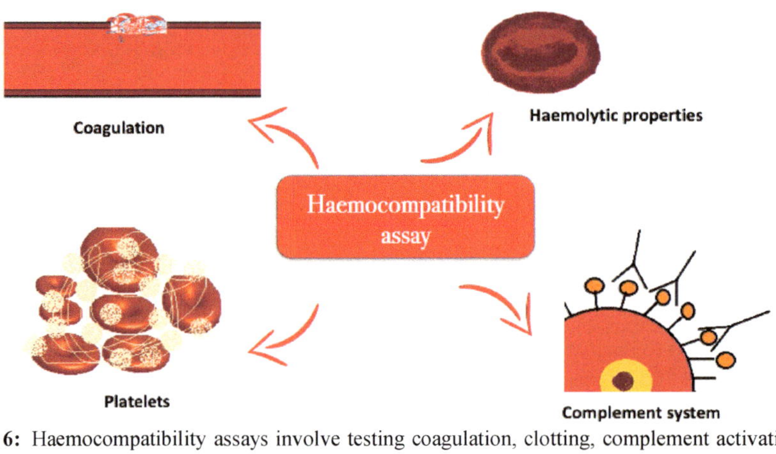

Figure 6: Haemocompatibility assays involve testing coagulation, clotting, complement activation and hemolysis. A biomaterial is considered haemocompatible if it shows <5% haemolysis.

C5a, Bb, iC3b, C4d, C3- or C5-convertase, and the C5b-9 complex in whole blood, along with the measurement of 50% complement-mediated hemolysis, are required by ISO 10993-4 (International Organization for Standardization, 2000) for the analysis of complement system activation (CH50). Anaphylatoxins C3a, C5a, and the C5b-9 complex concentrations are most typically assessed using ELISA. In the latest research, Endgberg and colleagues further proposed that a potential predictor for assessing hemocompatibility could involve examining the balance between the binding of C4 and its inhibitor C4BP (Figure 5; Weber et al. 2018).

4. Haemolytic Properties: The hemolysis test is targeted to ascertain the hemolytic activity of a medical device's components that have direct or indirect exposure to blood and satisfies the standards for hematologic testing for the selection of tests for blood-related interactions (Nelson's lab; hemocompatibility testing).

8. Future Prospects

There have been several recent breakthroughs in blood-interfacing materials as a result of the growing significance of biomaterials in our society over the past few decades. Medical devices, including arterial implants, renal dialysis devices, central venous access devices, and artificial cardiac valves, are responsible for a large portion of the industrialized society's improved quality of life. Meanwhile, sophisticated biomaterials have largely supplanted many conventional biomaterials, offering biologically inspired design and responsive properties. These issues must be taken into consideration in efforts to make innovative biomaterials, medical devices, and prostheses available for clinical use. According to ISO 10993, the specific uses of biomaterials in the particular medical device under consideration call for the *in vivo* verification of *in vitro* results. To correctly and sufficiently detect biological reactions to nanodevices and tissue-created scaffolds, which may contain a variety of components, including novel materials, cells, and proteins, new biological response assays must be developed. It is obvious that new and creative biological response test systems must be created in order to fully and properly evaluate nanodevices and tissue-generated scaffolds due to this substantial hurdle.

References

Aslantürk, Ö. S. 2018. *In vitro* cytotoxicity and cell viability assays: principles, advantages, and disadvantages. Genotoxicity-A Predictable Risk to our Actual World 2: 64–80.

Avelar-Freitas, B. A., V. G. Almeida, M. C. X. Pinto et al. 2014. Trypan blue exclusion assay by flow cytometry. Brazilian Journal of Medical and Biological Research 47: 307–315.

Baino, F., G. Novajra and C. Vitale-Brovarone. 2015. Bioceramics and scaffolds: a winning combination for tissue engineering. Frontiers in Bioengineering and Biotechnology 3: 202.

Braselmann, H., A. Michna, J. Heß et al. 2015. CFAssay: statistical analysis of the colony formation assay. Radiation oncology (London, England), 10: 223. https://doi.org/10.1186/s13014-015-0529-y.

Coagulation. *In*: Wikipedia. 2023 [cited 2023 Apr 11]. Available from: Https://en.Wikipedia.Org/w/index. Php?Title=Coagulationandoldid=1144076824.

Chraniuk, M., M. Panasiuk, L. Hovhannisyan et al. 2022. The preliminary assessment of new biomaterials necessitates a comparison of direct and indirect cytotoxicity methodological approaches. Polymers 14(21): 4522.

Ciapetti, G., D. Granchi, E. Verri et al. 1996. Application of a combination of neutral red and amido black staining for rapid, reliable cytotoxicity testing of biomaterials. Biomaterials 17(13): 1259–1264.

De Jong, W. H., J. W. Carraway and R. E. Geertsma. 2020. *In vivo* and *in vitro* testing for the biological safety evaluation of biomaterials and medical devices. In Biocompatibility and Performance of Medical Devices (pp. 123–166). Woodhead Publishing.

Dekaris, D., S. Smerdel and B Veselić.. 1971. Inhibition of macrophage spreading by upernatants of antigen-stimulated sensitized lymphocytes. European Journal of Immunology 1(5): 402–404.

Eloy, R. 2012. Challenges in biocompatibility and failure of biomaterials. In Biocompatibility and Performance of Medical Devices (pp. 18–29). Woodhead Publishing.

Fauve, R. M. and D. Dekaris. 1968. Macrophage spreading: inhibition in delayed hypersensitivity. Science 160(3829): 795–796.

Fini, M. and R. Giardino. 2003. *In vitro* and *in vivo* tests for the biological evaluation of candidate orthopedic materials: Benefits and limits. Journal of Applied Biomaterials and Biomechanics 1(3) 155–163.

Franken, N. A., H. M. Rodermond, J. Stap et al. 2006. Clonogenic assay of cells *in vitro*. Nature Protocols 1(5): 2315–2319. https://doi.org/10.1038/nprot.2006.339.

Freeberg, M. A., J. G. Kallenbach and H. A. Awad. 2019. Assessment of Cellular Responses of Tissue Constructs *in vitro* in Regenerative Engineering. Encyclopedia of Biomedical Engineering.

Groth, T., P. Falck and R. R. Miethke. 1995. Cytotoxicity of biomaterials—basic mechanisms and in vitro test methods: a review. Alternatives to Laboratory Animals 23(6): 790–799.

Hemocompatibility Testing. Nelson Labs. [cited 2023 Apr 11]. Available from: Https://www.Nelsonlabs. Com/testing/hemocompatibility/

Hudzicki, J. and K. B. D. D. S. Test. 2009. Protocol; American Society for Microbiology: Washington. DC, USA.

Inayat-Hussain, S., N. F. Rajab and E. L. Siew. 2009. *In vitro* testing of biomaterials toxicity and biocompatibility. In Cellular response to biomaterials (pp. 508–537). Woodhead Publishing.

Jung, F., S. Braune and A. Lendlein. 2013. Haemocompatibility testing of biomaterials using human platelets. Clinical Hemorheology and Microcirculation 53(1-2): 97–115.

Jung, O., R. Smeets, D. Porchetta et al. 2015. Optimized *in vitro* procedure for assessing the cytocompatibility of magnesium-based biomaterials. Acta Biomaterialia 23: 354–363.

Kaja, S., A. J. Payne, T. Singh et al. 2015. An optimized lactate dehydrogenase release assay for screening of drug candidates in neuroscience. Journal of Pharmacological and Toxicological Methods 73: 1–6

Kumar, P., A. Nagarajan and P. D. Uchil. 2018. Analysis of cell viability by the alamarBlue assay. Cold Spring Harb. Protoc. 2018(6.)

Levine, B. B. 1965. Studies on delayed hypersensitivity: I. Inferences on the comparative binding affinities of antibodies mediating delayed and immediate hypersensitivity reactions in the guinea pig. The Journal of Experimental Medicine 121(6): 873.

Li, W., J. Zhou and Y. Xu. 2015. Study of the *in vitro* cytotoxicity testing of medical devices. Biomedical Reports 3(5): 617–620.

McGaw, L. J., E. E. Elgorashi and J. N. Eloff. 2014. Cytotoxicity of African medicinal plants against normal animal and human cells. In Toxicological survey of African medicinal plants (pp. 181–233). Elsevier.

Mulinti, P., J. E. Brooks, B. Lervickal. 2018. Strategies to improve the hemocompatibility of biodegradable biomaterials. In Hemocompatibility of biomaterials for clinical applications (pp. 253–278). Woodhead Publishing.

Navarro, M., A. Michiardi, O. Castaño et al. 2008. Review: Biomaterials in orthopaedics. Journal of the Royal Society Interface 5(27): 1137–1158.

Paretsis, N. F., V. Gonçalves Junior, N. G. T. de Q. Hazarbassanov et al. 2021. *In vitro* evaluation of hydroxyapatite, chitosan, and carbon nanotube composite biomaterial to support bone healing. Brazilian Journal of Veterinary Research and Animal Science 58: e179885. https://doi. org/10.11606/issn.1678-4456.bjvras.2021.179885.

Paull, K. D., R. H. Shoemaker, M. R. Boyd et al. 1988. The synthesis of XTT: A new tetrazolium reagent that is bioreducible to a water-soluble formazan. Journal of Heterocyclic Chemistry 25(3): 911–914.

Pieters, R., D. R. Huismans, A. Leyva et al. 1989. Comparison of the rapid automated MTT-assay with a dye exclusion assay for chemosensitivity testing in childhood leukaemia. British Journal of Cancer 59(2): 217.

Repetto, G., A. Del Peso and J. L. Zurita. 2008. Neutral red uptake assay for the estimation of cell viability/ cytotoxicity. Nature Protocols 3(7): 1125–1131.

Scarcello, E., A. Lambremont, R. Vanbever et al. 2020. Mind your assays: Misleading cytotoxicity with the WST-1 assay in the presence of manganese. PloS one 15(4): e0231634.

Shafiee, M. A. M., M. A. M. Asri and S. S. S. Alwi. 2021. Review on the *in vitro* cytotoxicity assessment in accordance to the International Organization for Standardization (ISO). Malaysian Journal of Medicine and Health Sciences, 17(2).

Srivastava, G. K., M. L. Alonso-Alonso, I. Fernandez-Bueno et al. 2018. Comparison between direct contact and extract exposure methods for PFO cytotoxicity evaluation. Scientific Reports 8(1): 1425. https://doi.org/10.1038/s41598-018-19428-5.

Stout, J. T. and C. T. Caskey. 1985. HPRT: gene structure, expression, and mutation. Annual Review of Genetics 19(1): 127–148.

Thrivikraman, G., G. Madras and B. Basu. 2014. *In vitro/in vivo* assessment and mechanisms of toxicity of bioceramic materials and its wear particulates. RSC Advances 4(25): 12763–12781.

Torricelli, P., N. N. Aldini and M. Fini. 2017. Biological assessment of bioceramics. In Advances in Ceramic Biomaterials (pp. 111–125). Woodhead Publishing.

van Tonder, A., A. M. Joubert and A. D. Cromarty. 2015. Limitations of the 3-(4,5-dimethylthiazol-2-yl)-2,5-diphenyl-2H-tetrazolium bromide (MTT) assay when compared to three commonly used cell enumeration assays. BMC Research Notes, 8: 47. https://doi.org/10.1186/s13104-015-1000-8.

Vasdev, N., M. Deshpande, P. Katare et al. 2022. New emerging technologies for genetic toxicity testing. Pharmacokinetics and Toxicokinetic Considerations, 175–219.

Veselić, B., D. Dekaris and K. M. Hršak. 1973. *In vitro* studies of delayed-type hypersensitivity: The time course of the primary immunological reaction in individual rats determined by macrophage spreading inhibition. Immunology 24(2): 375.

Vichai, V. and K. Kirtikara. 2006. Sulforhodamine B colorimetric assay for cytotoxicity screening. Nature Protocols 1(3):1112–1116

Vijay, U., S. Gupta, P. Mathur et al. 2018. Microbial mutagenicity assay: Ames test. Bio-protocol. 8(6): e2763–e2763.

Voigt, W. 2005. Sulforhodamine B assay and chemosensitivity (pp. 39–48). Humana Press.

Wang, M. O., J. M. Etheridge, J. A. Thompson et al. 2013. Evaluation of the *in vitro* cytotoxicity of cross-linked biomaterials. Biomacromolecules 14(5): 1321–1329. https://doi.org/10.1021/bm301962f.

Weber, M., H. Steinle, S. Golombek et al. 2018. Blood-contacting biomaterials: *in vitro* evaluation of the hemocompatibility. Front. Bioeng. Biotechnol. 6, 99.

Weisenthal, L. M., P. L. Dill, N. B. Kurnick et al. 1983. Comparison of dye exclusion assays with a clonogenic assay in the determination of drug-induced cytotoxicity. Cancer Research 43(1): 258–264.

Weisenthal, L. M., J. A. Marsden, P. L. Dill et al. 1983. A novel dye exclusion method for testing *in vitro* chemosensitivity of human tumors. Cancer Research 43(2): 749.

Wieslander, A. P., M. K. Nordin, B. Hansson et al. 1993. *In vitro* toxicity of biomaterials determined with cell density, total protein, cell cycle distribution and adenine nucleotides. Biomaterials, Artificial Cells and Immobilization Biotechnology 21(1): 63–70.

Wu, Y., C. Lu, N. Pan et al. 2021. Serum lactate dehydrogenase activities as systems biomarkers for 48 types of human diseases. Scientific Reports 11(1): 12997.

14

Rheological Characterization of Biomaterials

Dharmesh R. Chejara,[1,*] *Amrutlal L. Prajapat,*[1] *Nilesh Pandya,*[1] *Anuraag Moturi*[1] and *Pranav Shah*[2]

1. Introduction

Thixotropic biopolymers or their modified conjugates are receiving significant consideration due to their wide-ranging applications in food, agricultural (Mishra et al. 2018), pharmaceutical (Silva and Sato 2017), biotechnological and cosmetic industries (Buruiana and Ioan 2018; Paul et al. 2017). Thixotropy is the characteristic of viscous matters or gelling systems wherein the matter turns to liquid upon applied stimuli, strain, or agitation, and the fluid displays a reversible structural change upon removing applied stimuli without getting the volume of the system changed (Lee et al. 2009). Several biopolymeric gelling systems become more fluid when subjected to such stimuli, stresses, or strain for a fixed time. It is reversible and regains its viscosity if left undisturbed.

A thixotropic can be defined as a phenomenon of fluid that takes a predictable time to achieve equilibrium viscosity while introduced to shear. Several fluids that get back to their original gel state almost instantly are called pseudoplastic fluids, e.g., ketchup. Others, like yogurt, take a longer period to become nearly solid. A number of gelling systems and colloidal fluids exhibit thixotropic behavior and a steady form on rest that becomes fluid upon shearing or agitation. Previously, temperature was the only significant stimulus that could cause such physical changes in gels, which allowed these gels to melt on heating and solidify on cooling (Barnes 1997). Figure 1 depicts the phenomenon of thixotropy of gel.

[1] Tarsadia Institute of Chemical Science, Uka Tarsadia University, Bardoli, Surat, India.
[2] Maliba Pharmacy College, Uka Tarsadia University, Bardoli, Surat, India.
* Corresponding author: dharmesh.chejara@utu.ac.in

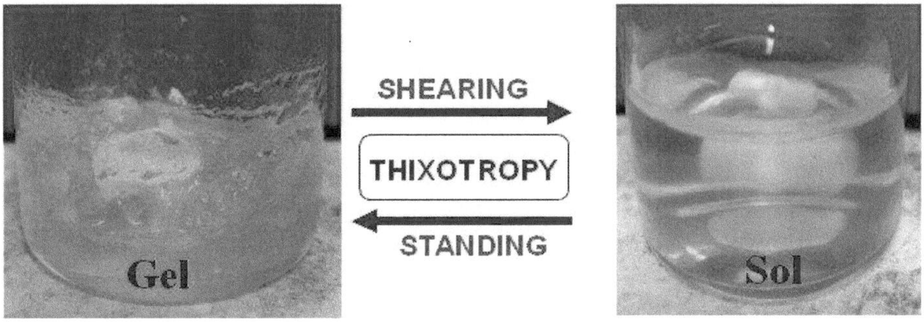

Figure 1: The phenomenon of thixotropy of gel (Chhatbar et al. 2012).

The field of biopolymer-based soft matter and its uses in biomedical engineering are fetching more attention and importance because of its availability at lower cost, environmentally friendly impact, and non-toxicity to humans. Some advanced materials can be designed by tuning these biopolymers in terms of physical or chemical modification. Many biopolymers have already been reported to have such properties. Chemically and physically modified or cross-linked gels showing thixotropic behavior attract great attention in the field of biomedical research due to their self-healing ability. These gels are especially suitable for injectable purposes.

Natural polymers-based biomaterials are widely applicable for designing stretchable gels due to their chemical or physical modification to introduce elasticity, thixotropy, etc. Such gelling matters have appeared as attractive candidates as matrixes for the regeneration and repair of various tissues expanding the field of tissue engineering (Li et al. 2021; Wang et al. 2019). The development of polymeric biomimetic materials for bone tissue engineering has also been studied and reported in the literature (Chahal et al. 2019). Some of the medical implantations exhibit some restrictions due to difficulties associated with injectability and their poor mechanical properties (Barbucci et al. 2008). With this in view, several attempts have been made to achieve biopolymers' smart-based gelling systems with better physico-mechanical properties to overcome these issues.

Polymers like cellulose and its analogs, guar gum, xanthan gum, and several seaweed polysaccharides—namely agarose, carrageenan, sodium alginate chitosan, and fucoidan—have been broadly studied and used in tissue engineering, regenerative medicines, biomedical implants, and other applications. Their favorable characteristics include biocompatibility, easy gelling nature, low toxicity, and non-immunogenicity (Lin et al. 2015; Lara et al. 2022; Kirtania et al. 2020).

Most of the research nowadays focuses on biopolymers-based gelling systems to design various formulations exhibiting potential utility to achieve targeted delivery systems. Delivery of the active pharmaceutical doses at the targeted place is achieved through biopolymeric gels for drug-delivery systems like buccal, oral, vaginal, rectal, topical, and nasal. Many of the biopolymeric gels show a great rheological transition that supports the physicho-mechanical characteristics of the designed drug-delivery systems. With this perspective in mind, this review highlights a few such studies based on several polysaccharide-based thixotropic gelling systems and their potential applications in the field of biomaterial research.

2. Rheological Studies of Thixotropic Gels

Thixotropic properties of materials can be evaluated using the following rheological experiments.

2.1 Dynamic Viscosities With Different Shear Rates

In this experiment, the flow property of the materials can be determined, in which dynamic viscosities at different shear rates are determined; the flow curve also represents shear stress against shear rate, which was used for the calculation using the Power Law model equation (1).

$$\tau = K \cdot \gamma^n \tag{1}$$

K is the consistency index, while 'n' represents the flow behavior index. τ and γ are the shear stress and shear rate, respectively.

2.2 Hysteresis Loop Area Experiment

This experiment is well known and one of the basic rheological experiments used for the determination of thixotrpicity. In this experiment, the material is exposed to varying shear rates against different shear stresses. The shape of the graph represents a rising sweep instantly, followed by a descending sweep. Thixotropic area (*Ar*) can be calculated by using the equation (2) as reported by Dolz et al. (2000).

$$Ar = (Aup - Adown/Aup) \text{ x } 100 \tag{2}$$

Here, *A*up = area of the upstream, *A*down = area of downstream data points, and (*Aup–Adown*) represents hysteresis or thixotropic loop area.

The hysteresis loop area is directly proportional to the thixotropicity of materials; in other words, the more the thixotropic area or hysteresis loop area, the more would be the thixotropicity.

2.3 Time-Dependent Thixotropic Property

This experiment is also used to determine the thixotropic properties of a material. The time-dependent thixotropic behavior of materials can be measured by employing time-dependent viscosities at various shear rates. For example, the experiment is to be conducted at shear rates 1s-1, 100s-1, and 1s-1. The time intervals were 25–45s (first 5 points), 46–96s (100 points), and 97–345s (500 points), respectively. This experiment shows the time-dependent viscosity of the material with varying shear rates.

2.4 Yield Stress Experiment

Yield stress is to be measured by a slow, constant shear rate applied to a sample, and the resulting stress is determined as a function of time. The determination of static yield stress is achieved by a stress sweep where the direct reversible deformation ends (Galindo-Rosales et al. 2011).

2.5 Creep-Recovery Experiment

The creep-recovery experiment is well-known and useful for the evaluation of the thixotropic properties. For this experiment, constant stress is to be applied, which is less than the measured yield stress of the material, and then the recovery is monitored for a specific time.

3. Biopolymeric Thixotropic Gels and Their Applications

Several polysaccharides, such as seaweed polysaccharides, plant polysaccharides, and animal-based polysaccharides, are to be modified to introduce thixotropic in their polymeric backbone to benefit from this property. However, few of these polysaccharides already exhibited such behavior, while some of their conjugates show interesting rheological transitions that made these polymers a candidate for modification to introduce these properties. Few such biopolymers belonging to each category have been highlighted in this section. Some of the examples of all types of polysaccharides, e.g., sodium alginate (seaweed polysaccharide), cellulose, guar gum (plant polysaccharide), and xanthan gum (bacterial polysaccharide), are highlighted in this section of the review.

3.1 Sodium Alginate as a Thixotropic Gel

Sodium alginate is one of the biopolymers that is extensively studied and used in drug-delivery systems and tissue engineering as well as in biomedical applications because of their biocompatible nature, gelling ability, non-immunogenicity and lower toxicity (George and Abraham 2006). Conventional uses of sodium alginate as an excipient in drugs are usually dependent on gel-forming and thickening properties (Tønnesen and Karlsen 2002). Modification of alginates has been reviewed by many authors (Cathell et al. 2010; Yang et al. 2011). Delivery of cells, insulin, and proteins in injectable mode are also reported using alginate as one of the prior materials (Silva et al. 2006; Silva and Sato 2017; Gombotz and Wee 1998). Unmodified alginate gels possess shear thinning behavior and low thixotropic properties. This property be induced by introducing various substrates on its backbone (Sasaki et al. 1956; "US Pat., 5,473,005, 1995," n.d.; "US Pat., 2003/0072803A1, 2003," n.d.; Voronko et al. 2002; Grassi et al. 2006; Pongjanyakul and Puttipipatkhachorn 2007; Daemi et al. 2013). A sprayable new soft gel based on sodium alginate with remarkable thixotropic behavior has been reported (Mahesh et al. 2012). Further, sodium alginate and 1-naphthylamine-based amide derivatives have also been reported to form soft thixotropic hydrogel (Chejara et al. 2013). Alginate-aminocaproic acid thixo-gel with multi-stimuli sensitivity as a bio-injectable gel is also reported that shows multi-stimuli responses with various applied stimuli, e.g., heat, stress, ultrasound, etc. (Chejara et al. 2016). Further, a microporous soft hydrogel derived from sodium alginate and aminosalicylic acid was studied for viscosupplementation application, wherein it was found to be a great alternative to any potential gelling viscosupplement for arthritis problems (Chejara et al. 2017).

3.2 Electrolyte Added Gelling Systems

Sodium alginate-based gelling systems with thixotropic behavior are reported in literature wherein these gels were obtained by the addition of significant amounts of several divalent ionic salts to sodium alginate solutions (Figure 2) (Sasaki et al. 1956). The thixotropic evaluation was carried out simply by determining the setting time detected when inverting the tube.

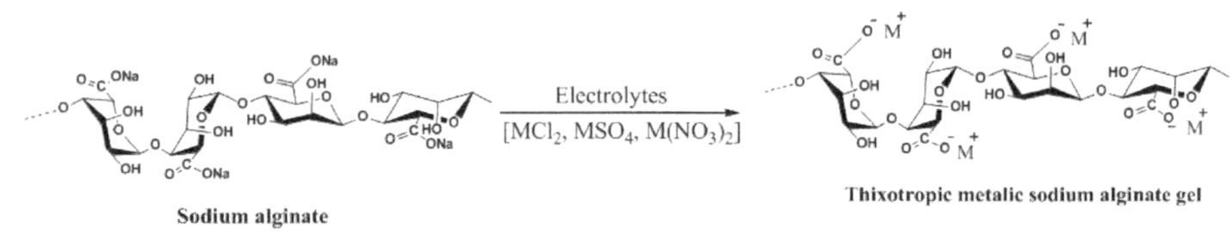

Where: MCl_2: cupric chloride, barium chloride, calcium chloride
MSO_4: cadmium sulfate, zinc sulfate, nickel sulfate
$M(NO_3)_2$: lead nitrate

Figure 2: Reaction mechanism of the addition of electrolyte in the sodium alginate solution.

Changes in the setting time of these gels were noted due to various factors such as the added different amounts of the electrolyte, the aging time of the gel, the degree of polymerization of the parent polymer, and the pH of the system.

Change in the thixotropic behavior was also seen with various volumes of the system. In addition, thixotropy was observed in total around 25–54% of alginate combined with cupric ions, while more than this percentage, the combination of alginate and cupric salt results in non-thixotropic behavior. Further, the thixotropic solution can also be obtained with extra added sodium alginate to the precipitated copper alginate.

3.3 Physical Blended Gelling Systems

The rheological behavior of gel obtained with sodium alginate and magnesium aluminum silicate (MAS) composite, in which various ratios of sodium alginate and magnesium aluminum silicate were studied. Adding MAS into the sodium alginate gels provides higher viscosity and change in rheological properties, and the flow behavior also gets changed from Newtonian to pseudoplastic with enhanced thixotropy (Pongjanyakul and Puttipipatkhachorn 2007). The change in rheological behavior caused by formation intermolecular hydrogen bonding and electrostatic force leads to a matrix structure of the synthesized composite gels (Figure 3).

The thixotropic property of composite gel was studied using the hysteresis loop area technique. This study suggested introducing MAS into the sodium alginate solutions change the flow behavior from Newtonian to pseudoplastic with enhanced thixotropy.

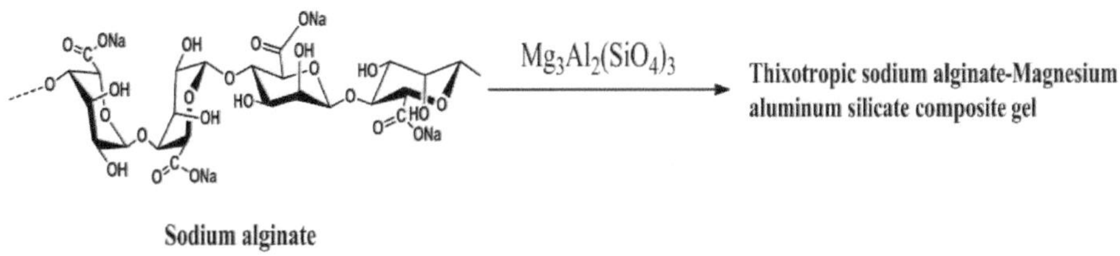

Figure 3: Reaction mechanism for sodium alginate-MAS composite.

3.4 Chemically Modified Gelling Systems

Sodium alginate-1-naphthylamine-based amide derivative (Figure 4) has been reported. The formed gel was efficient enough to hold water at the concentration of 4% w/v (Chejara et al. 2013). These gels behave like viscous sol on gentle stirring and come back to their original network upon removal of applied stirring at room temperature. This phenomenon appeared to be reversible manners. These types of thixotropic materials are claimed to have potential uses in biomedical and sensory applications. Sodium alginate and amino caproic acid-based new hydrogels cum thixo-gel system is stated in reports (Chejara et al. 2016). The designed soft thixotropic hydrogel was investigated for rheological studies, which resulted in thixotropic nature of the gel. This gelling system was converted to a flowable liquid when applying the stress, and its rigidity recovered upon removal of the applied stress. This gel was recommended for biomedical applications.

A new hydrogel with thixotropic behavior was designed and developed from alginate and 4-aminosalicylic acid which was found to have drug release profile as well on applied external stimuli. Gel displayed significant gel formation ability and

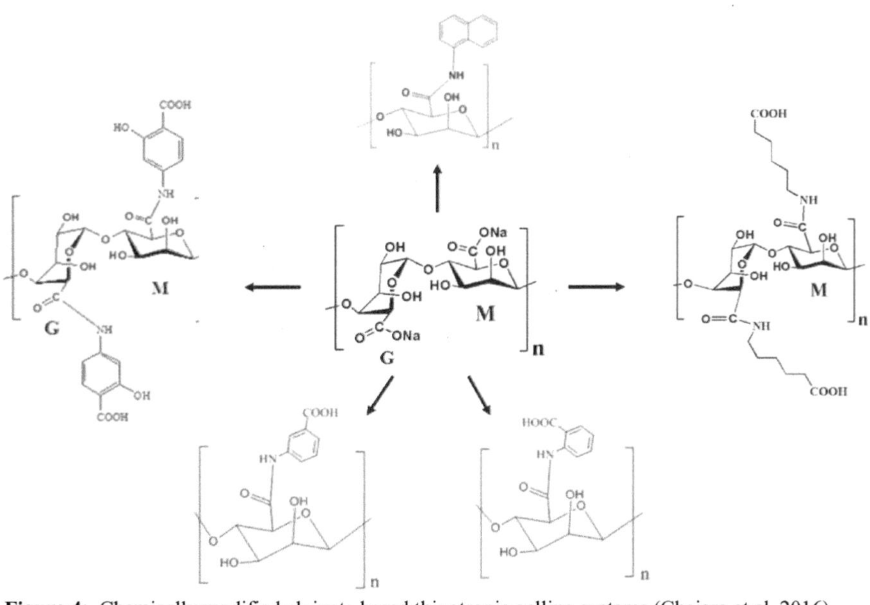

Figure 4: Chemically modified alginate based thixotropic gelling systems (Chejara et al. 2016).

was also found to have structure deformation on applied stress and recovery behavior on the removal of this stimulus, which was studied through rheological studies. The modified gelling system was recommended to be used as a viscosupplementation for injuries in the joints and several arthritis-induced conditions.

4. Guar Gum as a Thixotropic Gel

Guar gum (GG) is a naturally occurring water-soluble polymer made up of poly galactomannan units found from the seeds of Cyamopsis tetragonolobus, a native India guar plant, having linear backbone chain of β (1-4)-linked D-mannose units (M) with randomly attached side chains of α (1-6)-linked galactose units(G) (Prajapat and Gogate 2015; McCleary et al. 1985; Gupta et al. 2009). Figure 5 depicts the general structural formula of guar gum polymer.

Guar gum solution possesses high viscosities even at low concentrations due to its water-absorbing capacity. Guar gum is the most popular water-soluble hydrocolloid polymer used as a stabilizing, gelling, emulsifying, and mostly thickening agent for applications including food industries, cosmetics, pharmaceutical sectors, etc. (Wang et al. 2016). Generally, polysaccharides can be easily converted into their derivatives using either chemical or biochemical methods due to the presence of reactive functional moieties on the molecular chain (Prabaharan 2011). In addition, polysaccharides (guar gum) are an abundant source in nature, along with non-toxicity, stability, hydrophilicity, and safe nature. Due to the extensive properties of polysaccharides, many studies have been carried out on their application as drug-delivery systems. The property of flow behavior plays a key role in the development of the pharmaceutical process at each step during mixing, filling, and packing. Therefore, it is important for pharmaceutical scientists to study the flow properties with their influencing factors and effects on the efficacy of different physical properties, especially on liquid drugs (Lee et al. 2009). A viscosity system has either a Newtonian or non-Newtonian flow behavior with the applied shear/composition of the liquid, as depicted in Figure 6 (Likavčan et al. 2014). Water, honey, castor oil, alcohol, and ethyl ether are all examples of Newtonian fluids, whereas maximum thixotropic fluids, like putties, cosmetics, ointments, toothpaste, clays, and pastes, are examples of non-Newtonian fluids.

Depending on the end-use applications, the parent polymer, guar gum, can be modified chemically with the objective of altering its intrinsic viscosity as well as

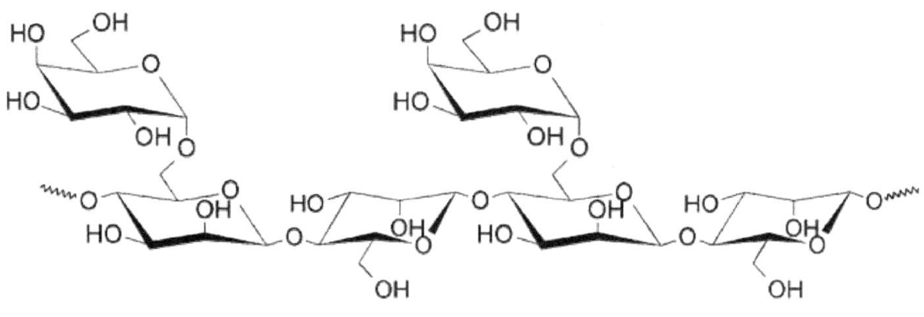

Figure 5: Structure of guar gum polymer.

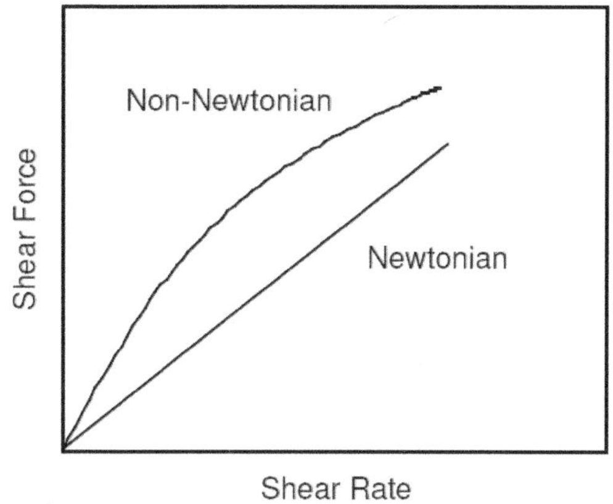

Figure 6: Schematic plot of shear force vs shear rate for Newtonian and non-Newtonian fluids (image sourced from an open internet source).

rheological behavior (Barbucci et al. 2008). However, physically modified or cross-linked conjugates are used with the aim of obtaining a water-insoluble compound. Wang et al. (2016) have studied the influence of pH on the rheological behavior of GG hydrogel and its derivative hydroxypropyl guar gum (HPG). The experimental results showed that the hydroxypropyl guar gum gel tests were performed in the range of pH of 7–14 and a temperature of 100°C. For the pH range of 7–12, it has been clearly observed that there was a sharp decrease in the viscosity in 5 minutes. While at pH 13, the viscosity of the gel decreased slowly. However, at pH 14, there was no change in the gel viscosity values.

The author also conducted a detailed study on pH levels ranging from 7–14 by calculating rheological parameters. It has been reported that at pH 8.5–12, the limiting viscosity with no shear rate (η_0) was exhibiting higher than 1,295.3 mPa·s, as the values compared with other pH. The side chain becomes negative due to the interrelate of GG molecules with OH^- ions in the alkaline pH range. With increasing concentrations of OH^- ions, there is an increase in negative electric charge density on guar gum polymer chains, which results in more negative zeta potential, and due to the electrostatic repulsion effect, guar gum polymer chains fall under stretching with relatively high entanglement density.

Barbucci et al. (2008) reported the experimental work on thixotropic guar gum hydrogel by chemically cross-linked. A new hydrogel was prepared in an alkaline condition with a 1:3 molar ratio of guar gum and polyethylene glycol diglycidylether (PEGDGE). The chemically cross-linked guar gum reaction was stirred for 15 hours in 0.1 M NaOH at 40°C. It has been observed from the experimental analysis data that GG hydrogel prepared using chemically cross-linked agents like glutaraldehyde and phosphate agents showed different thixotropic properties as compared with GG with PEGDGE crosslinking agent.

Li et al. (2008) synthesized and studied the thermal responses of guar gum-poly(N-isopropyl acrylamide) hydrogels, showing Interpenetrating Polymeric Networks (IPN) have the potential of specific colonic drug release profiles for the controlled drug-delivery system. In this research, experiments were carried out to prepare the thermoresposive hydrogel of guar gum-poly(N-isopropylacrylamide) IPN at various response rates. The experimental results showed faster deswelling rates by these hydrogels with reversible characteristics compared to native poly(N-isopropyl acrylamide) hydrogels and a much lower water absorption capacity of around 15 wt % GG contents. Hence, it has been concluded from result that the addition of guar gum polymer with IPN scheme can enhance the heat sensitivity and permeability of hydrogel as well. Similarly, the GG/poly(acrylic acid) IPN hydrogels approach has also been studied for swelling kinetics and water transportation tools as a function of solution pH for swelling and chemical composition for hydrogel (Li et al. 2006). It was observed that the swelling properties were highly dependent on solution pH along with formulation of hydrogels.

The rheological and fracturing properties of the sulfur-containing derivative of guar gum, i.e., sulfonated hydroxypropyl guar gum (SHG), were synthesized with different concentrations of sodium 3-chloro-2-hydroxy propane sulfonate (CHP) via sulphonation process by Qiu et al. (2018) so as to study the flow or rheological properties. Sulfonated hydroxypropyl guar gum was synthesized by first hydration of GG in NaOH solution for the period of 1 hour followed by the addition of sodium 3-chloro-2-hydroxy propane sulfonate for 20 min and further proceed for 3 hours at 60°C in a water bath. Two different fluids of SHG were prepared using various concentrations of NaCl and zirconium crosslinker. It has been reported that the increases in the degree of substitution (DS) of SHG solution decrease the thixotropic ability and strong anti-salinity capability. While SHG fracturing fluids showed excellent and shear, temperature resistance, and salt tolerance behavior. In addition, the gel-breaking properties of SHG exhibit lower interfacial surface tension and small core permeability damage. The experimental study reveals that SHG applications are mainly useful in the oil-extracting field.

5. Conclusion

Herein, in the present section of the book, the rheological characteristics of biomaterials, i.e., sodium alginate and guar gum-based thixotropic gelling systems, have been demonstrated. Both polymers show a significant improvement in their flow behavior after chemical or physical modification onto their polymeric backbone. The studied rheological properties open-up many doors for an individual researcher working on biopolymeric-based materials to be used for many potential applications. Such studies and modified conjugates may have potential use in biomedical applications, regenerative medicines, and other drug-delivery systems. Also, introducing thixotropy into these biopolymers creates other options for designing a wide variety of biopolymeric materials, with special reference to polymers such as agarose, carrageenan, xanthan gum, gum arabic, tamarind gum, etc.

Acknowledgment

The authors acknowledge the entire Department of Chemistry, Uka Tarsadia University, for their valuable input during the writeup of this review.

References

Barbucci, Rolando, Daniela Pasqui, Roberto Favaloro et al. 2008. A Thixotropic hydrogel from chemically cross-linked guar gum: synthesis. Characterization and Rheological Behaviour. Carbohydrate Research 343(18): 3058–65. https://doi.org/10.1016/j.carres.2008.08.029.

Barnes, H. A. 1997. Thixotropy—A review. Journal of Non-Newtonian Fluid Mechanics 70: 1–33.

Buruiana, L. I. and S. Ioan. 2018. Polymer Gel composites for bio applications. pp. 111–123. *In*: Vijay Kumar, T., Manju T. Kumari and Stefan V. Ioan (Eds.). In Polymer Gels. Springer Publication.

Cathell, Matthew D., Janah C. Szewczyk and Caroline L. Schauer. 2010. Organic modification of the polysaccharide alginate. Mini-Reviews in Organic Chemistry 7(1): 61–67.

Chahal, Sugandha, Anuj Kumar and Fathima Shahitha Jahir Hussian. 2019. Development of biomimetic electrospun polymeric biomaterials for bone tissue engineering. A review. Journal of Biomaterials Science, Polymer Edition 30(14): 1308–55. https://doi.org/10.1080/09205063.2019.1630699.

Chejara, Dharmesh R., Stalin Kondaveeti, Kamalesh Prasad et al. 2013. Studies on the structure-property relationship of sodium alginate based thixotropic hydrogels. RSC Advances 3(36): 15744–51. https://doi.org/10.1039/c3ra43070g.

Chejara, Dharmesh R., Mostafa Mabrouk, Ravindra V. Badhe et al. 2016. A Bio-injectable algin-aminocaproic acid thixogel with tri-stimuli responsiveness. Carbohydrate Polymers 135: 324–33. https://doi.org/10.1016/j.carbpol.2015.08.097.

Chejara, Dharmesh R., Mostafa Mabrouk, Pradeep Kumar et al. 2017. Synthesis and evaluation of a sodium alginate-4-aminosalicylic acid based microporous hydrogel for potential viscosupplementation for joint injuries and arthritis-induced conditions. Marine Drugs 15(8): 1–14. https://doi.org/10.3390/md15080257.

Chhatbar, Mahesh U., Kamalesh Prasad, Dharmesh R. Chejara et al. 2012. Synthesis of sodium alginate based sprayable new soft gel system. Soft Matter 8(6): 1837–44. https://doi.org/10.1039/c2sm06947d.

Daemi, Hamed, Mehdi Barikani and Mohammad Barmar. 2013. Highly stretchable nanoalginate based polyurethane elastomers. Carbohydrate Polymers 95(2): 630–36. https://doi.org/10.1016/j.carbpol.2013.03.039.

Dolz, M., F. González, J. Delegido et al. 2000. A time-dependent expression for thixotropic areas. Application to Aerosil 200 Hydrogels. J. Pharm. Sci. 89(6): 790–97.

Galindo-Rosales, F. J., F. J. Rubio-Hernández and A. Sevilla. 2011. An apparent viscosity function for shear thickening fluids. Journal of Non-Newtonian Fluid Mechanics 166(5-6): 321–25. https://doi.org/10.1016/j.jnnfm.2011.01.001.

George, Meera and T. Emilia Abraham. 2006. Polyionic hydrocolloids for the intestinal delivery of protein drugs: alginate and chitosan—A review. Journal of Controlled Release 114(1): 1–14. https://doi.org/10.1016/j.jconrel.2006.04.017.

Gombotz, Wayne R. and Siow Fong Wee. 1998. Protein release from alginate matrices. Advanced Drug Delivery Reviews 31(3): 267–85. https://doi.org/10.1016/S0169-409X(97)00124-5.

Grassi, G., A. Crevatin, R. Farra et al. 2006. Rheological properties of aqueous pluronic-alginate systems containing liposomes. Journal of Colloid and Interface Science 301(1): 282–90. https://doi.org/10.1016/j.jcis.2006.04.068.

Gupta, Sumit, Bhavini Shah, Bhaskar Sanyal et al. 2009. Role of initial apparent viscosity and moisture content on post irradiation rheological properties of guar gum. Food Hydrocolloids 23(7): 1785–91.

Kirtania, Moumita Das, Nancy Kahali and Arindam Maity. 2020. Biopolymeric Gels in Drug Delivery. Springer International Publishing. https://doi.org/10.1007/978-3-030-46923-8_3.

Lara, L. Reys, Simone S. Silva, Diana Soares da Costa et al. 2022. Fucoidan-based hydrogels particles as versatile carriers for diabetes treatment strategies. Journal of Biomaterials Science, Polymer Edition.

Lee, C. H., V. Moturi and Y. Lee. 2009. Thixotropic property in pharmaceutical formulations. Journal of Controlled Release 136(2): 88–98.

Lee, Chi H., Venkat Moturi and Yugyung Lee. 2009. Thixotropic property in pharmaceutical formulations. Journal of Controlled Release 136(2): 88–98. https://doi.org/10.1016/j.jconrel.2009.02.013.

Li, X., H. F. Lv, R. Zhao et al. 2021. Recent developments in bio-scaffold materials as delivery strategies for therapeutics for endometrium regeneration. Materials Today Bio. 11: 100101. https://doi.org/10.1016/j.mtbio.2021.100101.

Li, Xiuyu, Wenhui Wu and Weiqi Liu. 2008. Synthesis and properties of thermo-responsive guar gum/poly(N-isopropylacrylamide) interpenetrating polymer network hydrogels. Carbohydrate Polymers 71(3): 394–402. https://doi.org/10.1016/j.carbpol.2007.06.005.

Li, Xiuyu, Wenhui Wu, Jianquan Wang et al. 2006. The swelling behavior and network parameters of guar gum/poly(acrylic acid) semi-interpenetrating polymer network hydrogels. Carbohydrate Polymers 66(4): 473–79. https://doi.org/10.1016/j.carbpol.2006.04.003.

Likavčan, Lukáš, Miroslav Košík, Jozef Bílik, and Maroš Martinkovič. 2014. Determination of apparent viscosity as function of shear rate and fibres fraction in polypropylene. Ijeit 4(5): 23–26.

Lin, Hsin Yi, Shin Hung Chen, Shih Hsin Chang et al. 2015. Tri-layered chitosan scaffold as a potential skin substitute. Journal of Biomaterials Science, Polymer Edition 26(13): 855–67. https://doi.org/10.1080/09205063.2015.1061350.

Mahesh, U. Chhatbar, Kamalesh Prasad, Dharmesh R. Chejara et al. 2012. Synthesis of sodium alginate based sprayable new soft gel system. Soft Matter 8: 1837–44.

McCleary, B. V., A. H. Clark, I. C. M. Dea et al. 1985. The fine structures of carob and guar galactomannans. Carbohydrate Research 139: 237–260.

Mishra, S., N. Thombare, M. Ali et al. 2018. Applications of biopolymeric gels in agricultural sector. pp. 185–228. *In*: Vijay Kumar, T., T. Manju Kumari and V. Stefan Ioan. (Eds.). In Polymer Gels. Springer Publication.

Paul, S. D., H. Sharma, G. Jeswani et al. 2017. Novel gels: Implications for drug delivery. pp. 379–412. *In*: Nanostructures for Drug Delivery.

Pongjanyakul, Thaned and Satit Puttipipatkhachorn. 2007. Sodium alginate-magnesium aluminum silicate composite gels: characterization of flow behavior, microviscosity, and drug diffusivity. AAPS PharmSciTech 8(3). https://doi.org/10.1208/pt0803051.

Prabaharan, M. 2011. International journal of biological macromolecules prospective of guar gum and its derivatives as controlled drug delivery systems. International Journal of Biological Macromolecules 49(2): 117–24. https://doi.org/10.1016/j.ijbiomac.2011.04.022.

Prajapat, Amrutlal L. and Parag R. Gogate. 2015. Depolymerization of guar gum solution using different approaches based on ultrasound and microwave irradiations. Chemical Engineering and Processing: Process Intensification 88(February): 1–9.

Qiu, Liewei, Yiding Shen, Tao Wang et al. 2018. International journal of biological macromolecules rheological and fracturing characteristics of a novel sulfonated hydroxypropyl guar gum. International Journal of Biological Macromolecules 117: 974–82. https://doi.org/10.1016/j.ijbiomac.2018.05.072.

Sasaki, T., S. Miyamoto and Y. Moringaga. 1956. Thixotropy of Alginate Gels 32: 35–39.

Silva, Catarina M., António J. Ribeiro, Domingos Ferreira et al. 2006. Insulin encapsulation in reinforced alginate microspheres prepared by internal gelation. European Journal of Pharmaceutical Sciences 29(2): 148–59. https://doi.org/10.1016/j.ejps.2006.06.008.

Silva, K. C. G. and A. C. K. Sato. 2017. Biopolymer gels containing fructooligosaccharides. Food Res. Int. 101: 88–95.

Tønnesen, Hanne Hjorth and Jan Karlsen. 2002. Alginate in drug delivery systems. Drug Development and Industrial Pharmacy 28(6): 621–30. https://doi.org/10.1081/DDC-120003853.

"US Pat. 2003/0072803A1, 2003.

"US Pat. 5,473,005, 1995.

Voronko, N., S. Derkach and V. Izmailova. 2002. Rheological properties of gels of gelatin with sodium alginate. Russian Journal of Applied Chemistry Journal of Applied Chemistry 75(5): 790–94.

Wang, Shibin, Hongbiao Tang, Jianchun Guo et al. 2016. Effect of PH on the rheological properties of borate crosslinked hydroxypropyl guar gum hydrogel and hydroxypropyl guar gum. Carbohydrate Polymers 147: 455–63. https://doi.org/10.1016/j.carbpol.2016.04.029.

Wang, Wei, Kong-jun Lu, Chao-heng Yu et al. 2019. Nano-drug delivery systems in wound treatment and skin regeneration. J. Nanobiotechnol. 17(1): 1–15.

Yang, Ji Sheng, Ying Jian Xie and Wen He. 2011. Research progress on chemical modification of alginate: A review. Carbohydrate Polymers 84(1): 33–39. https://doi.org/10.1016/j.carbpol.2010.11.048.

15

In Vivo Testing of Biomaterials in Animal Models

Isha Zafar,[1] *Vishal K. Sahu,*[1] *Bommana Kavya,*[1] *Shuchi Nagar,*[1] *Jyotirmoi Aich*[2] and *Soumya Basu*[1,*]

1. Introduction

A biomaterial is an engineered substance with organic and inorganic origins specifically designed to replicate and replace lost tissues without causing damage to the surrounding environment. They have been used since ancient times, and their evolution has made them more potent, versatile, and simple to use in clinical settings. Keeping in mind that an ideal biomaterial must be biocompatible, bioinert, bioactive, bioabsorbable, and bioadaptable to be considered suitable for implantation (Figure 1); the classification of biomaterials is based on: (i) origin, i.e., natural or artificial; (ii) chemical nature, i.e., ceramic, metal, or polymer; and (iii) physical properties, i.e., roughness and porosity. Biomaterials play a crucial role in biological interactions, for which they are thoroughly assessed for their biocompatibility (Birajdar et al. 2021). Before heading toward a vibrant assessment of *in vivo* testing of biomaterials, it is imperative to understand the basic steps. First, *in vitro* studies, i.e., cell culture assays, are carried out by growing the cells on biomaterials. If favorable results are achieved, animal models are further used to assess the organ-implant interactions and the nature of the histological response induced by the implant is compared with that induced by a reference biomaterial for *in vivo* evaluation. However, the point-to-point association between *in vitro* and *in vivo* studies has

[1] Cancer and Translational Research Laboratory, Dr. D.Y. Patil Biotechnology and Bioinformatics Institute, Dr. D. Y. Patil Vidyapeeth, Pune – 411033, Maharashtra, India.

[2] School of Biotechnology and Bioinformatics, DY Patil Deemed-to-Be-University, CBD Belapur, Navi Mumbai, Maharashtra – 400614.

* Corresponding author: soumya.bs@gmail.com; ORCID ID: orcid.org/0000-0002-1726-0154

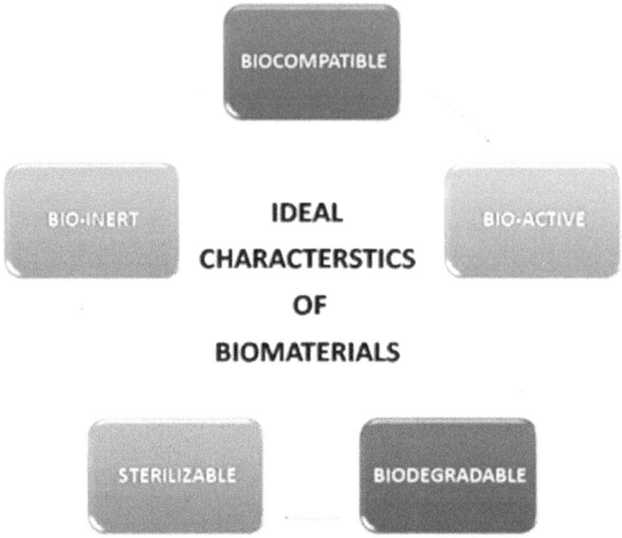

Figure 1: Ideal characteristics of biomaterials.

been observed to be quite low and could imply that these biomaterials may promote cellular adhesion and proliferation *in vitro* while failing to do the same *in vivo* and vice versa. This failure is strongly associated with the sophistication of *in vivo* tissue organization, which entails elaborated cell communication, mechanical load, and continuous flow; yet, none of these conditions recapitulate in standard cell cultures (Atif et al. 2021). This disengagement between *in vitro* and *in vivo* studies causes time delays, exorbitant expenditures, and the inordinate employment of animal experimentation to characterize biomaterials. Therefore, improved cell culture techniques that integrate with appropriate animal models enhance the predictive outcomes of implanting a biomaterial in the body. The choice of animal species and disease models (Ribitsch et al. 2020) may potentially improve the way we study *in vivo* biomaterial testing and thus facilitate addressing significant biological questions that have so far arisen in such studies.

Several natural and synthetic biomaterials are examined for their application as scaffolds in tissue engineering procedures for creating more enhanced, sophisticated, and reliable 3D microenvironments so as to promote the development and survival of various cell types. Furthermore, metal and its alloys are extensively used as scaffolds, bone plates, and bone screws for human tissue repair and regeneration (Pushp et al. 2021). The selection of an optimal biomaterial depends on material-driven cellular response and its ability to provide an environment for the proliferation of target cells and requires the incorporation of crucial initial steps in the design of biomimetic systems that aid in the investigation of various organ structures and functions (Ozgun et al. 2022). This highlights that the potential evolution of biomaterials necessitates a thorough perception of fundamental properties of materialistic behavior concerning the bio-environment (Paul 2019). Biomaterials play an indispensable function in addressing many medical complications with magnified precision and reflecting amazing progress in the context of orthopedics, cardiovascular, ophthalmology, neurology, etc. Their application in healthcare is undisputable and, in fact, has proven to be highly resourceful in the advancement of tissue engineering methodologies for therapeutic treatment (Amato and Ezzell 2014).

2. *In Vivo* Animal Models Used for Biomaterials Testing and Their Significance

The English scientists William Rusell and Rex Burch were the first to voice their concern for experimental animals in their publication, *The Principles of Humane Experimental Technique*. They proposed the principles of the "three Rs", which fulfilled the objectives of research scientists: (1) substitution of animals by use of *in vitro* methods or by using phylogenetically lower animals; (2) reduction of the number of animals required; and (3) improvement of experimental methods to lower the ethical costs of painful or stressful procedures (Mukherjee et al. 2022). Biomaterial testing is an intermediate step between preclinical research and human clinical trials. As a result, a suitable *in vivo animal* defect model for testing biomaterial substitutes for defect repair must be developed or adopted. It is critical to focus on *assessing* a clinically relevant *in vivo* testing model for analyzing biomaterials' biocompatibility, mechanical properties, degradation, and interactions with host tissues to evaluate biomaterials for development as organ substitutes for organ defect repair (de Lacerda et al. 2020). Every animal model possesses its advantages and disadvantages (Table 7), so it necessitates the consideration of several factors when selecting a species of animal for a particular model. The selection of animal species is based on the acknowledged research question (Muhamed et al. 2015). The factors that influence the selection of animal models include expenses to acquire and maintain animals, accessibility, social acceptability, forbearance to confinement, the convenience of housing and handling, inter-animal homogeneity, supporting staff with adequate facilities, biological characteristics analogous to humans, and an informational biological database on species. Additionally, the chosen species' lifespan should be adequate for the course of the experiment or research. Finally, the animal's size must be critical to ensure that it is appropriate for the number and size of implants chosen. According to Hazzard et al. (1992), the point of view captured *in vivo* biomaterial testing research, no single animal model can be used for all purposes, nor can a model be disregarded as unsuitable for all purposes. Furthermore, multiple model systems are expected to completely recapitulate a wide range of information (Pearce et al. 2007).

3. Selection of Animal Models

3.1 Rats

Rats are often preferred as a suitable rodent model for cardiovascular research, where their larger size facilitates surgical procedures and other types of biomaterial testing (Table 1). Different unique strains of rats have been generated to mimic the complex nature of human diseases such as obesity, cardiovascular defects, and diabetes. Rats of age 9–12 weeks have been used for the evaluation, which encompasses the assessment of the safety profile and degradation rate of biomaterials. Various experimental studies reveal the duration of implants is usually 8–12 weeks (Bryda 2013; Meng et al. 2020).

Table 1: Applications of rat models in biomaterial testing.

Sr. no.	Applications	References
1.	The function of hydroxyapatite and collagen in the regeneration of fractures in the fibula of rats was investigated. The biomaterials demonstrated biocompatibility and osteoconductive potential capable of inducing osteogenesis.	(Shirane et al. 2010)
2.	Poly (4-hydroxybutyrate), a biodegradable fibrous membrane as a potential dermal replacement, was constructed to evaluate its healing capacity in rat models. P4HB membranes demonstrated good air permeability and wound healing for full-thickness skin defects *in vivo*.	(Yuan et al. 2021a)
3.	The swim bladder was developed as a new biomaterial with anti-calcification capabilities by decellularization and glutaraldehyde cross-linking. The biomaterial was implanted in the abdominal artery of a rat model. *In vivo* analyses revealed that the biomaterial showed excellent hemocompatibility, mechanical strength and stability, and cytocompatibility with lower calcification.	(Liu et al. 2020)

3.2 Rabbits

Rabbits are the most frequently used animal models for screening implant material (Table 2). They are best for studying human bone defects. Rabbits are convenient animal models because they reach skeletal maturity around six months after sexual maturity. These animals are genotypically analogous to humans, comparatively economical, and easy to manage, which enables the use of a significantly large number of animals in a single experiment. Rabbits are the most commonly used animals in advanced dental implant research (Mapara et al. 2012).

Table 2: Applications of rabbit models in biomaterial testing.

Sr. no	Applications	References
1.	A biodegradable novel hybrid scaffold comprised of methacrylate chondroitin sulfate (CSMA) poly (ethylene) glycol methyl ether-ε-caprolactone-acryloyl chloride with grapheme oxide was examined for potential use in cartilage tissue engineering. The scaffold was found to be biodegradable with a favorable degradation rate. The CSMA/PECA/GO scaffolds demonstrated improved chondrocyte morphology, integration, persistent subchondral bone, and considerably denser newly generated cartilage.	(Liao et al. 2015)
2.	A silk fibroin/chitosan nano-hydroxyapatite biphase composite scaffold was created to evaluate its potential in tissue engineering for rabbit femoral lesions. The biphase scaffold implant was able to achieve chondral and subchondral repair, permitting both chondrogenesis and osteogenesis.	(Ruan et al. 2017)

3.3 Dogs

Dogs are large animal species most frequently used in biomedical research for musculoskeletal and orthodontic implants (Table 3). Dogs' highly tractable behavior can be advantageous during the post-operative recovery phase, where they can be

Table 3: Applications of dog models in biomaterial testing.

Sr. no	Application	References
1.	Two different hydroxyapatites (natural and synthetic) and bioactive glass were implanted in the dental sockets of six adult dogs to evaluate soft and hard tissue reactions. All biomaterials behaved in the same way. The bovine bone mineral exhibited an increased number of particles coated in the osseous tissue in comparison to synthetic hydroxyapatite and bioactive glass. All of the biomaterials affected the socket repair process.	(Santos et al. 2010)
2.	A Ti/Al-alloy device covered with biogenic hydroxyapatite was transplanted in the femurs of dogs to evaluate its toxicity and biocompatibility. The implant was found to be biocompatible and exhibited high osteoconductivity, which enables a total synergy between new and existing bone material and the development of new bone.	(Mahmoud et al. 2021)
3.	Octacalcium phosphate/weakly denatured collagen scaffold was fabricated to examine its osteoregenerative effects on skull defects in canine models. The histopathological analysis revealed that the novel scaffolds promoted osteoregeneration and osteoblast infiltration in skull defects, making this approach clinically applicable for bone reconstruction.	(Kaneko et al. 2021)

trained to participate actively in recuperative protocols. However, because of their status as the best companion animal, there are growing ethical concerns about using dogs in biomedical research. Joint disorders, particularly osteoarthritis, have long been studied using canine models. New reproductive technologies *based* on cell therapy in canine and feline medicine will have a remarkable impact on human cardiomyopathy treatment (Ribitsch et al. 2020).

3.4 Pigs

Porcine models *bear a close resemblance* to humans in terms of their physiology, anatomy, and metabolism. The onset of puberty in pigs usually occurs earlier (*at an average age of five to seven months*) than other large animals, have a short reproduction time, and have a substantial litter size. Due to their close physical and physiological similarities with humans, pigs are extensively employed as an

Table 4: Applications of pig models in biomaterial testing.

Sr. no	Applications	References
1.	Implants comprised of commercially refined Ti or Ti alloy were treated with three different roughening procedures (sandblasted acid-etched, anodized, and double acid-etched surfaces) for the assessment of osseointegration in mini-pigs. The osteogenicity of sandblasted acid-etched substrates was more than that of the anodized or dual acid-etched ones. The osseointegration showed a positive correlation with surface roughness.	(Del Fabbro et al. 2017)
2.	Using a sinus augmentation model for mini-pigs, researchers carried out an assessment of localized bone development and dental implant osteointegration with subsequent implantation of rhBMP-2/ACS coupled with ceramic bone biomaterial. rhBMP-2/ACS appears to be an efficient substitute for sinus augmentation treatments since it promotes significant bone formation/osteointegration.	(Susin et al. 2018)

appropriate animal model in organ transplantation and other surgical operations (Table 4). They also closely mimic the gastrointestinal tract, cardiovascular system, and skin, making them a suitable animal model for successful tissue engineering, regenerative medicine, and biomechanical research. Porcine models have often been used to investigate the biomechanics of specific joints, such as the knee and temporomandibular joint, as well as specific tissues, such as bone, cartilage, and ligaments. Its widespread presence and use in academia allowed it to be used in a variety of lung, liver, heart, and kidney transplantation models (Pearce et al. 2007).

3.5 Goats

While goats are used in 8.2% of animal experiments, they are most frequently used in cartilage, meniscal, and ligamentous repair studies. Their bodies are large enough to accommodate multiple human implants and prostheses (Table 5). Goats are considered a good animal model for testing and screening biomaterials because they have a similar metabolic bone remodeling rate as humans (Lamboni et al. 2016).

Table 5: Applications of goat models in biomaterial testing.

Sr. no	Applications	References
1.	A novel aragonite-hyaluronate biphasic scaffold was fabricated to assess osteochondral regeneration in the caprine model. The development of hyaline cartilage and regeneration of subchondral bone occur after the implantation of the Ar-HA scaffold in critical osteochondral defects.	(Kon et al. 2015)
2.	Tissue-engineered heart valves were created using modified PEG hydrogels, an acellular biomaterial scaffold, and autologous mesenchymal stem cells. The results demonstrated enhanced confluence of seeding cells and scaffold. In systemic circulation, it can successfully prevent BMSC development and differentiation.	(Zhang et al. 2010)

3.6 Sheep

Domestic sheep are predominantly used as animal models in experimental and preclinical research due to their easy accessibility, low costs, and social acceptance as research animals (Table 6). Their strong flocking and following behavior tends

Table 6: Applications of sheep models in biomaterial testing.

Sr. no	Applications	References
1.	Wool-derived keratin hydrogel was evaluated for its role in enhancing osseointegration in dental implants using an osseous bone model in sheep. The improved osseointegration observed in this prototype indicates that keratin hydrogel can enhance the osseointegration of dental implants implanted in low-quality bones.	(Duncan et al. 2018)
2.	A tissue-engineered hydroxyapatite bone scaffold embedded with osteoprogenitor cells was constructed as a substitute to stimulate bone regeneration in the sheep model.	(Osório et al. 2021)

Table 7: Animal models, their advantages and disadvantages, and their applications in biomedical research (Amato and Ezzell 2014; de Lacerda Schickert et al. 2020; Jiang et al. 2020; Kim et al. 2016).

Types of Animal Models	Advantages	Disadvantages	Applications
Rats	The most commonly used animal models, easy access, genetically modifiable, used in cardiovascular research, wound healing/skin tissue engineering, bone tissue engineering, disease mechanisms, and pathophysiology, are similar to humans.	Bodyweight, high cost, and organ structure vary.	Wound healing/ skin tissue engineering, bone tissue engineering, and cardiovascular conditions.
Rabbits	Placid and non-aggressive; easy to manage and observe; widely bred and affordable; brief vital cycles (gestation, lactation, and puberty).	Lack of well-equipped animal shelters, experienced animal handlers, and post- and intraoperative medications.	Wound healing/ skin tissue engineering, bone tissue engineering, and cardiovascular conditions.
Dogs	Size variations range from neonatal human size to adult size; longitudinal measurements are feasible. Large patient population (pets) is available when there is a clinical need.	Ethical concerns –costs, specialized housing, surgery, imaging, and necropsy facilities are required.	Dental implants, bone tissue engineering, skin tissue engineering, etc.
Pigs	Multiple and longitudinal measurements possible; moderate genetic variation within breeds and between breeds (CFR human population) (naturally occurring diseases); validated scoring methods and imaging are available (especially for orthopedics).	Ethical issues, lack of specialized housing, surgery, imaging, and necropsy facilities, and expensive.	Cardiovascular conditions, skin tissue engineering, bone tissue engineering, etc.
Goats	Easy access; inexpensive; easy to handle; bone structure and morphology are similar to humans; bone remodeling and metabolism rates are similar in humans.	High research costs; require skilled persons to perform experiments.	Bone tissue engineering and cardiovascular conditions.
Sheep	Multiple significant (critical size) defects may exist along with serial sampling. Size resembles humans more than horses or small animal models. Availability and social acceptance of research animals.	Requirements for specialized housing, surgery, imaging, necropsy, and technological expertise.	Bone tissue engineering.

to make sheep gentle, easy to manage, and relatively inexpensive, which drives the force for our research to pave the way to house and feed them. Clinically, their body weight (50–90 kg) is closer to that of humans than that of small animal models. The anatomical and biomechanical characteristics of sheep and humans had the closest resemblance (osseous composition, body weight, joint structure, and architecture), enabling a good simulation of bone or cartilage tissue healing and remodeling. Since their nerve size and regenerative behavior are similar, the ovine species is also a satisfactory and competent model for testing the regeneration of peripheral nerves. Due to their comparable valve architecture and size to humans, sheep have frequently been employed as a model for cardiovascular research, particularly for evaluating heart valves (Pearce et al. 2007).

4. Types of Biomaterials

Biomaterials have numerous advantages in medical therapies, allowing for a quicker response to clinical issues. Because of their characteristics in cell adhesion, proliferation, and compatibility with a wide range of drug classes and chemical structures, biopolymers are an important asset for biomedical applications (Osório et al. 2021).

4.1 Naturally-Derived Biomaterials

Plants, animals, insects, and even humans are sources of natural polymers. Since their chemical and mechanical characteristics are similar to those of natural tissues, these materials make reliable scaffolds.

4.1.1 Matrigel

The matrigel is made from the sarcoma of the Engelbreth-Holm-Swarm mouse, which generates a lot of extracellular matrix (ECM). Its composition promotes cell adhesion, including laminin (60%), type IV collagen (30%), entactin (8%), and heparin sulfate proteoglycan perlecan (2–3%). These proteins enable the presence

Table 8: *In vivo* testing of matrigel biomaterial and its applications.

Sr. no	Application	References
1.	Multicellular tumor spheroids (MCTS) are crucial tools for translating cancer research as they represent three-dimensional aggregates of cells with tumor-like traits like medication resistance oxygen gradients. Matrigel was utilized to create and test MCTS for therapeutic translatability. Matrigel closely mimics the ECM of tumor cells and can often recapitulate *in vivo* interactions, inducing MCTS formation.	(Kamatar et al. 2020)
2.	Human endometrial stem cells produced from matrigel, adipose-derived stem cells, and dental pulp stem cells have been identified as promising options for the therapy of pancreatic disorders.	(Xu et al. 2020)
3.	Growth factor reduced matrigel was used as a suitable scaffold that promotes differentiation and survival of neural progenitor cells for the treatment of neurological disorders.	(Uemura et al. 2010)

of soluble growth factors, such as transforming growth factor (TGF) family peptides (for example, TGF) and fibroblast growth factors (FGFs), as well as hormones and other major molecules, in the microenvironments (Table 8).

4.1.2 Collagen

Collagen is a predominant form of ECM proteins, which account for 30% of the proteins within a mammal. *Cross-linked collagen fibers* contribute to the

Table 9: *In vivo* testing of collagen biomaterial and its applications.

Sr. no	Application	References
1.	Collagen-based wound dressings have been used to heal burns and ulcers.	(Doillon and Silver 1986; Peters 1980)
2.	Collagen nanoparticles or nanospheres are thermally stable, non-immunogenic, and biocompatible particles that exhibit great efficacy when used as drug formulations to deliver antibacterial and anticancer medications.	(Arun et al. 2021; Peters 1980)
3.	Collagen biomaterial, due to its ideal physical and chemical characteristics, has been widely utilized to increase the efficacy of stem cells and bioactive cells for heart regeneration and repair (cardiac tissue engineering).	(Jiang et al. 2020; Wu et al. 2019)

extracellular scaffolding of natural tissue. Collagen influences cell morphology, adhesion, and differentiation and has been shown to have less immunogenicity, greater biodegradability, high biocompatibility, and permeability when compared to other proteins (Table 9).

4.1.3 Chitosan

Chitosan is an amino polysaccharide extracted as a deacetylated derivative of chitin. It is predominantly found in the exoskeletons of insects and crustaceans. The structure of chitosan is similar to that of glycosaminoglycans and has been widely used in the pharmaceutical industry because of its exceptional properties,

Table 10: *In vivo* testing of chitosan biomaterial and its applications.

Sr. no	Application	References
1.	Chitosan-based wound dressings exhibited effective antibacterial and procoagulant properties, promoting wound healing by creating an appropriate environment for favorable microbiota.	(Wang et al. 2021)
2.	Applicability of chitosan in the production of bio-dental materials and its utilization in oral drug delivery, treatment of periodontitis, and dental pulp revival via bone tissue engineering.	(Fakhri et al. 2020)
3.	Chitosan and hydroxyapatite composites possess outstanding properties such as biocompatibility, biofunctionality, and non-immunogenicity, indicating that they are feasible and ideal materials for the treatment of musculoskeletal disorders and hard tissue regeneration.	(Pighinelli and Kucharska 2013)

such as biocompatibility, biodegradability, and antibacterial and antifungal activity. It is a polycationic polymer with enhanced permeation activity (Table 11). Chitosan hydrogels offer low surface tension with water and biological fluids and can absorb water without compromising structural integrity. The charge and degree of cross-linking are correlated with the biomechanical properties of the biomaterial.

Table 11: *In vivo* testing of alginate biomaterial and its applications.

Sr. no	Application	References
1.	Exosomal-loaded alginate hydrogel exhibits improved healing, collagen production, and vascular development in the wound region, promoting tissue regeneration with full thickness.	(Shafei et al. 2020)
2.	The Alginate-based hydrogels possess potential advantages such as biocompatibility, biodegradability, safety, water absorption, and moisture retention, making them an appropriate biomaterial for homeostatic applications.	(Xie et al. 2022)
3.	Alginate-based scaffolds are widely used biomaterials for repairing and regenerating cartilage because of their ability to deliver cells, drugs, and sequences.	(Liu et al. 2022)

4.1.4 Alginate

Alginate is an example of an anionic polysaccharide biomolecule obtained from brown algae. The mechanical and chemical characteristics of alginate can differ depending on the source's seasonal growing conditions. G blocks are predominantly found in algae derivatives, whereas M blocks are abundant in bacteria derivatives. Over 200 alginates are currently being produced. *In vitro* and *in vivo* studies revealed that it is biocompatible with human tissues (Table 10).

Table 12: *In vivo* testing of cellulose biomaterial and its applications.

Sr. no	Application	References
1.	Silk sericin functionalized bacterial cellulose represents a potential biomaterial for the wound healing process, promoting keratinocyte growth and fibroblast proliferation with improved ECM production.	(Lamboni et al. 2016)
2.	Cellulose-based woven scaffolds embedded with silver nanoparticles (AgNPs) and bioglass were used as fabric scaffolds for bone cell proliferation. The scaffolds exhibited remarkable biomineralization capability, making them a suitable choice for bone tissue engineering.	(Ali et al. 2021)
3.	A new dental glass ionomer mechanically augmented with cellulose nanocrystals exhibited satisfying biocompatibility and advanced tissue repair.	(Menezes-Silva et al. 2017)

Table 13: *In vivo* testing of gelatin biomaterial and its applications.

Sr. no	Application	References
1.	*In vivo* implantation of an electrospun tubular construct based on gelatin-poly trimethylene carbonate, the blend has shown macrophage reduction within 15 days, demonstrating both its potential for host integration and biocompatibility. This makes it a viable biomaterial for vascular tissue engineering.	(Joy et al. 2020)
2.	Endogenic bone marrow cells combined with functionally and structurally optimized fibroin-gelatin scaffold prove to be a viable biomaterial for the repair and regeneration of joint cartilage. The dual-optimized scaffold works as a physical impediment for blood clots, offering mechanical stability prior to the production of neocartilage and an appropriate environment for endogenic bone cells for proliferation, differentiation, and ECM production.	(Shi et al. 2017)
3.	*In vivo* implantation of epinephrine-entrapped chitosan nanoparticles enveloped in gelatin nanofibers, a bilayer nano-biomaterial, proved to be an efficient and reliable homeostatic agent for hemorrhage control.	(Atashgahi et al. 2021)

4.1.5 Cellulose

Cellulose is a **β**-(1,4)-linked polymer of D-glucose that serves as a structural load-bearing component in plant cell walls. Nanofibrillated cellulose generates a network with excellent mechanical strength, water retention, and hydrophilicity in three-dimensional scaffolds, promoting cellular development and proliferation (Table 12).

4.1.6 Gelatin

Animal skin is hydrolyzed and denaturized to produce gelatin. It is biocompatible, biodegradable, and non-immunogenic due to its origin (Table 13). It has high hydrophilicity and can be easily acquired from commercial sources to simulate human tissue. While gelatin has a similar composition to collagen, it is a less expensive polymer and also less antigenic.

Table 14: *In vivo* testing of poly-(ε-caprolactone) biomaterial and its applications.

Sr. no	Application	References
1.	Electrospunpolycaprolactone-gelatin membranes were used as sealing patches to prevent post-operative anastomotic leakage and promote tissue repair.	(Joo et al. 2021)
2.	*In vivo* implantation of bilayered oxidized regenerated cellulose/ polycaprolactone composites promoted dural closure and induced osteogenesis.	(Chumnanvej et al. 2020)
3.	*In vivo* implantation of nanoscale demineralized bone matrix layered with PCL/β tricalcium phosphate scaffold demonstrated enhanced adhesion, proliferation, and survival of bone cells, promoting bone regeneration inside and surrounding the scaffolds.	(Yuan et al. 2021b)

4.2 Synthetic Biomaterials

4.2.1 Poly-(ε-caprolactone) – PCL

Poly-(ε-caprolactone) is a low-cost linear synthetic biodegradable polyester with a broad range of applications for scaffold development. However, PCL, as approved by the Food and Drug Administration (FDA), exhibits inherently ideal mechanical and biological properties; both of these properties limit its usefulness for biomedical and tissue engineering applications (Table 14). PCL biodegradability has been studied and found to vary greatly depending on a variety of factors (i.e., material properties, scaffold architecture, and environmental conditions) (Yuan et al. 2021c).

Table 15: *In vivo* testing of PDMS biomaterial and its applications.

Sr. no	Application	References
1.	A novel antibacterial compound, silicone-cationic steroid, was integrated into a polydimethylsiloxane membrane to create a novel active-release coating to combat infections caused by implants. The findings revealed that in all of the animals given biofilm treatment, the new coating was successful in avoiding infection.	(Williams et al. 2012)
2.	A 3-D printed silicone (PDMS) immunomodulatory hydrogel coated with a cocktail of cytokines consisting of interleukin-10 and prostaglandin-E2 was created to reduce unfavorable immunological responses while also promoting tissue integration.	(Barthes et al. 2021)
3.	Parylene-caulked PDMS was utilized as a platform for neural implantable electrodes. These electrodes demonstrated improved electrical durability in a damp environment and maintained the biomechanical properties of PDMS, making them appropriate for long-term *in vivo* applications.	(Jeong et al. 2016)

4.2.2 Poly-(dimethyl-Siloxane) – PDMS

PDMS is a silicon-based organic polymer that offers novel and versatile approaches in the constructs of microfluidic chips and scaffolds to address its excellent properties in cell biocompatibility, elastomeric properties, gas exchange, and optical transparency (Table 15). Small molecules and drugs are known to be absorbed by PDMS, which is either advantageous or disadvantageous based on its applications. The thermal and chemical stability of PDMS is due to the Si-O bond.

Table 16: *In vivo* testing of PLA biomaterial and its applications.

Sr. no	Application	References
1.	A new medicinal adhesive was created by combining PLA and poly(trimethyl) carbonate in ethyl acetate. PLA-PTMC was able to stop further infections effectively and has enormous promise as a medicinal adhesive. It had a beneficial influence on homeostasis and no sensitivity to wounds.	(Zhang et al. 2017)
2.	*In vivo* implantation of electrospun poly-(L-lactic) acid fiber-based membranes, incorporating calcium phosphate ormoglass particles, demonstrated enhanced vasculogenesis and bone regeneration.	(Oliveira et al. 2017)
3.	A ring polymerization strategy was applied to synthesize poly-(L-lactic) acid biomaterial. The biomaterial demonstrated excellent biocompatibility with improved tissue regeneration and osteoregeneration at the site of injury.	(Xavier et al. 2021)

4.2.3 Polylactic Acid – PLA

Polylactic acid (PLA) can be derived by the *fermentation* process of the carbohydrate as well as through the chemical synthesis of acetaldehyde. The molecular mass of PLA influences the mechanical, biological, and degradation rates. A longer polymerization time and azeotropic distillation can be used to modify these parameters. PLA can be soluble in dioxane, oxolane, benzene, and other solvents. It is an example of a thermoplastic polymer that is widely used because it can be quickly molded into desired thin fibers and films. PLA exhibits various mechanical properties depending on the scaffolding techniques used while manufacturing the biomaterial (Table 16).

Table 17: *In vivo* testing of PEG biomaterial and its applications.

Sr. no	Application	References
1.	Administration of a novel thermosensitive and locally injected, 4-aminopyridine PLGA-PEG has improved functional recovery after severe peripheral nerve damage.	(Manto et al. 2021)
2.	*In vivo* evaluation of osteogenic clinoptilolite/PCL-PEG-PCL scaffolds revealed enhanced regeneration of bone in a dose-dependent manner.	(Pazarçeviren et al. 2017)
3.	The rhein-PEG-nano-hydroxyapatite conjugate was used as a drug delivery vehicle to administer doxorubicin and phosphorus-32 concurrently for improved cancer chemo-radiotherapy. DOX/^{32}P@Rhein-PEG-nHA was able to suppress the growth of bone metastasis in breast cancer.	(Yang et al. 2020)

4.2.4 Polyethylene Glycol – PEG

Hydrogels made of PEGs are becoming more biocompatible with human tissues (Table 17). Biocompatibility is achieved by manipulating its mechanical properties by adjusting weight percentage, molecular chain length, and cross-linking density. The cross-linked PEG forms a porous hydrogel that closely recapitulates the ECM of human tissues. In the presence of cell suspensions, the liquid-to-solid mechanism of

Table 18: *In vivo* testing of PGA biomaterial and its applications.

Sr. no	Application	References
1.	The PCL-PGA scaffolds with β-tricalcium phosphate were considered prospective options for bone tissue engineering applications, particularly for bone defect restoration, where strength and biodegradability are crucial features.	(Kumar et al. 2021)
2.	PGA-coordinated hydroxyapatite nanoparticles were utilized to perform dual functions, i.e., a DOX delivery system and prolonged calcium flow supply to accomplish synergistic, tumor-specific therapy. The *in vivo* tests also confirmed that the current system inhibited tumors with high selectivity and low heart toxicity, making it an effective anti-tumor platform.	(Xiaoyu et al. 2019)
3.	A mixture of HA and pluronic F127 was combined with PGA to create a formulation for a tissue barrier. The *in vivo* investigations showed that the F-127 HA and PGA combination produced improved stability and a strong *in vivo* anti-tissue adhesion activity.	(Kim et al. 2016)

gelation can be fine-tuned and regulated. It is a low-cost scaffold obtained from the living anionic ring-opening polymerization of ethylene oxide. It is soluble in water, and its molecular weight ranges from 0.4 to 100 kDa.

4.2.5 Polyglycolic Acid – PGA

Polyglycolic acid is non-toxic, non-immunogenic, biodegradable, and biocompatible. It is a linear aliphatic polyester approved by the FDA. The PGA can be derived by the process of fermentation and polymerization of natural resources such as rice, wheat, and sweet potato, which makes it a substitute for animal-derived scaffolds. Because of their tunable degradation rate and ease of processing, these scaffolds are widely used. The degradation of PGA occurs by non-enzymatic hydrolysis, which offers the advantage of producing non-toxic metabolites (Table 18). The tensile strength of PGA is 57 MPa, and Young's modulus ranges between 6 and 7 GPa.

Table 19: *In vivo* testing of PU biomaterial and its applications.

Sr. no	Application	References
1.	Propolis-coated polyurethane wound dressings were prepared using the salt-leaching method. The wound dressings exhibited enhanced *in vitro* biocompatibility and *in vivo* wound healing, proving their potential as an appropriate biomaterial for preclinical applications.	(Khodabakhshi et al. 2019)
2.	Polyurethane and tannin-modified hydroxyapatite (THA) composites were tested on critical-size calvarial defects in rats. THA was able to promote osteogenesis and angiogenesis and is a better bone composite than pure hydroxyapatite.	(Tian et al. 2020)
3.	The *in vivo* tissue reactions and functional performance of electrospun membranes made of gelatin and polyurethane used as biomimetic films on glucose biosensors were assessed. The research found that electrospun membranes can help implantable glucose biosensors achieve a long *in vivo* sensing lifetime.	(Burugapalli et al. 2018)

4.2.6 Polyurethane – PU

Polyutherane scaffolds are gaining popularity in various biomedical applications due to their biocompatibility and favorable mechanical properties (Table 19). The segmented structure of the synthetic polymer allows for the manipulation of a broad spectrum of parameters. The important variables, such as hydrophilicity and degradation rate, are strongly affected by several important parameters, such as the composition, the preparation method, and the manufacture of the polymer.

5. Recent Advances in Biomaterials Testing—Future Prospects

In recent years, many human body tissues, such as osseous, denticles, sinew, ligaments, and other load-bearing constructs, have been successfully replaced in the human body. For biomedical applications, traditional remedies, such as wood and rubber, and synthetic remedies, such as metals and glasses, have been utilized since ancient times. Commonly used biomaterials such as alloys, ceramics, and polymers have a unique atomic geometry, resulting in distinctive physical, mechanical, and chemical properties. Considering these, biomaterials have been gaining momentum as an alternative application for various body parts implantation. Over the last few years, metals have been used for the majority of load-bearing applications in the biomedical system due to their substantial mechanical advantages, including strength, stiffness, and fatigue life (Paul 2019). Despite the fact that many metals and alloys have been manufactured so far, only a few will be benchmark materials for implantation materials in the future, such as stainless steel, cobalt-chromium alloys, nickel, pure titanium and its alloys (Ti-6Al-4V and Ti-6Al-7Nb), magnesium, and so on. Besides these, the highly porous ceramics assist with prosthetic stabilization and bone ingrowth. They have, however, been exploited for reduced-loaded or unloaded bearing applications due to their porous properties and minimum mechanical strength. Recently, three basic types of ceramics, such as bioactive, bioinert, and bioresorbable, have become significantly more promising biomaterials for biomedical applications. The most prominent behavior of the first generation (the 1960s and 1970s) of ceramic biomaterials was their "inertness" (alumina and zirconia), reported in Hench's classification; they drastically reduced the foreign particle response as well as the immunogenic response (Pearce et al. 2007). In an instance, Merola et al. recreated the articulation functionality in the replacement surgery of the hip joint by substituting the joint with high-density, highly pure alumina (Al_2O_3). In this aspect, ceramics act as a potential alternative due to their limitless wear and corrosion resistance, along with their high strength and good biocompatibility (Mayer et al. 2018). Further, second-generation (1980 and 2000) became more apparent in the literature, supporting the notion that ceramics exhibited dual properties in terms of "bioactive (hydroxyapatite glass ceramics) as well as bioresorbable (calcium phosphate ceramics). Bioactive materials stimulate a biological response in the body, such as chemical bond formation with soft tissues, and reinforce them toward particular responses and behavioral patterns (Yadav and Gangwar 2018). The principles of bioactivity and biodegradability are combined in the new generation of ceramics, and the combination of material properties with their ability to obtain efficient signaling and activate specific cellular mechanisms and behaviors. Because of their outstanding properties, natural and synthetic polymers are suitable for a wide range of biomedical applications. Natural polymers with good bioactivity and cytocompatibility, such as collagen, fibronectin, silk fibroin, fibrin, chitosan, and many others, whose microstructure, degradation rate, mechanical properties, porosity, and others can be tailored based on the nature of the polymers, have been extensively utilized in tissue engineering applications.

6. Conclusion

Many lines of evidence in this chapter showed the *in vivo* methods for evaluating biomaterial biological compatibility. The methods were presented from two viewpoints: (1) international *in vivo* assays oriented around preclinical biosafety evaluation and potential risk assessments, and (2) more specific material and application-specific *in vivo* assays for specific cell- and tissue-based responses. In particular, important methods for gathering extensive information were narrated in concomitant case examples because the latter area is complex and diverse. These examples show how well-designed *in vivo* assays are essential for advancing new biomaterial therapies into clinical practice. In today's proof-of-concept studies, a comprehensive biocompatibility test plan developed by engineers and material scientists is more likely to endow vast areas of prototypes made of well-characterized biocompatible materials with a long history of appropriate utilization. The challenges faced by most production engineers, material scientists, regulatory bodies, and standard developers will undoubtedly increase with the range of biomaterials and the expansion of their applications. Nonetheless, the progress of such a functional framework for the development and application of *in vivo* assays that can accommodate novel therapies, improved medical materials, and advances in medicine will undoubtedly continue.

References

Ali, M.A., N. M. Aly, M. Mabrouk et al. 2021. A novel synthetic approach to produce cellulose-based woven scaffolds impregnated with bioactive glass for bone regeneration. Int. J. Biol. Macromol. 181: 905–918.

Amato, S. F. and R. M. Ezzell Jr. 2014. Regulatory affairs for biomaterials and medical devices. Elsevier.

Arun, A., P. Malrautu, A. Laha et al. 2021. Collagen nanoparticles in drug delivery systems and tissue engineering. Appl. Sci. 11: 11369.

Atashgahi, M., B. Ghaemi, A. Valizadeh et al. 2021. Epinephrine-entrapped chitosan nanoparticles covered by gelatin nanofibers: A bi-layer nano-biomaterial for rapid hemostasis. Int. J. Pharm. 608: 121074.

Atif, A. R., M. Pujari-Palmer, M. Tenje et al. 2021. A microfluidics-based method for culturing osteoblasts on biomimetic hydroxyapatite. Acta Biomater. 127: 327–337.

Barthes, J., P. Lagarrigue, V. Riabov et al. 2021. Biofunctionalization of 3D-printed silicone implants with immunomodulatory hydrogels for controlling the innate immune response: An *in vivo* model of tracheal defect repair. Biomaterials 268: 120549.

Birajdar, M. S., H. Joo, W.-G. Koh et al. 2021. Natural bio-based monomers for biomedical applications: a review. Biomater. Res. 25: 8.

Bryda, E. C. 2013. The Mighty Mouse: the impact of rodents on advances in biomedical research. Mo. Med. 110: 207.

Burugapalli, K., S. Wijesuriya, N. Wang et al. 2018. Biomimetic electrospun coatings increase the *in vivo* sensitivity of implantable glucose biosensors. J. Biomed. Mater. Res. Part A 106: 1072–1081.

Chumnanvej, S., T. Luangwattanawilai, V. Rawiwet et al. 2020. *In vivo* evaluation of bilayer ORC/PCL composites in a rabbit model for using as a dural substitute. Neurol. Res. 42: 879–889.

de Lacerda Schickert, S., J. J. J. P. van den Beucken, S. C. G. Leeuwenburgh et al. 2020. Pre-clinical evaluation of biological bone substitute materials for application in highly loaded skeletal sites. Biomolecules 10: 883.

Del Fabbro, M., S. Taschieri, E. Canciani et al. 2017. Osseointegration of titanium implants with different rough surfaces: a histologic and histomorphometric study in an adult minipig model. Implant Dent. 26: 357–366.

Doillon, C. J. and F. H. Silver. 1986. Collagen-based wound dressing: Effects of hyaluronic acid and firponectin on wound healing. Biomaterials 7: 3–8.

Duncan, W. J., P. F. C. Greer, M. Lee et al. 2018. Wool-derived keratin hydrogel enhances implant osseointegration in cancellous bone. J. Biomed. Mater. Res. Part B Appl. Biomater. 106: 2447–2454.

Fakhri, E., H. Eslami, P. Maroufi et al. 2020. Chitosan biomaterials application in dentistry. Int. J. Biol. Macromol. 162: 956–974.

Jeong, J., N. Chou and S. Kim. 2016. Long-term characterization of neural electrodes based on parylene-caulked polydimethylsiloxane substrate. Biomed. Microdevices 18: 1–9.

Jiang, H., G. Liang, M. Dai et al. 2020. Preparation of doxorubicin-loaded collagen-PAPBA nanoparticles and their anticancer efficacy in ovarian cancer. Ann. Transl. Med. 8.

Joo, G., T. Sultana, S. Rahaman e al. 2021. Polycaprolactone-gelatin membrane as a sealant biomaterial efficiently prevents postoperative anastomotic leakage with promoting tissue repair. J. Biomater. Sci. Polym. Ed. 32: 1530–1547.

Joy, J., R. Aid-Launais, J. Pereira et al. 2020. Gelatin-polytrimethylene carbonate blend based electrospun tubular construct as a potential vascular biomaterial. Mater. Sci. Eng. C 106: 110178.

Kamatar, A., G. Gunay and H. Acar. 2020. Natural and synthetic biomaterials for engineering multicellular tumor spheroids. Polymers (Basel). 12: 2506.

Kaneko, M., K. Shigeno, M. Wakatsuki et al. 2021. Bone regeneration utilizing a newly developed octacalcium phosphate/weakly denatured collagen scaffold. Int. J. Artif. Organs 44: 139–145.

Khodabakhshi, D., A. Eskandarinia, A. Kefayat et al. 2019. *In vitro* and *in vivo* performance of a propolis-coated polyurethane wound dressing with high porosity and antibacterial efficacy. Colloids surfaces B Biointerfaces 178: 177–184.

Kim, M., Y. Hwang and G. Tae. 2016. The enhanced anti-tissue adhesive effect of injectable pluronic-HA hydrogel by poly (γ-glutamic acid). Int. J. Biol. Macromol. 93: 1603–1611.

Kon, E., G. Filardo, J. Shani et al. 2015. Osteochondral regeneration with a novel aragonite-hyaluronate biphasic scaffold: up to 12-month follow-up study in a goat model. J. Orthop. Surg. Res. 10: 1–17.

Kumar, A., S. M. Mir, I. Aldulijan et al. 2021. Load-bearing biodegradable PCL-PGA-beta TCP scaffolds for bone tissue regeneration. J. Biomed. Mater. Res. Part B Appl. Biomater. 109: 193–200.

Lamboni, L., Y. Li, J. Liu et al. 2016. Silk sericin-functionalized bacterial cellulose as a potential wound-healing biomaterial. Biomacromolecules 17: 3076–3084.

Liao, J., Y. Qu, B. Chu et al. 2015. Biodegradable CSMA/PECA/graphene porous hybrid scaffold for cartilage tissue engineering. Sci. Rep. 5: 9879.

Liu, J., B. Li, H. Jing et al. 2020. Swim bladder as a novel biomaterial for cardiovascular materials with anti-calcification properties. Adv. Healthc. Mater. 9: 1901154.

Liu, W., H. Madry and M. Cucchiarini. 2022. Application of alginate hydrogels for next-generation articular cartilage regeneration. Int. J. Mol. Sci. 23: 1147.

Mahmoud, E. M., M. Sayed, M. Awaad et al. 2021. Evaluation of Ti/Al alloy coated with biogenic hydroxyapatite as an implant device in dogs' femur bones. J. Mater. Sci. Mater. Med. 32: 1–15.

Manto, K. M., P. K. Govindappa, D. Parisi et al. 2021. (4-Aminopyridine)–PLGA–PEG as a novel thermosensitive and locally injectable treatment for acute peripheral nerve injury. ACS Appl. Bio. Mater. 4: 4140–4151.

Mapara, M., B. S. Thomas and K. M. Bhat. 2012. Rabbit as an animal model for experimental research. Dent. Res. J. (Isfahan). 9: 111.

Mayer, Y., O. Ginesin, A. Khutaba et al. 2018. Biocompatibility and osteoconductivity of PLCL coated and noncoated xenografts: An *in vitro* and preclinical trial. Clin. Implant Dent. Relat. Res. 20: 294–299.

Menezes-Silva, R., F. V. Pereira, M. H. Santos et al. 2017. Biocompatibility of a new dental glass ionomer cement with cellulose microfibers and cellulose nanocrystals. Braz. Dent. J. 28: 172–178.

Meng, X., R. Ziadlou, S. Grad et al. 2020. Animal models of osteochondral defect for testing biomaterials. Biochem. Res. Int. 2020.

Merola, M. and S. Affatato. 2019. Materials for Hip Prostheses: A Review of Wear and Loading Considerations. Materials (Basel, Switzerland), 12(3): 495. https://doi.org/10.3390/ma12030495.

Muhamed, J., D. Revi, A. Rajan et al. 2015. Biocompatibility and immunophenotypic characterization of a porcine cholecyst–derived scaffold implanted in rats. Toxicol. Pathol. 43: 536–545.

Mukherjee, P., S. Roy, D. Ghosh et al. 2022. Role of animal models in biomedical research: a review. Lab. Anim. Res. 38: 18.

Oliveira, H., S. Catros, O. Castano et al. 2017. The proangiogenic potential of a novel calcium releasing composite biomaterial: Orthotopic *in vivo* evaluation. Acta Biomater. 54: 377–385.

Osório, L. A., E. Silva and R. E. Mackay. 2021. A review of biomaterials and scaffold fabrication for organ-on-a-chip (OOAC) systems. Bioengineering 8: 113.

Ozgun, A., D. Lomboni, H. Arnott et al. 2022. Biomaterials-based strategies for *in vitro* neural models. Biomater. Sci. 10: 1134–1165.

Paul, S. 2019. Biomedical engineering and its applications in healthcare. Springer.

Pazarçeviren, E., Ö. Erdemli, D. Keskin et al. 2017. Clinoptilolite/PCL–PEG–PCL composite scaffolds for bone tissue engineering applications. J. Biomater. Appl. 31: 1148–1168.

Pearce, A. I., R. G. Richards, S. Milz et al. 2007. Animal models for implant biomaterial research in bone: a review. Eur Cell Mater 13: 1–10.

Peters, W. J. 1980. Biological dressings in burns—a review. Ann. Plast. Surg. 4: 133–137.

Pighinelli, L. and M. Kucharska. 2013. Chitosan–hydroxyapatite composites. Carbohydr. Polym. 93: 256–262.

Pushp, P., R. Bhaskar, S. Kelkar et al. 2021. Plasticized poly (vinylalcohol) and poly (vinylpyrrolidone) based patches with tunable mechanical properties for cardiac tissue engineering applications. Biotechnol. Bioeng. 118: 2312–2325.

Ribitsch, I., P. M. Baptista, A. Lange-Consiglio et al. 2020. Large animal models in regenerative medicine and tissue engineering: to do or not to do. Front. Bioeng. Biotechnol. 8: 972.

Ruan, S., L. Yan, J. Deng et al. 2017. Preparation of a biphase composite scaffold and its application in tissue engineering for femoral osteochondral defects in rabbits. Int. Orthop. 41: 1899–1908.

Santos, F. A., M. T. Pochapski, M. C. Martins et al. 2010. Comparison of biomaterial implants in the dental socket: histological analysis in dogs. Clin. Implant Dent. Relat. Res. 12: 18–25.

Shafei, S., M. Khanmohammadi, R. Heidari et al. 2020. Exosome loaded alginate hydrogel promotes tissue regeneration in full-thickness skin wounds: An *in vivo* study. J. Biomed. Mater. Res. Part A 108: 545–556.

Shi, W., M. Sun, X. Hu et al. 2017. Structurally and functionally optimized silk-fibroin–gelatin scaffold using 3D printing to repair cartilage injury *in vitro* and *in vivo*. Adv. Mater. 29: 1701089.

Shirane, H. Y., D. Y. Oda, T. C. Pinheiro et al. 2010. Biomaterial implants in bone fractures produced in rats fibulas. Rev. Bras. Ortop. (English Ed.) 45: 478–482.

Susin, C., J. Lee, T. Fiorini et al. 2018. Sinus augmentation using rh BMP-2/ACS in a mini-pig model: Influence of an adjunctive ceramic bone biomaterial. J. Clin. Periodontol. 45: 1005–1013.

Tian, X., X. Yuan, D. Feng et al. 2020. *In vivo* study of polyurethane and tannin-modified hydroxyapatite composites for calvarial regeneration. J. Tissue Eng. 11: 2041731420968030.

Uemura, M., M. M. Refaat, M. Shinoyama et al. 2010. Matrigel supports survival and neuronal differentiation of grafted embryonic stem cell-derived neural precursor cells. J. Neurosci. Res. 88: 542–551.

Wang, C.-H., J.-H. Cherng, C.-C. Liu et al. 2021. Procoagulant and antimicrobial effects of chitosan in wound healing. Int. J. Mol. Sci. 22: 7067.

Williams, D. L., B. S. Haymond, J. P. Beck et al. 2012. *In vivo* efficacy of a silicone–cationic steroid antimicrobial coating to prevent implant-related infection. Biomaterials 33: 8641–8656.

Wu, W., S. Peng, Z. Song et al. 2019. Collagen biomaterial for the treatment of myocardial infarction: an update on cardiac tissue engineering and myocardial regeneration. Drug Deliv. Transl. Res. 9: 920–934.

Xavier, M., N. Farez, P. L. Salvatierra et al. 2021. Biological performance of a bioabsorbable Poly (L-Lactic Acid) produced in polymerization unit: *in vivo* studies. F1000Research 10.

Xiaoyu, M., D. Xiuling, Z. Chunyu et al. 2019. Polyglutamic acid-coordinated assembly of hydroxyapatite nanoparticles for synergistic tumor-specific therapy. Nanoscale. 11: 15312–15325.

Xie, Y., P. Gao, F. He et al. 2022. Application of alginate-based hydrogels in hemostasis. Gels 8: 109.

Xu, B., F.-Z. Yuan, L. Lin et al. 2020. The higher inherent therapeutic potential of biomaterial-based hDPSCs and hEnSCs for pancreas diseases. Front. Bioeng. Biotechnol. 8: 636.

Yadav, S. and S. Gangwar. 2018. An Overview on Recent progresses and future perspective of biomaterials, in: IOP Conference Series: Materials Science and Engineering. IOP Publishing, pp. 12013.

Yang, X., D. Zhai, J. Song et al. 2020. Rhein-PEG-nHA conjugate as a bone targeted drug delivery vehicle for enhanced cancer chemoradiotherapy. Nanomedicine Nanotechnology, Biol. Med. 27: 102196.

Yuan, B., Z. Wang, Y. Zhao et al. 2021. *In vitro* and *in vivo* study of a novel nanoscale demineralized bone matrix coated PCL/β-TCP scaffold for bone regeneration. Macromol. Biosci. 21: 2000336.

Yuan, S., X. Sun, Y. Shen et al. 2021. Bioabsorbable poly (4-hydroxybutyrate)(P4HB) fibrous membranes as a potential dermal substitute. J. Mater. Chem. B 9: 8074–8080.

Zhang, D. Q., Q. Yang and C. M. Jiang. 2010. Zhonghua wai ke za zhi [Chinese Journal of Surgery]. Zhonghua Wai Ke Za Zhi 48: 1637–1641.

Zhang, S., H. Li, Mingwei Yuan et al. 2017. Poly (lactic acid) blends with poly (trimethylene carbonate) as biodegradable medical adhesive material. Int. J. Mol. Sci. 18: 2041.

Index

Editors' Biography

Dr. Arvind K. Singh Chandel is a researcher and educator specializing in Pharmaceutical Biotechnology, Polymer Chemistry, and Tissue Engineering. He completed his B. Pharmacy at Barkatullah University, Bhopal, in 2011 and his M.S. Pharm. at NIPER, Hajipur, in 2013. He earned his PhD in Biological Sciences from AcSIR-CSIR-CSMCRI, Bhavnagar, in 2018. Dr. Chandel received the Japan Society for the Promotion of Science (JSPS) postdoctoral fellowship and conducted research at the University of Tokyo. His work focuses on the design of polymeric materials for biomedical applications, advanced drug delivery systems, and water desalination. Currently an Assistant Professor at the University of Tokyo, Dr. Chandel has published 35 peer-reviewed papers, holds eight patents, and has contributed to 20 book chapters. He has received awards like the Marie Skłodowska-Curie Individual Postdoctoral Fellowship and the M.K. Bhan Young Researcher Fellowship. His research is known for its interdisciplinary approach, advancing innovation in healthcare.

Arpana Parihar is currently working as a research associate at the Department of Translational Medicine, All India Institute of Medical Sciences Bhopal, MP, India. She did her Ph.D. at the Raja Ramanna Centre for Advanced Technology, Indore, under the aegis of HBNI, Mumbai. Her doctoral research involves the evaluation of the tumor selectivity and photodynamic therapy (PDT) efficacy of chlorin p6 through receptor-mediated targeted drug delivery. She has more than 9 years of research and teaching experience at prestigious institutes including BARC, RRCAT, IITD, CSIR-AMPRI, etc. She has been awarded many prestigious fellowships including DST-WOS-B, DBT-RA, CSIR-NET, GATE, etc. Her current research interests include 3D bioprinting, micro/nano fabrications of biosensors for early diagnosis of diseases, infectious diseases, stem cell biology, tissue engineering, photodynamic therapy of cancer, receptor-mediated targeting of tumors, 3D cell culture, expression of receptors and cytokines, molecular docking, molecular dynamic simulations, and immunoinformatics. She has published ~50 articles, 30 Book chapters, and edited 11 books with renowned publisher.

Dr. Raju Khan, currently works as a Senior Principal Scientist and Professor at CSIR—Advanced Materials Processes and Research Institute, Bhopal, Madhya Pradesh, India. He earned an MSc Chemistry and PhD Chemistry from Jamia Millia Islamia, Central University, New Delhi, India. He has worked as a Postdoctorate at the University of the Western Cape, Cape Town, South Africa, Young Scientist at CSIR—National Physical Laboratory, New Delhi, India, and Visiting Scientist at the University of Texas at San Antonio (UTSA), USA. He has been awarded Fellow of Royal Society of Chemistry (FRSC) and also a reputed BOYSCAST Fellowship, Government of India. His achievements include Scientific Advisor for FSSAI, Government of India, a Life Member of the Materials Research Society of India, and Member of the International Advisory Committee, DST, Government of India. His work has also been highlighted in Nature India. He has published more than 180 articles with a high citation score of more than 5000 in several renowned journals including 45 book chapters by esteemed publishers, editing 25 books with various reputed publishers and several patents filed. He has completed several projects and also have ongoing projects, including national and international scientific collaborations with the USA, Czech Republic and Russia.